Pacific Islands Guestworkers in Australia

"Guestworkers are not immigrants. They are workers who are not accorded full citizenship rights and are subject to eventual removal. While the earnings of guestworkers may be remitted to help family in their own country, their status in the new country is always secondary. This means they are vulnerable to exploitation. The difference between an immigrant and a guestworker is that a guestworker is expected not to assimilate because at the end of their work visa they will be required to leave. Guestworkers become a society within a society. We have already had an unfortunate experience with guestworkers, for example the Kanakas brought to the North Queensland sugar industry."

—Peter Costello, *Australian Federal Treasurer (July 2006)*

"[Pacific island countries] will continue to survive ... because of large aid assistance from Australia ... and because many of their workers come here and pick our fruit, pick our fruit with hard Australian enterprise and endeavour and we welcome them and we always will .... But the fact is we're not going to be hijacked into doing something that will shut down an industry...and I'm only talking coal, let alone all our other resources."

—Michael McCormack, *Australian Deputy Prime Minister (August 2019)*

"[The aged and disability care] workforce who, whilst they do the work of angels, are effectively treated like fruit pickers."

—Bill Shorten, *Shadow Minister for Government Services (August 2020)*

Kirstie Petrou • John Connell

Pacific Islands Guestworkers in Australia

The New Blackbirds?

palgrave
macmillan

Kirstie Petrou
Griffith University
Brisbane, QLD, Australia

John Connell
University of Sydney
Sydney, NSW, Australia

ISBN 978-981-19-5386-6 ISBN 978-981-19-5387-3 (eBook)
https://doi.org/10.1007/978-981-19-5387-3

Cover illustration: kenkuza_shutterstock.com

This Palgrave Macmillan imprint is published by the registered company Springer Nature Singapore Pte Ltd.
The registered company address is: 152 Beach Road, #21-01/04 Gateway East, Singapore 189721, Singapore

Preface

This is the first book to examine contemporary seasonal migration from Pacific island states to Australia primarily to work in the agricultural sector. It reflects on the new age of guestworkers from both ends—the source Pacific island states and Australian destinations—from a broad social, economic and cultural perspective; and, as the opening quotations suggest, from a political perspective. It seeks to offer a critical perspective on different phases of managed labour migration. We first trace the antecedents of present policies and practices in nineteenth-century 'blackbirding' that brought thousands of Pacific islanders, mainly from the New Hebrides (Vanuatu) and Solomon Islands, to work in Queensland sugar plantations. After a century-long hiatus, labour migration from Pacific island states to Australia began again in the 2000s (modelled on temporary migration schemes from the Caribbean to Canada, and from the Pacific to Aotearoa-New Zealand). In a normal season about 10,000 agricultural workers arrive in Australia from the Pacific island states (especially Vanuatu and Tonga), alongside some from Timor-Leste, working in various agricultural activities, mainly in Queensland and New South Wales. These numbers may expand significantly in the future as Australia negotiates new strategies with island states, as one approach to supporting island economies, and in response to China taking a more active interest in the region. That also depends on responses to and outcomes of the COVID-19 pandemic. This book examines why and how these policies and programmes have developed, what impact this has had in Australia and for the people, villages and islands of the sending states. It particularly focuses on Vanuatu, the main source of labour, based on studies in Australia and in Vanuatu. It therefore traces new patterns of migration, with intriguing economic and social consequences, that are restructuring parts of rural and regional Australia, in response to labour demands from agriculture and an evolving regional geopolitics.

Adelaide, SA, Australia — Kirstie Petrou
Sydney, NSW, Australia — John Connell

Acknowledgements

We conducted fieldwork in Vanuatu and in Australia at various times and for various durations through the 2010s (and early but mainly indirectly in 2020). Conversations were conducted in Bislama, Vanuatu's lingua franca, and English. Our Bislama and Tok Pisin helped in gaining trust and the greater ability to converse with workers and their families. All the translations from Bislama were done by Kirstie. One or both of us visited the Pacific island states that were involved in Australia's guestworker schemes (with the exception of Nauru and Solomon Islands) at least once during the decade. COVID-19 constrained fieldwork we would like to have conducted, resulting in occasional over-dependence on secondary sources and on generous colleagues. Limited research possibilities have meant some dependence on key informants, particularly loquacious workers and social media, which occasionally necessitated 'educated guesses' and 'valiant assumptions', and the all-too-frequent use of 'some', 'many' and other unhelpful adjectives. We have sought to move somewhat away from the basic but limited quantitative data to provide fragments of the everyday lives and narratives of the workers and their families, and those who have helped them, that shed some light on their attitudes, objectives, feelings and emotions. Workers and others directly quoted have been given pseudonyms. Unless otherwise indicated currency is demarcated in Australian dollars. A few direct quotes from seasonal workers that appear throughout this book have been used in our previous publications as listed in the reference list. This includes our 2019 paper in the *Journal of Australian Political Economy* and our 2018 paper published in *Rural Society*.

Many people have contributed to the long process of writing this 'quick' book. Above all we thank the many seasonal workers who shared their experiences with us. We are indebted to Richard Curtain and Richard Brown for their generous and helpful comments on an earlier version of this manuscript. Others, located across several countries, kindly responded to our queries and probing emails, notably Michelle Carnegie and Sarah James. And we are particularly grateful to our kith and kin for their patience and support.

Adelaide and Sydney
October 2021

Contents

Abbreviations

ABC	Australian Broadcasting Commission
AE	Approved Employer
AOST	Add On Skills Training
APTC	Australia Pacific Technical College/Australia Pacific Training Coalition
AWU	Australian Workers Union
DAMA	Designated Area Migration Agreement
DESE	Department of Education, Skills and Employment
DFAT	Australian Department of Foreign Affairs and Trade
ESU	Vanuatu Employment Services Unit
FWO	Fair Work Ombudsman
GDP	Gross Domestic Product
ILO	International Labour Organisation
KANI	Kiribati-Australia Nursing Initiative
LMAP	Labour Mobility Assistance Program
MDGs	Millennium Development Goals
MIRAB	Migration, Remittances, Aid and Bureaucracy
NAWPP	Northern Australia Worker Pilot Program
NFF	National Farmers Federation
NSW	New South Wales
NUW	National Union of Workers
OECD	Organisation for Economic Cooperation and Development
PACER	Pacific Agreement on Closer Economic Relations

PALM	Pacific Australia Labour Mobility
PIC	Pacific Island Country
PICSA	Pacific Islands Council of South Australia
PICTA	Pacific Island Countries Trade Agreement
PIF	Pacific Islands Forum
PIFS	Pacific Islands Forum Secretariat
PIR	Pacific Islands Report
PLM	Pacific Labour Mobility
PLMAM	Pacific Labour Mobility Annual Meeting
PLF	Pacific Labour Facility
PLS	Pacific Labour Scheme
PNG	Papua New Guinea
PSWPS	Pacific Seasonal Worker Pilot Scheme
RNZ	Radio New Zealand
RSE	Recognised Seasonal Employer
SAWP	Seasonal Agricultural Worker Program (Canada)
SDGs	Sustainable Development Goals
SIDS	Small Island Developing States
SPC	Pacific Community (formerly South Pacific Commission)
SWP	Seasonal Worker Programme
TASWA	Tonga Australia Seasonal Workers Association
VNSO	Vanuatu National Statistic Office
WHM	Working Holiday Maker

List of Figures

List of Tables

1

Introduction: A New Age of Temporary Migration?

This book seeks to examine the new age of guestworkers in Australasia, in both the source Pacific Island Countries (PICs) and the destinations, especially Australia. In so doing, it takes a broad social, economic, political and cultural perspective. This chapter first examines the background and history of labour migration, and the multiple rationales for and evolution of guestworker schemes. First, we look at the history of temporary labour migration globally, and in the next chapter focus on Australia and the Pacific, and the very similar twenty-first-century scheme in Aotearoa-New Zealand from where the present Australian employment scheme drew its inspiration. Ultimately temporary employment programmes in Australia are merely modern versions of similar programmes that have existed in many countries and continents for over a hundred years. We thus recount the recent rise in Australia of what may seem a new phenomenon: seasonal migrant workers from relatively poor countries in the Pacific, who come for periods of several months at a time to work in agricultural industries where few Australians wish to work. But its novelty is an illusion. Temporary migrant workers from some of the same countries, notably Vanuatu, were crucial for the sugar cane industry in nineteenth-century Queensland. That story is told in the next chapter.

K. Petrou, J. Connell, *Pacific Islands Guestworkers in Australia*,
https://doi.org/10.1007/978-981-19-5387-3_1

These contemporary schemes raise basic questions of why guestworker migration has occurred in many places in different eras, and whether trends in Australia are different or distinctive from elsewhere in the world. This chapter briefly considers the extent to which early schemes have met the needs of those who participated and how that has worked out. Do such schemes benefit people from the sending states, or are they designed to benefit the hosts? Which countries and which people are involved in these schemes and what do they get out of it? We first ask what history suggests of guestworker schemes. We then review contemporary schemes and how they have responded to employment needs in Australia, economic challenges in the PICs, and finally to geopolitical issues. Typically, the schemes have been touted as 'triple win' programmes that benefit the sending and receiving countries and the participants, but it is rarely ever quite so neat. Questions over who the schemes were designed to benefit, and why, and to what extent this has been achieved, are recurrent themes throughout the book.

History

Temporary migrant agricultural 'guestworker' schemes have existed for almost a century in America, but the concept of guestworkers is usually associated with post-war European migration of unskilled workers especially from Turkey into Germany (from where the word—*gastarbeiter*—comes). But these were not at all the first schemes: Prussian farmers had brought seasonal workers from Poland as early as the 1860s (Hahamovitch, 2011; Surak, 2015) and before that Irish workers were similarly employed seasonally in England (Johnson, 1967). However, during the 1950s and 1960s, West Germany signed bilateral recruitment agreements with several southern European and North African countries, mostly for work in the booming construction and industrial sectors rather than in agriculture. Turkey particularly pressed Germany to admit its citizens as guestworkers, and Germany justified the scheme as a form of development aid. It was expected that guestworkers would learn useful skills which after return could help them build their own livelihoods and nations. While employment was centred on the idea that guestworkers would eventually return home, many stayed after the programme ended in 1973, and people of Turkish descent are now a prominent component of the German population (Chin, 2007). At much the same time Vietnamese were being recruited as guestworkers in East Germany, and similar programmes existed in the Netherlands and Belgium. The term *gastarbeiter* survives in several east European states, but more generally means 'expatriate', and in Russia it is used, with a somewhat negative connotation, to denote workers from former Soviet republics, notably Ukraine, Moldova and Tajikistan.

Rather earlier, before the post-war European *gastarbeiter* schemes, in 1942 the United States introduced a Bracero Program, initially created to alleviate wartime labour shortages, but which lasted until 1964, bringing 4.5 million Mexican workers into the United States during its lifespan. A second key labour source was the United States Caribbean island territory of Puerto Rico (Meléndez, 2017). The Bracero Program was followed by an H-2 Visa Program that has brought smaller numbers of both agricultural and non-agricultural guestworkers into the United States, and which remains in place. Rather later, Canada introduced a Seasonal Agricultural Worker Program (SAWP) in 1966 to respond to agricultural labour shortages. Around 18,000 farm workers from Caribbean island states and Mexico arrive in Canada annually to work in fields, orchards and greenhouses. The scheme first linked Canada and Jamaica but subsequently expanded to include Trinidad and Tobago and Barbados in 1968, the smaller Eastern Caribbean states in 1976, and finally Mexico in 1974 (Preibisch & Binford, 2007). The SAWP enabled Canadian farmers to hire Caribbean workers, initially and still mainly in Ontario and to a lesser extent in Quebec and British Columbia, when local workers could not be found. Since many workers came from small Caribbean island states Canada offers the closest comparisons with the much later Pacific schemes, and the SAWP eventually became a partial model for the Aotearoa-New Zealand and Australian schemes (see Chap. 4). The SAWP is bilateral, based on agreements between countries that allow the source countries to have some influence on the shape and structure of the programme.

Many European countries have had versions of one or other of these schemes. The United Kingdom had a Seasonal Agricultural Worker

Scheme (SAWS) between 2005 and 2018 that was then formally replaced by a new version, as Brexit meant that labour shortages became a key concern. The new scheme allowed employers to take on migrants for a six-month period each year and was expected to bring an annual 2500 workers into the country. Several European countries, from Scandinavia to Spain, have similar schemes. In fact, most developed countries have experienced shortages of agricultural workers in the past half century and have chosen to resolve this problem with short-term temporary migrants. Present schemes in Australia have multiple parallels and precedents.

In post-war Europe temporary employment schemes never faded away, as they had been expected to do, but broadened focus in both destinations and origins of the workers, moving eastwards in Europe and then beyond. As the EU expanded, increased migration to England eventually became a partial catalyst for Brexit. Seemingly paradoxically some of the regions that were most dependent on eastern European guestworkers, such as Lincolnshire, voted most strongly in favour of Brexit. Politics underpinned migration. Previously, migrant numbers had been falling and in 1986 the distinguished migration scholar, Stephen Castles, argued that the age of guestworkers in Europe was over: an obituary was required. More permanent migration streams were expected to dominate. The prominent International Labour Organisation (ILO) economist, Roger Böhning (1984, p. 162), argued that 'temporary worker programmes for non-temporary jobs are incompatible with the fundamental tenets of Western democracy, the charter of the United Nations and the Universal Declaration of Human Rights …Temporary worker programmes and restrictions are not only morally offensive but politically less and less tenable in Western plural societies'. It made no difference.

Just two decades later, Castles (2006) was forced to conclude that a 'new age' of temporary migration had begun in Europe. Alongside Germany and the UK, seasonal schemes emerged in Spain, Italy, Norway and elsewhere, particularly focused on migrants from Eastern Europe, with the same objectives of securing cheap agricultural labour in the absence of local workers and with the same flaws as earlier schemes (e.g. Plewa & Miller, 2005). This matched parallel trends in North America, Asia and elsewhere: 'Thai labourers toil in the fields of Israel [and pick berries in Sweden and Finland], Moroccans tend the tomatoes in Spain [alongside others from Colombia and Romania] and Chinese workers pluck apples in Japan' (World Bank 2006, p. 116). In few, if any, places was there any obvious decline, so that by the end of the 2010s there were more guestworker programmes and more guestworkers in existence than ever before (Binford, 2019). In New Zealand and Australia the 'new age' of proliferation was being replicated in the southern hemisphere, as both countries developed temporary migration schemes for agricultural workers from PICs.

The Pacific Island Countries

Around 500 inhabited Pacific islands are spread across 30 million km^2 of the tropical Pacific Ocean, a third of the world's surface, divided among 14 independent countries and 12 dependent territories. With the exception of Papua New Guinea (PNG), which accounts for half the land area and three-quarters of the population of the Pacific islands, the PICs have limited natural resources, small populations, are distant from global markets and somewhat isolated from global trade, all circumstances that have led to limited economic growth, hampered sustainable development, and delayed or discouraged independence (see Chap. 2). In some contexts, Timor-Leste (East Timor)—the small nation north of Australia that fought to gain its independence from Indonesia in 1999 (more than a decade later than most PICs)—is linked in with PICs, and is part of the contemporary seasonal work schemes. We have however, focused this book on the PICs, and especially on Vanuatu, despite passing references to Timor-Leste.

The history of the Pacific is one of mobility. The high islands of Melanesia, notably New Guinea itself, were populated at least 40,000 years ago with people gradually spreading eastwards, but migrants only reached some of the smaller Polynesian and Micronesian islands in the last 2000 years. Most islands have been characterised by some form of mobility and migration in the more recent past. Islanders moved in response to environmental stresses, the quest for resources, social tensions and even adventure, so that the Tongan anthropologist, Epeli Hau'ofa (1993), has spoken of the Pacific being rather less a vast ocean of tiny isolated islands but a region bound together through oceanic networks, loosely integrated through trading and cultural exchange: at least metaphorically a blue continent. Such linkages shaped and reshaped the population geography of the Pacific, and this pre-colonial mobility has some resonance in contemporary movements in the quest for employment, education and resources.

In colonial times, from around 1800 onwards, newer patterns of migration emerged, some of them forced. This era of mobility followed European expansion of commercial and colonial empires—as traders sought everything profitable, from whales and bêche-de-mer (sea cucumbers) to copra, phosphate and sandalwood. In the second half of the century Melanesians went from the nearby New Hebrides (Vanuatu) and Solomon Islands primarily to work as labourers on Australian sugar cane plantations (see Chap. 3). That labour trade ended early in the twentieth century. Migration however never ended, and Pacific islanders, especially in the second half of the twentieth century, increasingly began to seek out 'modern' economic and social opportunities, and began migrating, many permanently, to the metropolitan nations around the Pacific, especially New Zealand, the United States and Australia.

At the beginning of the twenty-first century, after a gap of almost 100 years, a new 'labour trade' began, with islanders moving initially to New Zealand and later to Australia, to work in commercial agriculture. This migration formally began in New Zealand as the Recognised Seasonal Employer (RSE) scheme in 2007 and, after an extended trial period, was followed in Australia in 2012 with the Seasonal Worker Program (SWP). Both schemes were encouraged by the World Bank and partly modelled on the Canadian SAWP. The SWP, in some respects a parallel to labour mobility of the nineteenth century, with its consequences for Australia and for the Pacific islands, is the subject of this book.

While the PICs vary enormously, this new pattern of migration unites them in one dimension, and indicates their shared reliance on Australia and New Zealand, for what in both these metropolitan states is seen as a form of development aid. Indeed in Australia an expanded SWP is regarded as one of the more significant components of a recent 'step-up' in involvement in the region, and in New Zealand the RSE forms part of a parallel 'reset', both responses to the new geopolitical significance of the

Pacific region where China is now thoroughly involved. It is reasonable to query whether the guestworker schemes, recently extending beyond agriculture, are centred on employment support for farmers, constitute development aid or are exercises in practical geopolitics and soft power.

Mobility in the Pacific is nothing new. After almost a century, the revival of labour migration to Australasia has meant that the region has experienced two distinct phases of formal guestworker migration with some structural similarities. The twenty-first-century agricultural labour migration schemes in New Zealand and Australia have presented new temporary migration opportunities, for some PICs for the first time in almost a century, and have begun to contribute to change and development in several states, most notably in Vanuatu, where participant numbers have been greatest. This book therefore sets out to examine this new phase of migration from the PICs, particularly focused on the newest recipient, Australia, and the impacts that migration has had on seasonal workers and their families and on the countries they have come from. It focuses especially on Vanuatu which has supplied the greatest number of workers and where the impact has been greatest. To its supporters the scheme has often been triumphed as a triple win, with benefits going to the sending countries, Australia and the workers themselves. It is this win-win-win that the book seeks to investigate. How did the schemes come about, how do they work and who and where are the beneficiaries? We now provide a background to the countries of the Pacific (Chap. 2), followed by an account of the diverse strands of migration in the region, with a particular focus on blackbirding (Chap. 3). That leads in to the first tentative phase of twentieth-century seasonal migration, and the early involvement of Vanuatu and Tonga (Chaps. 4 and 5). The following five chapters—the core of the book—examine the present scheme, its organisation and expansion and its wide-ranging social and economic impacts in Australia and in the island states (Chaps. 6, 7, 8, 9 and 10), before we seek to make conclusions about whether it really has constituted a triple win, how that might be nuanced, what that means for development and what economic and geopolitical shifts might mean for its future (Chaps. 11 and 12).

Bibliography

Binford, A. (2019). Assessing temporary foreign worker programs through the prism of Canada's Seasonal Agricultural Worker Program: Can they be reformed or should they be eliminated? *Dialectical Anthropology, 43*(4), 347–366.

Böhning, W. (1984). *Studies in international labour migration*. Macmillan.

Castles, S. (2006). Guestworkers in Europe. A resurrection? *International Migration Review, 40*(4), 741–766.

Chin, R. (2007). *The guest worker question in postwar Germany*. Cambridge University Press.

Hahamovitch, C. (2011). *No man's land. Jamaican Guestworkers in America and the Global History of Deportable Labor*. Princeton University Press.

Hau'ofa, E. (1993). Our Sea of Islands. In E. Waddell, V. Naidu, & E. Hau'ofa (Eds.), *A New Oceania: Rediscovering Our Sea of Islands* (pp. 2–16). University of the South Pacific.

Johnson, J. (1967). Harvest migration from nineteenth-century Ireland. *Transactions of the Institute of British Geographers, 41*, 97–112.

Meléndez, E. (2017). *Sponsored migration. The state and Puerto Rican Migration to the United States*. Ohio University Press.

Plewa, P., & Miller, M. (2005). Postwar and post-cold war generations of European Temporary Foreign Worker Policies: Implications from Spain. *Migraciones Internacionales, 3*(2), 58–83.

Preibisch, K., & Binford, L. (2007). Interrogating racialized global labour supply: An exploration of the racial/national replacement of foreign agricultural workers in Canada. *Canadian Review of Sociology and Anthropology, 44*(1), 5–36.

Surak, K. (2015). Guestworker regimes globally: A historical comparison. In L. Talani & S. McMahon (Eds.), *Handbook of the International Political Economy of Migration* (pp. 167–184). Edward Elgar.

2

The Pacific Island Countries

This chapter provides some background to the Pacific Island Countries (PICs) involved in seasonal labour migration to Australia, by first analysing their economies, and suggesting that economic issues have often pointed towards one rationale for migration. That is taken up in detail in the following chapter which focuses on the history of Pacific migration and examines why it has become an attractive option for both countries and households.

Conventionally the Pacific has been subdivided into three great cultural and geographical realms: Melanesia, Micronesia and Polynesia. Although often illusory, these divisions have some basis in the history of colonisation. The much larger Melanesian states (Papua New Guinea [PNG], Solomon Islands, Vanuatu and Fiji) form an arc northeast of Australia, while the more distant Polynesian states (Tonga, Samoa and Tuvalu) are in the central Pacific, with the small Micronesian islands (Kiribati and Nauru) to the north. All these independent states are regarded as small island developing states (SIDS). Despite being classified as a SIDS, PNG is much the largest state in the region both in size (nearly half a million km^2) and in population (with 9 million people). No other Pacific island state has a population greater than 900,000, and most are very much smaller (Table 2.1). Cultural differences are regarded as

K. Petrou, J. Connell, *Pacific Islands Guestworkers in Australia*,
https://doi.org/10.1007/978-981-19-5387-3_2

Table 2.1 Pacific Island Countries and Australia

	Population (estimate, mid-2020)	Land area (sq. km)	Population density (people/sq. km)	Urban population (%)	Youth unemployment	GDP per capita 2019	Population below poverty line (%) 2019	Remittances (USD) 2019, millions
Melanesia								
Fiji	895,000	18,333	56	58	15	13,634	30	287
PNG	8,935,000	462,840	19	13	4	4355	37	3
Solomon Islands	712,100	28,230	20	25	1	2663	13	25
Vanuatu	294,700	12,281	25	26	12	3153	13	35
Micronesia								
Kiribati	118,700	811	146	57	17	2272	..	20
Nauru	11,700	21	557	..	27	11,583	..	..
Polynesia								
Samoa	198,700	2934	68	19	32	6521	20	147
Tonga	99,800	746	133	23	9	6389	23	183
Tuvalu	10,600	26	408	63	21	4281	26	..
Australia	25,800,000	7,632,300	3	86	14	49,854	13	1754

Source: Based on Secretariat of the Pacific Community, *Pacific Island Populations 2020* and CIA, *The World Factbook 2021*. Many numbers in the table, especially on employment and economic issues, are unreliable—based on local measures (or guesses)—but are included simply for crude comparative purposes

distinguishing the small-scale and localised societies of Melanesia, where leadership is contested between generations, and multiple languages exist (more than 800 in PNG), and the larger scale, chiefly societies of Polynesia, where single languages cover a wide area and leaders may be seen as hereditary 'kings'. Fiji sits between these two realms with a single language covering the entire country and leaders also being hereditary and of wider significance. Micronesia (and also Tuvalu) consists of mainly small islands, most being coral atolls, but often with significant hereditary chiefly leadership (Fig. 2.1).

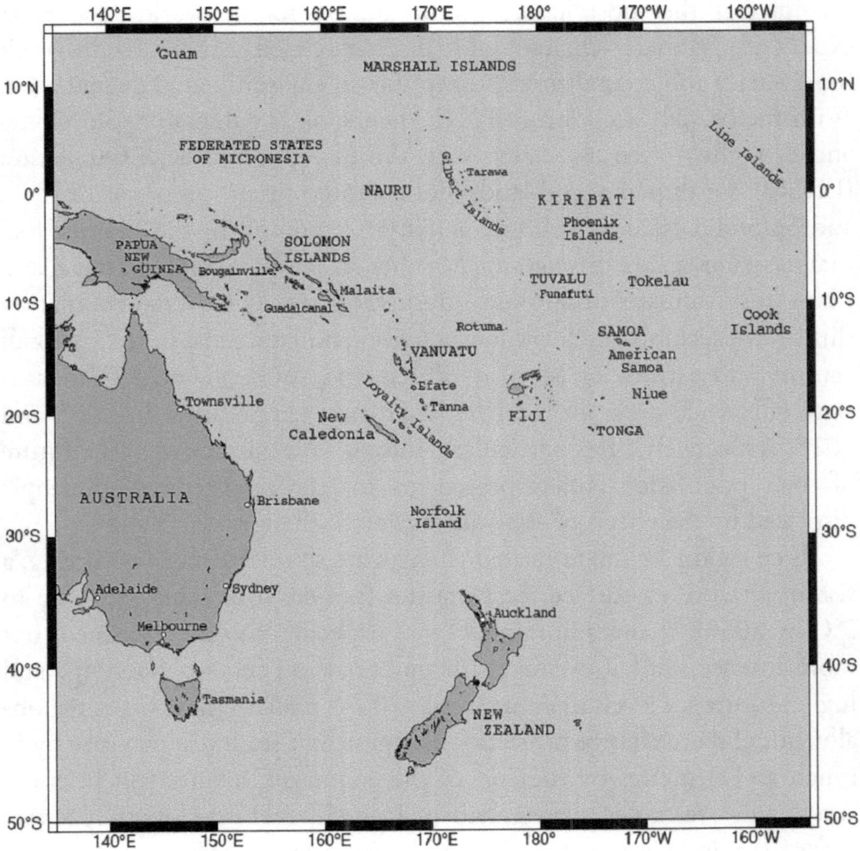

Fig. 2.1 Pacific Island countries. (Source: Map drawn by Xuesong Ding)

The larger island states—PNG, Solomon Islands, Vanuatu, Samoa and Fiji—are mountainous (high island) countries. By contrast Kiribati and Tuvalu are composed of coral atolls, none more than 2 metres above sea level. Nauru is a raised atoll. While Nauru is a single island, most PICs consist of many islands scattered over very large areas. Kiribati, with a population of just 120,000 on 24 populated atolls, crosses both the equator and the date line, and extends over 3.5 million km^2. Outer islands everywhere are disadvantaged. All the countries have extensive coastlines—Kiribati and Tuvalu are exclusively coastlines—and are subject to problems emerging from environmental hazards. The Pacific includes some particularly vulnerable islands and countries; Vanuatu has been classified as the most hazard-prone country in the world. The PICs experience climate change and disaster-related shocks, with their biodiversity and coastal zones susceptible to environmental degradation. With the partial exception of PNG, the region is vulnerable to cyclones and droughts, often associated with the El Niño Southern Oscillation (ENSO). Earthquakes and landslides are intermittent problems in PNG and Solomon Islands. Volcanic activity is common in the 'ring of fire' that describes an arc through the Melanesian states, including Tonga, and PNG and Vanuatu host some unusually active volcanoes. Population shifts have accompanied volcanic activity (and tsunamis) in the past half century; the islands of Manam (PNG) and Ambae (Aoba) (Vanuatu) were effectively depopulated (the latter only temporarily) in 2005 and 2018, respectively, after particularly intense volcanic activity. The future of their populations offers precedents for the resettlement of people displaced by the effects of slow onset climate change.

Three regional climate-related changes are simultaneously occurring: a rise in air and sea-surface temperatures (projected to increase by up to 2°C by 2050), changes in rainfall (with expectations of less frequent but more intense rainfall events, with potential impacts on flooding) and higher future sea levels (Mimura et al., 2007). Each of these varies regionally, but climate change projections suggest that sea levels may rise by as much as 1.4 metres by the end of the century. Cyclones will be more severe and over a wider area, evident in such recent cyclone impacts as that of Tropical Cyclone Harold that devastated parts of Solomon Islands,

Vanuatu, Tonga and Fiji in April 2020. Climate change threatens agricultural and fisheries production and thus livelihoods and incomes.

From around the time of independence, mostly in the 1970s, the coastal zones of PICs have experienced changes in land use, from being primarily agricultural to becoming increasingly urban and tourism orientated. This has resulted in a 'coastal squeeze' with the concentration of towns, people, infrastructure and relative wealth in low-lying coastal areas, prone to multiple natural hazards, notably flooding, whose impacts are exacerbated by climate change. Fijian towns, including Nadi and Ba, have been badly flooded and, in the Rewa river delta, increased flooding has resulted in mangroves suffering dieback, reducing riverine productivity (Kinch et al., 2019). Many of the region's coastlines have been affected by deforestation, sea level rise, tidal and wave action, storms, sand mining and housing, wharf and causeway construction. So far human impacts have had a greater influence on localised erosion and coastline changes than more natural processes, with erosion greatest in urbanised areas.

On the north coast of New Guinea, and on several islands in Fiji, shoreline recession and erosion have displaced villages from established sites (Lipset, 2011; Charan et al., 2017; Neef et al., 2018; McMichael et al., 2019). Migration from the atolls of Bougainville (PNG) and small islands in Solomon Islands has occurred in advance of substantial climate change (Connell, 2016; Albert et al., 2018). The PICs of Kiribati and Tuvalu are particularly at risk. The unprecedented impacts of Cyclone Winston and Cyclone Yasa in Fiji (2016 and 2020), Cyclone Pam and Cyclone Harold in Vanuatu (2015 and 2020), and Cyclone Gita in Tonga (2018) provided acute reminders of how disasters can derail development prospects by devastating agricultural activities, housing and transport infrastructure.

Late Independence and Development

Colonialism was belated in the Pacific. In Melanesia especially it barely existed until towards the end of the nineteenth century. In some cases it was even reluctant; the Gilbert and Ellice Islands, later to become Kiribati and Tuvalu, were an unwelcome British trophy, described as the 'cinderellas of empire' (Macdonald, 1982) and only acquired in the late 1800s. In the twentieth century Britain sought to devolve the New Hebrides (Vanuatu) to Australia as it had done with New Guinea. For most island groups, colonialism lasted little more than a century, and colonial rule sometimes had limited influence beyond administrative capitals, partly because of archipelagic PICs, the distance from the colonial powers and the limited resources apparently available there. Missionaries, traders, planters and colonial administrations enforced pacification, enabled plantations and mines, and provided very basic infrastructure. Samoa became independent in 1962 and Nauru in 1966 (on account of its mineral wealth). It was not until 1975 that PNG followed, and Vanuatu somewhat reluctantly went to independence in 1980 with military force being used to put down secessionist opposition. In Melanesia cultural diversity and island distinctiveness have resulted in several attempts at secessionism, most obviously in Bougainville (PNG), where a 2019 referendum brought a vote of 98 per cent in favour of independence but which has not been ratified.

Average annual population increase in the PICs is around 1.9 per cent but this varies widely. It is greatest in Melanesia, with higher fertility rates, reaching 2.1 per cent in Vanuatu and 2.2 per cent in PNG, and lowest in the larger Polynesian states. High-fertility rates have resulted in large numbers of young people—a 'youth bulge'—who require education, training and job opportunities, while high teenage fertility rates are linked to poor education and unemployment. In most PICs, more than half the population are aged below 24. In Vanuatu especially there is a strong sense of young urban men simply wandering around, *kilim taem* (killing time), because of lack of work (see below). In the last half-century, there have been both accelerated international migration from the Polynesian states and ubiquitous rural-urban migration, with declining populations in national peripheries and growing high-density urban concentrations. Emigration is so substantial in Tonga and Tuvalu that the present population is probably declining, and it is barely growing in Samoa.

One characteristic of the PICs is the persistence of broadly democratic institutions, but stability has not always followed, with persistent votes of no-confidence disrupting continuity in the Melanesian states and coups in Fiji creating international concern and destabilised development. In Solomon Islands instability resulted in external intervention, on two separate occasions, as also in PNG, when a civil war ensued after the dramatic closure of the massive Panguna gold-copper mine on Bougainville island. In the early 2000s parts of the region became known to some as the 'arc of instability' (Ayson, 2007). After its phosphate resources were exhausted at the end of the twentieth century, without any other resources in an environmentally devastated landscape, and with numerous changes of government, Nauru was regarded as the first 'failed state' in the region (Connell, 2006). Political turbulences, and rapid changes in government, have hampered economic growth in several PICs, but not in Polynesia where chiefly, even royal, leadership extends into 'modern' parliaments.

Indicative of the small size of island states and the challenges of development have been the several instances of direct external involvement in the region, whether involving military intervention—that enabled the decolonisation of Vanuatu as a single state, and contributed to peace in Bougainville and Solomon Islands—or direct involvement in the police, judiciaries and public services, where financial support and local human resources have been limited. The vast size and huge distances between islands have meant that the larger states have accommodated provincial governments. Since independence, and to some extent before it, PICs have been involved in regional politics, through institutions like the South Pacific Commission (later the Pacific Community) and the Pacific Islands Forum Secretariat (PIFS). Regionalism has long been heralded as one partial solution to problems of small size. The Pacific Plan Review of the early 2010s, for example, argued that:

> Pacific economies, with small populations and tax bases on which to draw, need to do all they can to leverage voice, influence and competitiveness, and to overcome their inherent geographical and demographic disadvantages. (PIFS, 2013, p. 15)

This tendency to see a wider region of many states as empowering and integrative was also present in the more evidently cultural vision of ties within an interconnected 'sea of islands' (Hau'ofa, 1993). Such visions of integration were frustrated early in 2021 when disputes over leadership led to the Micronesian states withdrawing from the PIFS. Nonetheless, regional empowerment hitherto subverted the discursive hegemony of colonial and subsequent development projects that belittled the region and its peoples. In a similar vein, SIDS have more recently been seen less as island microstates but more as 'big ocean states' within a wider blue economy. The significance of climate change has given PICs particular prominence in global organisations. In coalition with other members of the Alliance of Small Island States, PICs have been leaders in global climate negotiations and, in coalition with other small states, become a recognised grouping in the multilateral sphere.

Economic Development

PICs face various constraints to standard notions of economic development: isolation and remoteness, archipelagic populations, few natural resources, susceptibility to extreme events and other external shocks. Broad socio-economic vulnerability is the outcome of population size and distribution, small open economies, limited human resources, poor and ageing infrastructure, distance from markets and few economies of scale. Island economies are further constrained by small domestic markets, poor connectivity to regional and international markets, and high costs of doing business (insurance and energy costs), and infrastructure, scarce local skills, problems of matching skills and jobs, and limited domestic capital. Entrepreneurialism is stifled by the lack of skills, capital and contacts, but also by scarce opportunities, beyond small-scale services. Public administration is inefficient, and in most PICs, provision of services—especially transport—to remote areas has worsened rather than improved since independence. Relatively generous overseas aid and remittances have not compensated for such cumulative disadvantages. These constraints have resulted in narrow export specialisation and market orientation, so that export prices and foreign exchange earnings can fluctuate considerably and alarmingly. Manufacturing is conspicuous by its absence and without backward linkages to the agricultural or fisheries sectors.

Significantly, progress in economic development has usually been conceptualised in standard Western, 'modern' terms that have had little to do with any local alternative perceptions or with the environmental footprint of development. Vanuatu briefly achieved fame when it topped the Happy Planet Index in 2016 that measured subjective wellbeing. It has also led the region in pioneering an 'Alternative Indicators of Wellbeing for Melanesia' initiative, that centred around three themes: access to customary land and natural resources, traditional knowledge and practice, and community vitality, offering a 'more holistic approach to development'. This resulted in policies around traditional medicine and knowledge, access to customary land and the use of indigenous languages becoming part of the *2016 Vanuatu National Sustainable Development Plan* and documented in *Wellbeing in Vanuatu* (VNSO, 2021a). However, such a focus is exceptional, is largely ignored by international agencies and conflicts with the aspirations of some Melanesians, including Vanuatu's Minister of Finance and Economic Management who simply argued that 'economic growth is the only realistic means to lift the poor out of extreme poverty' (quoted in UNDP, 2013).

In the more or less half-century since independence, domestic development strategies have generally been based on agriculture and fisheries, but with tourism dominant in Vanuatu and Fiji, and mining of considerable value in PNG. Agriculture has produced the bulk of exports for most PICs but in this century has proved more difficult under a regime of increasingly free trade. Fisheries offer existing and future prospects, dependent on international negotiations. Tourism offers options for some, but the tyranny of distance and intervening opportunities have resulted in tourism being trivial in the atoll states, where alternative livelihood opportunities are particularly scarce.

Wide-ranging economic, social, environmental and political challenges impeded PIC achievement of many Millennium Development Goals (MDGs). Some were successful in reducing child mortality and in achieving universal primary education, but few made gains in reducing poverty or accomplishing gender equity. Health and education outcomes are generally superior in Polynesia PICs are currently facing a non-communicable disease (NCD) crisis, with NCDs the leading cause of death in all the states except PNG, where infectious diseases remain significant. In most PICs income inequalities have increased in this century, with poverty emerging in both disadvantaged remote areas and more central towns. Dwarfing all of these challenges in the early 2020s was the need to combat COVID-19 and establish a normal development trajectory.

Agriculture, Forests and Fish

PICs generally combine subsistence and cash economies, although Nauru has only vestiges of agriculture. Most communities, especially in Melanesia, remain rural and depend on agricultural and marine ecosystems for their livelihoods and sustenance. More than two-thirds of Melanesians live on their traditional lands, growing food crops and harvesting forest and marine resources for personal consumption, exchange and market sales. Subsistence agriculture in Melanesia is based on a diminishing range of root crops, dominated by sweet potato and taro, and complemented by such crops as coconuts, breadfruit and bananas. Beyond ubiquitous chickens and pigs, the latter having a valuable social role, livestock are of limited significance.

Food security was once primarily achieved through subsistence farming, fishing and market trade and exchange (Bourke, 2019), but now involves a mix of local and imported food: the ability to purchase food has become crucial. Increased dependence on store-bought food, especially rice, sugar, noodles and flour, has introduced a new kind of food insecurity and exposed PICs both to the uncertainties of global food production, supply chains and costs, and to NCDs. Diabetes and obesity are major problems in the region, especially in the Micronesian and Polynesian PICs (Connell, 2020). In Melanesia food consumption is more likely to be dominated by local foods and greater resilience has been retained, evident in several Vanuatu islands (D'Arcy, 2011; Wilson, 2013; Allen, 2020).

Particularly in Melanesia there is something of a gendered division of labour in agriculture with men primarily engaged in cash crop production, and women in more laborious and regular subsistence food crop production. By contrast, in Micronesia and Polynesia, men are more likely to be involved in both food and cash crop production. The cash economy is critically important as people seek to earn income for children's education, transport and various household needs. However, the narrow economic base and small markets in most PICs have resulted in a limited range of commodities, mainly the 'traditional' plantation crops of cocoa, coffee and copra, and more recently oil palm. Coffee dominates the highlands of PNG just as cocoa dominates the lowlands. Both are grown more sporadically in Vanuatu and Solomon Islands.

Commodity exports are the main sources of national and household revenue, but export incomes have declined over time with deteriorating terms of trade. Declining copra prices have meant the coconut plantations that played a key role in 'opening up' much of the Pacific to capitalism, and once dominated all PICs have almost entirely disappeared. Sugar production, once the mainstay of Fiji, has declined since trade regulations mean it is no longer subsidised, factories are ageing and leases are more difficult to negotiate. With the expiry of preferential sugar price arrangements in 2017, the industry is no longer competitive but it still provides a livelihood for about 15 per cent of Fiji's population. By contrast oil palm development has expanded massively in this century and enabled resettlement in some lowland areas of PNG from more densely populated Sepik and highland areas, becoming the single-most important agricultural export from the country, more valuable than the old plantation staples of cocoa and coffee combined. In the atoll economies of Kiribati and Tuvalu village, copra production remains important, even subsidised in Kiribati, since little else can be grown on poor coralline atoll soils. One outcome is poor nutritional status. Attempts at developing niche commodities, such as ginger, vanilla and chillies, have been limited over time and space. Kava, briefly referred to as 'green gold', boomed at the end of the last century (Murray, 2000) and is again of new export and domestic importance mainly for Fiji, Vanuatu and Tonga. Marijuana has found a clandestine domestic economic niche in several PICs. Other temporary success stories, such as squash and vanilla in Tonga, have floundered on marketing, taste, environmental stresses and intervening opportunities.

Food crops supply flourishing urban and rural markets, but especially in a few fertile coastal areas, such as the Guadalcanal plains (Solomon

Islands) and the Sigatoka Valley (Fiji) with good access to markets, so empowering many female market vendors. In PNG one of the most valuable locally marketed commodities is betel nut (Sharp, 2013; Sharp & Busse, 2019). Vegetables from the cool highlands are marketed in coastal towns, while betel nuts and coconuts travel in the opposite direction. With the exceptions of Kiribati, which has a tiny urban market, and Tuvalu, which has none, urban markets are huge; Port Vila's main market usually has at least 250 vendors six days a week, and spills over into handicraft, flower, fish and cooked food sections.

More than half of PNG is still covered by forest. Much of this is illegally logged and exported especially to China, with PNG being China's single largest supplier of tropical logs. Solomon Islands is the next biggest exporter of logs to China, where they are processed and sold as tropical hardwood products. In both countries, the economic benefits have been widely perceived as highly uneven, and a basis for corruption, and far less than the social and environmental costs, with landowners gaining little and environmental degradation considerable (West, 2016). Outside PNG and Solomon Islands, forestry is of limited significance. In Fiji timber resources are now largely made up of introduced pine and mahogany plantings, but it has had some success with downstream processing and furniture manufacture. Vanuatu once possessed valuable stands of sandalwood that have largely disappeared, but efforts are being made to redevelop them, and have been successful on a few islands such as Aniwa (Wilson, 2013). Otherwise, Pacific timber industries are trivial.

The PICs have exclusive economic zones (EEZs), covering large ocean areas with considerable fishing potential, and more distant possibilities of deep-sea mining. However, Pacific states have no tradition of deep-water fishing, and EEZs are leased to the vessels of distant nations, with low rentals characteristically no more than about 10 per cent of the value of the tuna taken. Consequently, the contribution of fishing to national GDPs has rarely been more than about 10 per cent, but it has created numerous onshore jobs, mainly in PNG (Barclay, 2010). One of the recent economic success stories in the region has been the emergence of the Parties to the Nauru Agreement, a consortium of ten PICs that has been successful in restricting fishing access in their EEZs, thus increasing the possibility of sustainability, and simultaneously bringing considerable national economic benefits. For Kiribati and Tuvalu, without other resources, that has been invaluable; Kiribati now has the largest productive tuna fishery in the region (Webb, 2020a). Kiribati raised nearly 90 per cent of its GDP in 2015 from tuna licence fees but these record receipts have subsequently declined. Coastal fishing for food production and local markets is valuable, especially in the atoll states, but rather less so in Melanesia. In coastal villages, in Paama (Vanuatu) for example, only half the households fished at all, and then usually infrequently (D'Arcy, 2011) whereas on Aniwa three-quarters of households fished, a function of a more impoverished agricultural economy on a rather smaller island (Wilson, 2013). Such patterns are replicated in PNG and Solomon Islands. By contrast, on most islands in Kiribati and Tuvalu, fishing is nearly ubiquitous and crucial to food consumption.

Mines

Mining has a long history in PNG, with alluvial gold mining opening up parts of New Guinea at the end of the nineteenth century. Mining boomed in PNG after the construction of the Panguna mine in Bougainville shortly before independence, which then provided more than half the nation's exports, until a civil war brought its closure. Few mines have been without problems, but the larger PNG mines, notably Ok Tedi and Porgera, have made substantial contributions to the national economy, while oil and gas from the Southern Highlands are two of the country's most valuable exports. Gas alone accounts for a third of the nation's exports; gold accounts for a quarter. The newest large mine, the Ramu nickel mine constructed in 2012, is operated by the Metallurgical Corporation of China. Much of the future of PNG depends on the ability to bring new mines online, as older mines are worked out. Re-opening the Panguna mine is a possibility, partly dependent on whether Bougainville secedes to become an independent state. Mining in Fiji has been as long-established as in PNG but no new mines have been developed in recent years and it remains largely a residual economic activity. The Vatukoula gold mine has operated since the 1930s, with something of a chequered history (Emberson-Bain, 1994). Repeated plans have been made for a copper mine at Namosi but low-grade ore has constantly thwarted development. The Gold Ridge mine in Solomon Islands, 30 kms from Honiara, began operations in 1998 but subsequently closed twice, after the 'tensions' (civil unrest) at the end of the century and later flooding of the site. It was eventually sold to a Chinese company in 2019 with the prospect of exports again being possible by 2022. Vanuatu had a manganese mine until 1978 but subsequent exploration licences have failed to find economic resources. Outside Melanesia there are no prospects for terrestrial mining. Deep-sea mining in the Pacific, especially around the New Guinea islands, has been anticipated for at least a quarter of a century but been delayed by technological, economic, environmental, legal and political complications. The general lack of fossil fuels in the Pacific, despite the wealth of resources in PNG, has increased costs of living and development.

Nauru and its smaller neighbour, Banaba, both had phosphate mines for much of the twentieth century until the resources were largely depleted. Mining was so profitable that at its peak Nauru had the highest per capita income in the world, which enabled the small state to become independent before much larger PICs. After mining ended the Nauru economy collapsed following corruption and mismanagement (Cox, 2009), resulting in economic dependence on small-scale residual mining, fishing rental fees and hosting Australia's offshore refugee processing centre for most of this century. The number of refugees in Nauru had fallen considerably by 2021, so there was some prospect of the centre finally closing (as a second centre on Manus (PNG) had already done). Nauru has never had a significant diaspora population to provide remittances, hence, without realistic alternatives, closure would leave Nauru largely dependent on overseas aid, much of which has traditionally come from Australia.

Tourism

Tourism offers some obvious comparative advantages for most PICs and has proved lucrative for a few PICs. In Fiji and Vanuatu, tourism is the largest economic sector. It is of lesser importance in Samoa and Tonga, where much of that tourism is Visiting Friends and Relatives (VFR). High transport costs, limited infrastructure and insecurity have discouraged tourism development in PNG, while equally high transport costs, remoteness and intervening opportunities have largely excluded Kiribati and Tuvalu. Tourism partly emerged through default, when other sectors floundered, but grew substantially from the 1970s, with cheaper airlines, higher incomes and longer vacations. In recent years Chinese tourists have diversified the regional tourism market, again particularly in Vanuatu and Fiji, and Chinese companies have constructed hotels. However, Australia and New Zealand remain the main tourist sources. Other than in Fiji, tourism is often highly centralised around the main urban area (as in Vanuatu) and its impact localised. Economic benefits are less than in larger states because of disproportionate imports of foods (as local food production is too uncertain and uneven) and other materials, and overseas ownership of tourism infrastructure (hotels, travel agents, airlines, rental cars etc.). That is exacerbated where tourism is foreign-owned and resort-based, but it nonetheless generates more income than commodity exports, and much employment. The rapid recent rise of cruise tourism in the Pacific brought unusual incomes to some relatively remote islands such as 'Mystery Island'/Aneityum (Vanuatu). Tourism is labour intensive, creating many jobs, particularly for women (though few are well-paid, and promotion prospects are poor), provides foreign exchange and has multiplier effects in the service sector. More than any other sector, tourism was devastated by COVID-19. Reviving the industry is a considerable challenge.

Employment, Incomes and Poverty

Creating employment is a regional priority. Weak economic growth has meant that the number of school leavers, and thus potential entrants to the workforce, invariably exceeds the number of jobs created annually. In Kiribati, for example, at the turn of the century, around 2000 school leavers entered the job market each year, but only around 500 found formal sector jobs. Generally, more people are applying for fewer jobs, and as levels of education have risen it has become more difficult to secure even unskilled jobs without a '*pepa*' (qualification certificate), a classic example of 'diploma disease'. In Vanuatu this phenomenon was immortalised in pop culture via the local string band hit song, *Class Six*. As one ni-Vanuatu in Port Vila observed: 'It's hard to find work in Vila. Now you need a certificate of education. You have to finish school in Year12 and have papers. When you finish in primary school it is hard. There is no work you can do' (quoted in Petrou, 2020, p. 165; Kraemer, 2013, p. 100). Extreme competition for jobs, and high levels of youth unemployment have contributed to social tensions in some urban centres and pressures for migration, especially where most school leavers seek formal employment, and disdain informal activities. Women's formal labour force participation rates are much lower than those of men. Public service employment is widely favoured because of its usually superior wage levels and conditions. By contrast agricultural work generates lower, more unstable and fluctuating incomes, and workers lack access to social protection. The possibility of 'fast money' from wages is increasingly preferred to the 'slow money' of agricultural work, although opportunities for urban 'fast money' are becoming more and more scarce. Relatively few people have post-school (tertiary) education and training, which leaves shortages for skilled positions, and potential migrants without the skills to access good jobs elsewhere. Vocational training is a contemporary priority.

Unemployment is mostly an urban issue and has risen in the PICs in recent years, but adequate measures are few. In Kiribati, unemployment was estimated to be 38 per cent in 2015, while the labour force grew by 16 per cent between 2010 and 2015, eclipsing a 3 per cent growth in employment (Webb, 2020a). In Fiji in 2018 the overall youth unemployment rate stood at 20 per cent (and was 30 per cent for females and 10 per cent for males) and had steadily grown during the century (Wainiqolo, 2019). Urban youth unemployment is particularly high, with growing numbers of unemployed and marginalised youths in most urban centres, such as Port Vila, Vanuatu (Mitchell 2004, 2011). In 2005–2006 the unemployment rate for 15–19-year-olds in Honiara, Solomon Islands, was estimated at 75 per cent and for 20–24-year-olds was 49 per cent (Russell 2009). By contrast the ILO in 2009 recorded a youth unemployment rate of 10.6 per cent, with young women less likely to find employment than men. By 2010 youth unemployment had seemingly passed 40 per cent in the PICs of Kiribati, Samoa, Solomon Islands and Tonga, but data are not always officially collected or reported (Table 2.1). Such numbers and percentages are at best indicative, and scarcely comparable between countries, since concepts and methodologies vary enormously, yet they crudely indicate that difficult labour market conditions have persisted and generally worsened in this century.

Urban informal settlements are widely perceived—not always accurately—as the locus of high unemployment rates and other forms of disadvantage. Urban areas more broadly can be seedbeds of discontent which may have contributed to urban riots in Nuku'alofa, Honiara and elsewhere. Typically, unemployed youth in Vila as elsewhere are *kilim taem*, bored, frustrated and without adequate opportunities, so sometimes contributing to crime, including domestic violence, and previously unsanctioned earning opportunities such as prostitution. In the extreme circumstances of COVID-19, with rising unemployment and declining incomes, as in Suva, prostitution has increased significantly (*Fiji Times*, 2 November 2020). But the informal sector often offers little pay and considerable insecurity. While for many youth urban life is frustrating and disappointing, they are not without aspirations for better lives, and that includes the possibility of seasonal migration. Although large numbers of PIC workers taking up employment overseas alleviates some of the pressure on domestic labour markets, where there is a bias towards more educated workers (as can be the case), this may not only constrain domestic growth by creating shortages of skilled workers but can also increase the frustrations of unemployed urban youth (see Chap. 6).

Average PIC incomes are low. In 2019, the national minimum hourly wage was A$1.26 in Fiji, A$0.91 in PNG, A$1.10 in Samoa and A$1.80 in Vanuatu. A fairly standard working week of around 40 hours, even with some overtime, produces little. While that is mediated by access to some services at low cost or free—such as healthcare and education—living costs can be considerable, especially in urban areas. An average Fijian household (of 4.5 people) is estimated to require about F$250 (A$160) per week to satisfy their needs but that would take approximately 108 hours to earn. Moreover, since many people cannot access formal employment, and incomes in the informal sector can be lower than in the formal sector, the outcome is some degree of hardship.

Until quite recently there was a widespread belief that poverty, however defined, did not exist in PICs, since urban and rural 'safety nets' existed, where extended families could and would support those who, for whatever reason, experienced temporary problems. Such safety nets are breaking down and it is no longer possible, if it once was, for urban people simply to return to their home villages and be supported by rural kin. Thus, in 2017 some 32 per cent of working Fijians were assessed as poor, rising to 52 per cent of those in the informal sector (Gangopadhyay, 2018). Despite relatively weak economies, other PICs are unlikely to be faring significantly worse since a higher proportion of the population live in rural areas, where access to some basic needs is easier and it is more feasible to live on the margins of the cash economy.

In most PICs, income inequalities have increased in this century, with poverty emerging in several places, from disadvantaged remote areas to urban settlements. As elsewhere people from poor places are moving to better-off places, seeking access to markets for their produce and health and education services for their families (Allen et al., 2005). In Fiji, where the most finely grained data exist, the major pockets of poverty are in urban informal settlements and in a few rural areas where the sugar industry has declined (Pabon et al., 2012). A similar pattern recurs in PNG, Solomon Islands and Vanuatu with poverty greatest in urban settlements where access to food is particularly costly; in Vanuatu 56 per cent of all expenditure is on food, but urban areas in both Port Vila and Luganville had the highest incidence of food deprivation (VNSO, 2012; UNDP, 2013; Deeming & Gubhaju, 2015). However, not everyone in urban settlements is poor. In Honiara, for example, many settlement residents worked in the informal sector and had average incomes more than double those in full-time formal employment (Maebuta & Maebuta, 2009). At least in the late 2000s, the two Vanuatu provinces most distant from the capital, Port Vila, Torba in the north and Tafea in the south, had much higher levels of poverty than urban centres or nearby provinces, in terms of income but also in access to adequate shelter, sanitation, water, education and health (Deeming & Gubhaju, 2015). In Fiji too, the rise in poverty has been most evident in outer islands and rural areas, whereas in Samoa it is more an urban and peri-urban phenomenon. Poverty in many outer islands is compounded by poorer access to basic services, especially health and education.

Typically, in three Fijian villages within 25 kilometres of the capital, incomes came mainly from selling produce (primarily fish and vegetables) and some local employment (at low wages) but almost every household could not afford living costs that centred on food, school fees, church tithes and village social functions. Some households sold almost all their fish, and consequently had an inadequate diet based on rice and cassava, could not afford school fees for some secondary school children and lived in bush material houses without electricity or water. In two of the villages average households lived in 'extreme poverty' on less than A$1 per capita per day (Zann & Zann, 2008). On Paama, average weekly household incomes were roughly A$22 and A$38 in the two villages of Liro Nesa and Liro, respectively, while weekly household expenditure averaged about A$62 and A$76 in these same locations. Villagers could not meet their income needs locally, and relied on remittances from urban kin to make ends meet (Petrou, 2020). Similar situations are true of many other PIC villages where households may have limited access to land (and cash crops) and to markets. Consequently, in Fiji, Tonga and probably other PICs, migrant remittances have had a positive effect on poverty alleviation and wealth creation, although the impacts on income distribution are mixed, especially during crises (Brown et al., 2014; Webb, 2020b). Migration and remittances play an increasingly valuable role in smaller PICs where domestic development opportunities are limited.

While income poverty is more easily measurable, a non-income component of poverty consists of people's capabilities, vulnerabilities and participation: broadly a poverty of opportunity, alongside inadequate access to services. Some households are poor primarily in the sense that they do not have enough food, clean water or access to adequate housing or basic education (Abbott 2007). In Vanuatu in 2015 just over half (53 per cent) of the population was using at least basic sanitation services. In Honiara as many as one-third of the urban population are below the national poverty line and perhaps half of these are in urban settlements, so that 'many families are mired in poverty [leading to] family breakdown, street children, sleeping out, sexual exploitation, alcohol and drug abuse, delinquency, violence and lawlessness' (Russell 2009, p. 16; Maebuta & Maebuta, 2009). Social disruption has followed poverty and domestic violence has increased, and in parts of Melanesia, notably PNG and Vanuatu, a resurgence of sorcery has also occurred (Rio 2011). However much poverty is also hidden in outer islands and remote areas, emphasised by a lack of 'voice' and limited access to educational and health facilities, and formal employment. Education can be inadequate to equip rural school leavers for wage employment. In Vanuatu, where there are not enough places for all children, most of those in secondary education are from medium- and high-income households, and live in the two urban areas.

Extreme poverty has declined in most PICs, since Fiji, Vanuatu and Samoa have benefited from a stronger service sector, notably tourism (Narain et al., 2020), but remains high in PNG and Solomon Islands. Between 2006 and 2010 there was some reduction in poverty in rural areas of Vanuatu, attributed to both government expenditure and internal remittances (VNSO, 2012; UNDP, 2013; World Bank, 2014). By 2019, the hardship rate was estimated at 15.9 per cent of the population, and rural areas were home to 97 per cent of people living in hardship (VNSO, 2021b). By 2010, Samoa, which a decade earlier had rejected any notion of poverty, now recorded 20 per cent of people below its own poverty line. In Fiji too, over a longer time period from 1975 to 1991, the extent of both poverty and inequality grew steadily, despite Fiji having achieved greater levels of economic growth than other PICs. In Tonga absolute poverty rose between 2001 and 2009. Meanwhile, the number of people living below the basic needs poverty line jumped from 16 per cent of the population in 2001 to 22 per cent in 2009, when the largest numbers in poverty were on the urban fringes of Nuku'alofa, many being migrants from outer islands (Tonga Ministry of Finance 2011). Recent assessments of poverty are at best fragmented but overall poverty rates in the PICs appear to have changed little in the 2010s. They remain high, have been a significant factor in rural-urban migration and are an incentive to international migration.

Trade

PICs experience significant trade deficits, with imports substantially greater than exports, made possible because of aid and remittances. Island economies have tended to be weakened through recent trade liberalisation, multinationalisation of production and capital deregulation, so reducing protection for traditional commodities, as in the sugar industry. That has also constrained industrialisation beyond limited import substitution; thus, a nascent Fijian clothing industry has virtually disappeared; industrial parks, as in Tonga, have effectively closed and the largest manufacturing industry in the region, the Yazaki Samoa plant that made wire harnesses for export to the Australian vehicle industry, and had once employed 300 people, closed after 25 years in 2017. For the past 50 years the rate of growth of exports from small states has been inferior to that of larger states, while open economies mean that even small trade shocks have a large impact, because of narrow export concentrations. PICs characteristically trade with distant metropolitan states rather than with neighbours (with similar exports). Lack of complementarity, and greater competition rather than cooperation between states in a more deregulated context, has even resulted in localised disputes such as the 'kava-biscuit war' between Vanuatu and Fiji (Connell, 2007). Diversification of exports is rarely an easy option, due to narrow resource bases and high transport costs.

More rapid growth and development, derived from free market reform, have been widely touted as justification for reducing trade barriers, but little empirical evidence exists of real advantages from a more open economy, and PICs are particularly vulnerable to fluctuating international trade regimes. Increasingly free trade, centred on notions of comparative advantage, threatens countries that primarily produce agricultural commodities, particularly where diversification from traditional export staples is difficult. Protection policies have had to be abandoned, while regional agreements, notably the Pacific Island Countries Trade Agreement (PICTA) and the Pacific Agreement on Closer Economic Relations (PACER), have sought economic integration to overcome systemic challenges (Morgan, 2018; Kautoke-Holani, 2017). The assumption that

PICs can produce tradeable goods of value elsewhere, and that this will outweigh the costs in terms of sovereign control over economic affairs, has yet to be demonstrated. Geographically disadvantaged SIDS cannot compete on a level playing field. Trade has emphasised asymmetry and uneven development.

Regional trade agreements have been negotiated between the 14 independent PICs (until the 2021 split, all the members of the Pacific Islands Forum) and Australia and New Zealand. The principal agreement is PACER Plus (with added features beyond trade), intended to provide a framework for future trade cooperation, that moves the region towards free trade but is considered by Australia and New Zealand to go beyond that in also constituting 'aid for trade'. The PICs stressed that they wanted migration to be part of the PACER Plus negotiations but Australia sought to focus only on trade. New Zealand and Australia have argued that PACER Plus would boost regional trade, but there were few indications that this would benefit the PICs rather than Australia and New Zealand. The final communique of the 41st Pacific Islands Forum in Vanuatu in 2010 stated that: 'Leaders noted ongoing developments on labour mobility in the region as well as parallel developments on Temporary Movements of Natural Persons-related activities and the labour mobility objectives of Smaller Island States under the auspices of PACER-Plus, PICTA, EPA and other trade negotiations'. At the time, there was little immediate movement towards any consideration of new or expanded forms of labour mobility, but in 2015 that resulted in the expansion of the SWP into additional non-agricultural employment, initially for Northern Australia, and subsequently for the entire country. It was the first time that any part of Australia had been singled out as being in particular need (see Chap. 10). At the 2015 meeting of Forum Trade Ministers, the Ministers noted that the opportunity to hold annual meetings of this Framework (Pacific Labour Mobility Annual Meeting [PLMAM])

> would enable enhanced cooperation including in the following areas: investments in skills development in the FICs [Forum Island Countries] to respond to domestic, regional and international demand for labour; efforts for better collection and analysis of data on intra-regional and international

labour mobility for targeted policy formulation; and enhanced dialogue with the private sector in ensuring that the region's employment needs are effectively addressed.

The inclusion of labour mobility issues in PACER Plus was a key reason for some PICs signing the agreement.

Most PICs (including Kiribati, Nauru, Samoa, Solomon Islands, Tonga, Tuvalu and Vanuatu) have signed up to PACER Plus, after negotiations formally concluded in 2017, and PACER Plus came into force in December 2020. However, the two largest states, PNG and Fiji, did not ratify the agreement, arguing that the trade agreement was lopsided: it reduced Pacific island tariffs on imports (which constitute 10–20 per cent of government revenue) and enabled fewer controls on foreign investment, so disadvantaging their own industries, hence their own needs were absent. PICs further argued that their main interest lay in climate change issues and not trade agreements, and sought more positive policy changes in that context.

Aid

PICs have gained considerable overseas aid from metropolitan states, notably Australia, New Zealand, Japan and the EU, and from international agencies and multilateral organisations, especially UN agencies, the World Bank and the Asian Development Bank. Benefiting from 'small country bias', PICs are the most heavily aid-assisted parts of the world on a per capita basis. In this century China has become a substantial aid donor, alongside a trade, investment and migration presence, perceiving the region as part of a maritime Silk Road, where the Pacific enables access to natural resources, in a potential sharing of prosperity (e.g. Szadziewski, 2020) and in competition with Taiwan for political recognition. By 2019 China had become the single largest lender to Tonga, Samoa and Vanuatu, and in the same year both Kiribati and Solomon Islands agreed to recognise China rather than Taiwan.

Aid exhibits urban bias and has spawned an urban bureaucracy arguably at some variance with local needs and national development strategies. Increasing demands from donors for consultation, accountability and engagement, alongside complicated adherence requirements, have created an 'inverse sovereignty effect' where PICs are losing control over their development strategies, policies and programmes, as these became politicised (Overton et al., 2018). That was complicated by the growing significance of aid from 'non-traditional' donors, notably China, that may be linked to natural resource use and infrastructure provision, rather than local priorities or poverty alleviation. China's aid has often been in the form of substantial loans, raising concerns over the ability of smaller states, such as Vanuatu, to repay these loans. That has raised fears of 'debt-trap diplomacy' where predatory lending practices may result in PICs having unsustainable debt burdens and to swapping 'debt forgiveness' for geopolitical interventions, ranging from port construction to natural resource concessions, yet such concerns may be unwarranted (Rajah et al., 2019). Despite commitments to harmonisation, the proliferation of aid donors, including many NGOs, and projects (especially related to climate change), has sapped the attention, energy and resources of recipient governments, and reduced the capacity for aid management, policy development and practical implementation, which in turn has resulted in donors focusing on 'good governance' (Dean et al., 2017). It has also contributed to the 'Pacific paradox' of high levels of aid but low economic growth.

Aid delivery has failed to effectively restructure Pacific economies in favour of either a more sustainable and equitable development, or one where economic growth has been achieved. Indeed aid has been rather less likely to directly target economic issues following a transition from its role in economic and social development to capacity building to prevent state fragility and failure, and enhancement of regional security. It is in this more obviously geopolitical context that both Australia and New Zealand have sought to 'step up' their engagement in the region and to argue that migration opportunities represent a valuable form of development aid.

Overcoming Tyrannies of Distance?

Generating modern economic activity, let alone economic growth in the PICs, has proved difficult for a range of reasons stemming from limited human and physical resources, inadequate infrastructure, high energy costs and low wages but, above all, location and the lack of economies of scale. Weak national economies are tied to global political economies. With most Pacific island populations still growing quickly, unemployment and poverty have become more evident markers of relative contemporary economic status, especially as agricultural stagnation and new aspirations foster rural-urban migration in search of increasingly elusive 'fast money' via wage employment. A growing sensitivity to environmental influences, as cyclones have gathered strength and the coastal squeeze reduced resilience, is compounded by most PICs being composed of many islands scattered over very large areas. As outlined above, broader socio-economic vulnerability is the outcome of population size and distribution, small open economies, remoteness, poor connectivity and limited resources. Conventional development has always been elusive.

In this century islanders have become increasingly familiar with uneven development and relative deprivation, as inequalities are more visible, but, in every past era, islanders travelled in search of new opportunities, sometimes pushed by environmental influences and population pressures that probably accounted for the settlement of most of Oceania. In early colonial times young men signed up on whaling ships or were enticed to work on distant mines and plantations. Excitement played a part but the belief that some form of development might be secured by migration was never far away. In more recent times urban life lured many to the seemingly superior prospects in towns and cities in the Pacific, and then to the great metropolises that bordered the region. Polynesians led the way followed by Micronesians and eventually Melanesians. In the next chapter we trace this history of mobility, and especially of labour migration, both voluntary and forced. In the past, migration was both 'required' for the sake of colonial development, as mines and plantation enterprises needed workers, as did, much later, urban bureaucracies and commerce. In the last century, island states trained workers for overseas employment. Such patterns of mobility were always linked to opportunities at home, and usually to the possibility of successful return, assisted by new wealth or routine remittances. Islanders increasingly exercised their own agency in search of development opportunities, while evolving aid delivery and international relations played their part in fostering mobility, whether for education and training or, in this century, the new guestworker schemes. National development became increasingly deterritorialised.

Bibliography

Abbott, D. (2007). Poverty and pro-poor policies for Pacific Island countries. *Asia-Pacific Population Journal, 22*(3), 59–74.

Albert, S., Bronen, R., Tooler, N., Leon, J., et al. (2018). Heading for the hills: Climate-driven community relocations in the Solomon Islands and Alaska provide insight for a 1.5°C future. *Regional Environmental Change, 18*(8), 2261–2272.

Allen, B., Bourke, M., & Gibson, J. (2005). Poor rural places in Papua New Guinea. *Asia Pacific Viewpoint, 46*(2), 201–217.

Allen, M. (2020). Framing food security in the Pacific Islands: Resilience in Malo, Vanuatu. In J. Connell & K. Lowitt (Eds.), *Food security in Small Island states* (pp. 91–110). Springer.

Ayson, R. (2007). The 'arc of instability' and Australia's strategic policy. *Australian Journal of International Affairs, 61*(2), 215–231.

Barclay, K. (2010). Impacts of tuna industries on coastal communities in Pacific Island countries. *Marine Policy, 34*, 406–413.

Bourke, R. (2019). Subsistence food production in Melanesia. In E. Hirsch & W. Rollason (Eds.), *The Melanesian World* (pp. 143–163). Routledge.

Brown, R., Connell, J., & Jimenez-Soto, E. (2014). Migrants' remittances, poverty and social protection in the South Pacific: Fiji and Tonga. *Population, Space and Place, 20*(5), 434–454.

Charan, D., Kur, M., & Singh, P. (2017). Customary land and climate change induces relocation—A case study of Vunidogoloa Village, Vanua Levu, Fiji. In W. Filho (Ed.), *Climate change adaptation in Pacific countries* (pp. 19–34). Springer.

Connell, J. (2006). Nauru: The first Pacific failed state? *The Round Table, 95*(383), 47–63.

Connell, J. (2007). Towards free trade in the Pacific? The genesis of the 'Kava-Biscuit War' between Fiji and Vanuatu. *Geographical Research, 45*, 1–12.

Connell, J. (2016). Last days in the Carteret Islands? Climate change, livelihoods and migration on coral atolls. *Asia Pacific Viewpoint, 57*(1), 3–15.

Connell, J. (2020). Lost roots? Fading food security in Micronesia. In J. Connell & K. Lowitt (Eds.), *Food security in Small Island states* (pp. 57–76). Springer.

Cox, J. (2009). The money pit: An analysis of Nauru's phosphate mining policy. *Pacific Economic Bulletin, 24*(1), 174–186.

D'Arcy, M. (2011). *Away from subsistence: Food security and resource use on Paama, Vanuatu*. Honours thesis, University of Sydney School of Geosciences.

Dean, A., Green, D., & Nunn, P. (2017). Too much sail for a small craft? Donor requirements, scale and capacity discourses in Kiribati. In E. Stratford (Ed.), *Island Geographies: Essays and conclusions* (pp. 54–77). Routledge.

Deeming, C., & Gubhaju, B. (2015). The mis-measurement of extreme global poverty: A case study in the Pacific Islands. *Journal of Sociology, 51*(3), 689–706.

Emberson-Bain, A. (1994). *Labour and Gold in Fiji*. Cambridge University Press.

Gangopadhyay, P. (2018, February 27). What Australia can learn from Fiji in reducing the working poor. *The Conversation*.

Hau'ofa, E. (1993). Our Sea of Islands. In E. Waddell, V. Naidu, & E. Hau'ofa (Eds.), *A New Oceania: Rediscovering Our Sea of Islands* (pp. 2–16). University of the South Pacific.

Holani, A. (2017). *Enhancing the development impact of labour mobility in Pacific Island countries: The role of the seasonal worker program and the PACER plus*. PhD thesis, University of Adelaide.

Kinch, J., Vitukawalu, B., Nalasi, U., Waqainabete, P., & Bermudes, M. (2019). Socioeconomic aspects of Oyster harvesting in the Rewa river delta area, Fiji. *SPC Fisheries Newsletter, 159*, 45–54.

Kraemer, D. (2013). *Planting roots, making place: An ethnography of young men in port Vila, Vanuatu*. PhD thesis, London School of Economics.

Lipset, D. (2011). The tides: Masculinity and climate change in coastal Papua New Guinea. *Journal of the Royal Anthropological Institute, 17*(1), 20–43.

Macdonald, B. (1982). *Cinderellas of the empire*. ANU Press.

Maebuta, H., & Maebuta, J. (2009). Generating livelihoods: A study of urban squatter settlements in Solomon Islands. *Pacific Economic Bulletin, 24*(3), 118–131.

McMichael, C., Katonivualiku, M., & Powell, T. (2019). Planned relocation and everyday agency in low-lying coastal villages in Fiji. *Geographical Journal, 185*(3), 325–337.

Mimura, N., Nurse, L., McLean, R., Agard, J. et al. (2007). Small islands. In M. Parry, O. Canziani, J. Palutikof, P. van der Linden & C. Hanson (Eds.), *Climate change 2007: Impacts, adaptation and vulnerability* (pp. 687–716). Cambridge: Cambridge University Press.

Mitchell, J. (2004). 'Killing time' in a postcolonial town: Young people and settlements in Port Vila, Vanuatu. In V. Lockwood (Ed.), *Globalization and culture change in the Pacific Islands* (pp. 358–376). Upper Saddle River: Pearson.

Mitchell, J. (2011). 'Operation restore public hope': Youth and the magic of modernity in Vanuatu. *Oceania, 81*, 36–50.

Morgan, W. (2018). Much lost, little gained? Contemporary Trade Agreements in the Pacific Islands. *Journal of Pacific History, 53*(3), 268–286.

Murray, W. (2000). Neoliberal Globalisation, 'exotic' agro-exports and local change in the Pacific islands: A study of the Fijian kava sector. *Singapore Journal of Tropical Geography, 21*, 355–373.

Narain, D., Rogoiruwai, K., & Singh, T. (2020). *Trade and poverty: Some considerations for Pacific Island countries.* Griffith Asia Institute Research Working Paper No. 15.

Neef, A., Bege, L., Boruff, B., Pauli, N., Weber, E., & Varea, R. (2018). Climate adaptation strategies in Fiji: The role of social norms. *World Development, 107*, 125–137.

Overton, J., Murray, W., Prinsen, G., Ulu, T., & Wrighton, N. (2018). *Aid, ownership and development : The inverse sovereignty effect in the Pacific Islands.* Routledge.

Pabon, L., Umapathi, N., & Waqavonovono, E. (2012). How geographically concentrated is poverty in Fiji? *Asia Pacific Viewpoint, 53*(2), 205–217.

Petrou, K. (2020). *If everyone returned, the Island Would Sink: Urbanisation and migration in Vanuatu.* Berghahn.

PIFS. (2013). *Framework for Pacific regionalism.* Suva: Pacific Islands Forum Secretariat.

Rajah, R., Dayant, A., & Pryke, J. (2019). *Ocean of debt? Belt and road and debt diplomacy in the Pacific.* Lowy Institute.

Rio, K. (2011). Policing the holy nation: The state and righteous violence in Vanuatu. *Oceania, 81*, 51–71.

Russell, L. (2009). *Stayin' alive. A report on urban livelihoods in Honiara, Solomon islands.* Sydney: APHEDA.

Sharp, T. (2013). Baias, Bisnis, and Betel nut: The place of traders in the making of a Melanesian market. In F. McCormack & K. Barclay (Eds.), *Engaging with capitalism: Cases from Oceania* (pp. 227–256). Emerald.

Sharp, T., & Busse, M. (2019). Cash crops and markets. In E. Hirsch & W. Rollason (Eds.), *The Melanesian World* (pp. 194–222). Routledge.

Szadziewski, H. (2020). Converging anticipatory geographies in Oceania: The Belt and Road Initiative and Look North in Fiji. *Political Geography, 77*, 102–119.

Tonga Ministry of Finance. (2011). *Second national millennium development goals report*. Nuku'alofa: Ministry of Finance.

UNDP. (2013). *Vanuatu Hardship and Poverty Report*. Vanuatu National Statistics Office and UNDP Pacific Centre.

Vanuatu National Statistics Office. (2012). *Vanuatu Household Income and Expenditure Survey 2010*. Vanuatu Government.

Vanuatu National Statistics Office. (2021a). *Wellbeing in Vanuatu. 2019–20 NSDP Baseline Survey*. Vanuatu Government.

Vanuatu National Statistics Office. (2021b). *Hardship in Vanuatu: 2019–20 NSDP Baseline Survey*. Vanuatu Government.

Wainiqolo, I. (2019, December). Labour productivity and youth unemployment in Fiji. *Pacific Economic Monitor*, 7–8.

Webb, J. (2020a). Kiribati economic survey: Oceans of opportunity. *Asia Pacific Policy Studies, 7*, 5–26.

Webb, J. (2020b, December). Samoa and Tonga: Opportunities in the storm?, *Pacific Economic Monitor*, 19–22.

West, P. (2016). *Dispossession and the environment: Rhetoric and inequality in Papua New Guinea*. Columbia University Press.

Wilson, K. (2013). Wan Laki Aelan? Diverse development strategies on Aniwa, Vanuatu. *Asia Pacific Viewpoint, 54*(2), 246–263.

World Bank. (2014). *Vanuatu Socio-Economic Atlas*. World Bank.

Zann, S., & Zann, L. (2008). Poverty in paradise? Issues in poverty and development in Fijian fishing villages. *SPC Women in Fisheries Information Bulletin, 18*, 36–41.

3

Two Centuries of Pacific Migration

This chapter traces the long history of labour migration in the Pacific region, and focuses especially on nineteenth-century 'blackbirding', the most significant phase of labour recruitment in the colonial era, and one with distinct repercussions for Australia. Remarkably these early years suggest intriguing similarities and also differences between the two phases of labour migration. Here we focus on Vanuatu, then the New Hebrides, which along with Solomon Islands was the main nineteenth-century source of labour. However, the Australian labour trade was simply one substantial part of an evolving diversity of Pacific migration, long before the present phase of temporary labour migration. We therefore also examine the more varied migration streams that have followed in the twentieth and early twenty-first centuries.

The earliest European explorers came to the Pacific looking for valuable resources that could generate profits on the other side of the world. Solomon Islands were optimistically named by sixteenth-century Spanish explorers in the vain hope that they would yield vast quantities of gold and become a new King Solomon's mine. Mid-nineteenth-century sandalwood traders exploited the forests of the southern New Hebrides (Shineberg, 1967). But in most islands only disappointment over

K. Petrou, J. Connell, *Pacific Islands Guestworkers in Australia*,
https://doi.org/10.1007/978-981-19-5387-3_3

resources awaited and, beyond Melanesia, many islands were too small for traders to even establish viable coconut plantations. The seas offered a little more and whaling was prevalent in the first half of the nineteenth century, centred first on Hawai'i and later around Fiji and the Gilbert Islands (now Kiribati). Bêche-de-mer too was widely traded until it was exploited nearly to extinction. By the end of the nineteenth century interest had returned to the land: mining was being established at several sites in Papua and New Guinea, New Caledonia and Nauru. Such trends calibrated the evolution of migration in the Pacific as it took on ever-evolving forms and directions.

Complex Labour Flows

If the Pacific provided few valuable resources to potential colonists, especially once whaling had ended, sandalwood and bêche-de-mer stocks were depleted, copra prices had fallen and cotton plantations in Fiji, Samoa and Tahiti failed, what it did provide was human labour. Labour was recruited for all these activities within the Pacific but especially for various plantations. By the middle of the nineteenth century, labour shortages in Australia and elsewhere had begun to transform once remote and often-isolated Pacific islands into new labour reserves. The larger populations of the more accessible Melanesian islands of the south-west Pacific especially acted as a subsequent and more significant labour pool for extensive German, French, British and, later, Australian agricultural and mining interests in the region.

Australia was far from alone in seeking island labour. In the 1860s, Polynesians and Micronesians were forcibly taken from such contemporary Pacific states as Kiribati, Tuvalu, Tokelau, Rapa Nui (Easter Island) and French Polynesia, to work in Chilean and Peruvian plantations and mines. Kidnapping achieved notoriety, some 'recruits' never arrived and few returned home hence several islands lost more than half their population in as few as two years (McCall, 1976; Maude, 1981). Even so, as late as the 1890s, Gilbertese (i-Kiribati) were recruited to work on coffee plantations in Mexico and Guatemala, where many died (Munro, 1992). On larger Pacific islands, like those of Hawai'i, Fiji and Samoa, plantations had been established by the mid-nineteenth century, and many islanders had gone to work in sugar and cotton production, especially from the Polynesian islands and the Gilbert Islands. Belatedly, the island of Tanna (New Hebrides) proved a valuable and popular labour source, resulting in competition between Fiji and Queensland for New Hebridean labour. Gradually the labour frontier was moving westwards and the larger islands of what are now Papua New Guinea (PNG), Solomon Islands and Vanuatu became a significant part of the plantation economy. Conspicuously absent from much of the labour trade was Fiji. Fijians were reluctant to work for outsiders so that even when the sugar industry became established there, they were uninterested in employment. Much as in the other sugar plantation colonies of Mauritius, Trinidad and Guyana, between 1879 and 1917 as many as 60,000 indentured Indian labourers (*girmitiya*) were brought in to work in the Fijian cane fields (Lal, 2004). Their descendants eventually made up half the population of Fiji.

Elsewhere the outcome of colonial expansion in the Pacific was multiple flows of workers traversing the Pacific and beyond in various directions in the second half of the nineteenth century. Few islands were spared or remained aloof from the incursions of the labour trade, involving complex flows for different reasons, especially in the south-west Pacific, graphically depicted in Edward Docker's early 1970 synthesis of just a part of the labour trade (Fig. 3.1). Gilbertese (i-Kiribati) workers were on British plantations in Samoa by the 1870s. Micronesians had ventured away on whaling ships. Polynesians had been taken to Peru. Solomon Islands provided more than 80 per cent of Fiji's indentured Melanesians (Corris, 1973). Melanesians, from as far away as Bougainville (PNG), worked in Samoa, some eventually remaining there. Those who returned to their homes brought back new goods that proved useful, such as plants (including a useful banana species in south Bougainville still known as *samoa*) and even, in parts of the New Hebrides and Bougainville, the boomerang as a ceremonial object (Thurnwald, 1932). Many left the island of Malaita, always a dominant source of migrants in Solomon Islands, while Tanna dominated the New Hebrides labour trade. Islanders moved for many other reasons. Pacification and missionisation meant some households and villages moved from island interiors to more

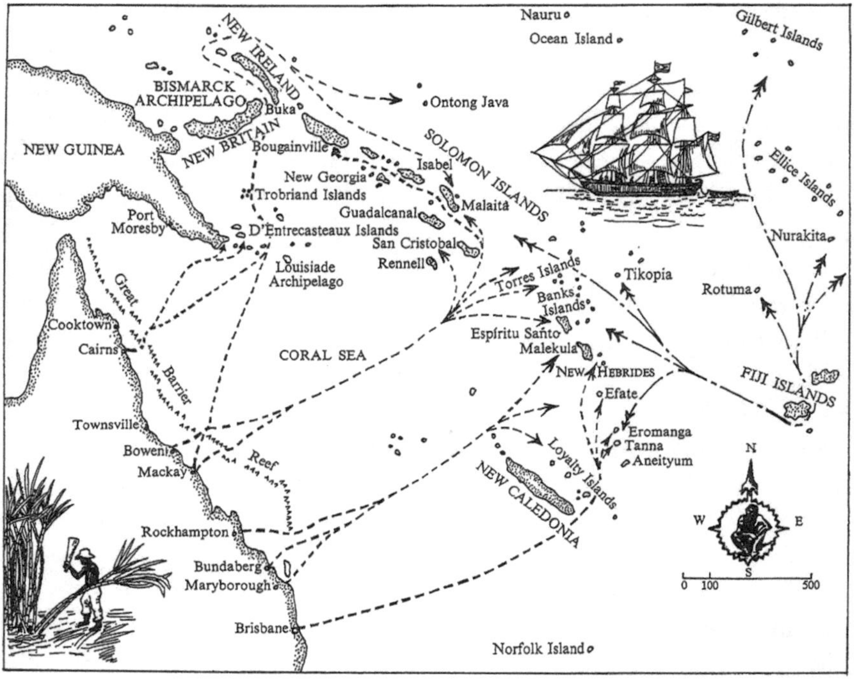

Fig. 3.1 The main recruiting routes for nineteenth century labour migration. (Source: Docker, 1970, p. 3)

accessible coastal locations and in some parts of Melanesia especially, notably on such islands as Erromango (Vanuatu), populations also fell substantially. By the end of the nineteenth century most Pacific islands had experienced labour migration, with some Pacific Island Countries (PICs) being both sources and recipients of labour, although it was those closest to Australia that were experiencing the greatest repercussions. Over four decades, 'blackbirding', the colloquial term for the early labour trade in the South Pacific, took Solomon Islanders, New Hebrideans and Loyalty Islanders to the cane fields of Queensland. More than at any time in the past, and more than for nearly another century, the late nineteenth century was a time of remarkable long-distance mobility, in many respects the forerunner of the contemporary migration of agricultural workers to Australia.

Blackbirding: The Early Labour Trade

In the nineteenth century 'blackbirding' took Pacific islanders from various colonies to work in many places, but above all in the canefields of Australia. It was the first substantial contact of Melanesians with Australia, and rather more than 60,000 people were involved over a period of 40 years. It might have begun even earlier; at the turn of the century, Lord Sydney had proposed importing 200 island women from New Caledonia, Tonga and elsewhere, as wives for the first convicts settled on Botany Bay, an idea quashed by the first Governor. The legacy of blackbirding lives on in the South Sea Islanders, who remain in Australia today, the descendants of those blackbirds who chose to stay on.

Blackbirding was initiated as illegal and uncontrolled recruitment, long after such practices had ended in Europe, the Americas and Africa. It accompanied the establishment of a new industrial plantation frontier in Australia and the Pacific. In the mid-nineteenth century Melanesians—the 'blackbirds'—went from Solomon Islands and the New Hebrides to work in the canefields of Queensland. At its worst, it was barely short of slavery, but over time most people moved relatively freely and had some idea of what would be involved.

This was not the first time that indentured labour had been brought into Australia. Convict labour, euphemistically 'assigned service', had been valuable early on in the colonies but the end of convict transportation to New South Wales in the 1830s also ended that source of cheap labour, creating a perceived labour crisis (Cullen, 2012). Although the Indigenous Aboriginal population were also used as unpaid labour in many sectors, including the pastoral industry, they could not easily be coerced into regular agricultural employment, and were liable to abscond, whereas migrants could be more easily controlled and sequestered. A few 'coolies' were brought to New South Wales from India in the 1840s where they worked in agriculture in exploitative conditions. Later in the decade, over 3600 Chinese coolies were brought in, mainly to Victoria. Their social and economic circumstances were so dire that the Victorian government formed a select committee to investigate, albeit to no effect. The Chinese indentured labourers were expected to be docile, live on low wages (much less than free labourers) consume cheap food and remain for five years (Loy-Wilson, 2017). By 1855, experiments with cheap Asian labour were over, although soon afterwards voluntary Chinese migration to Australia boomed in the wake of the gold rushes.

The first major Pacific blackbirding event in Australia occurred in 1847 when a ship chartered by Benjamin Boyd, a Scottish settler and former stockbroker on the south coast of the colony of New South Wales, brought almost 200 Melanesian workers (mainly from the New Hebrides) to work on his pastoral leases (and perhaps be involved in his plans to create a thriving port, build an elegant hotel and construct a huge tower for spotting whales). It was an ill-fated venture. Workers soon absconded, probably due to harsh conditions, some were later to be found starving and destitute in Sydney, while a group marched on Sydney to draw attention to their exploitation (Banivanua-Mar, 2019). Very much later, in this century, an increasing clamour of calls were being made that Boyd's name be removed from the Ben Boyd National Park and other New South Wales sites (McKinnon, 2019) and steps were in place to do so late in 2021 (Barlass, 2021). The experiment of employing Melanesian labour was briefly abandoned, until Robert Towns, who had previously been a sandalwood and whaling merchant, and was already regarded as 'Australia's most notorious coolie trader' (Loy-Wilson, 2017), restarted the practice in Queensland in the early 1860s, bringing Melanesians to work on his cotton plantation near Brisbane.

Around that time, the establishment of cotton and sugar plantations in Queensland began the blackbirding era. As the American Civil War loomed, a new source of cotton was required, so Queensland was promoted as a cotton-growing colony and the 'future cotton field of England'. Plantation owners migrated, many from Louisiana, to establish the industry in Queensland, and profits from American slavery were reinvested in Queensland's new tropical plantation economy. Most however failed by 1866 due to flooding and crop disease, and owners reinvested in sugar (Gleeson, 2020a). In due course, much like Boyd, Towns too was accused of kidnapping (and, in this century, South Sea Islanders have asked for his statue to be removed from the centre of Townsville).

From 1863 to 1906 the labour trade brought Melanesians from many islands, initially to work on Queensland cotton plantations and later on the sugar cane plantations. Others were employed as shepherds and bullock drivers. They played a large part in developing the North Queensland economy and establishing the Australian sugar industry. Most came from the Loyalty Islands (New Caledonia), the New Hebrides (especially Tanna) and Solomon Islands. Recruitment was easier where there had been persistent droughts, as in the southern Gilbert Islands, or where a combination of an impoverished ecology and high population densities existed, as in the Lau lagoon area of Malaita (Solomon Islands) (Graves, 1986). Youthful adventure and escape from social tensions attracted others. A moving labour frontier began in the New Hebrides and the Loyalty Islands in the 1860s, stretched northwards to Solomon Islands in the 1870s and had reached islands off eastern New Guinea by the 1880s (Fig. 3.1). In the New Hebrides, the major source of labour, it contributed to population declines in the southern islands.

Initially labourers were coerced, kidnapped or promised great wealth (and defrauded), until the growth of widespread opposition in various Western countries, often following missionary protests. The missions themselves were only then becoming established in the islands, and both trading and missions preceded an administrative presence. Fearing the loss of innocence of their potential congregations, missionaries were usually sympathetic to pressure from islanders who had been exploited. One such missionary, H. A. Robertson, remarked on the plight of an Erromango man, returning from overseas in the late 1870s, whose wife had in the meantime been acquired by another man. He regarded the labour trade as 'little else than a slave trade' and 'nothing but a curse to our island' (1902, pp. 297, 419–420). Another, Maurice Frater, writing in 1922, stated: 'The horrors and misdemeanours of the sandalwood traders who invaded Erromanga [*sic*] with drink and firearms form a black chapter in the history of Britain's dealings with the islands [alongside] the misconduct of unscrupulous traders and labour agents' (1922, p. 137). None were more scathing about the inequities of the labour trade than John Paton, who spent more than 30 years in and around Tanna and produced a substantial document 'The Kanaka Labour Traffic; Or, Slavery in the South Seas' describing it as 'thinly veiled slavery' and railing at violent recruiting, exploitative wages and resultant depopulation (Paton, 1891, pp. 454–467). His son Frank, a missionary in his own right, took up the same themes, blaming the labour trade, the accompanying trade in firearms and alcohol, and the spread of European diseases through returned recruits as the main causes of depopulation: 'the labour traffic sucks out the young lifeblood of the community, breaks up family ties and everywhere leaves desolation and bad feeling behind it ... [and] the rapid dying out of our people in the New Hebrides' (Paton, 1903, pp. 47–48).

Early on, intimidation and force meant that it was no more than a 'colonial fantasy' (Breman, 1990) to maintain that employment, let alone contracts, were entered into voluntarily. In 1867 news of kidnappings and the sale of recently arrived labourers for as little as £2 each prompted concern in Australia and the United Kingdom over what appeared a new slave trade. Reports of violence, kidnap and murder in the 'recruitment' process resulted in occasional inquiries but no action was ever taken. After 1870, under pressure from the British government, Queensland appointed government agents to regulate the trade, by ensuring that people engaged of their own free will, that appropriate food and clothing were issued to them and that they were returned to the islands they had come from. However, most of the agents were a 'poor lot' on low wages and had difficulty in retaining independence; many were nominated by traders with a financial interest in the recruiting, were easily bought off and exercised minimal scrutiny (Corris, 1973). Nonetheless, in 1872, after a decade of the labour trade, forcible blackbirding had effectively been replaced by more legal recruitment.

Diverse interpretations of blackbirding attach varied significance to the degree of kidnapping, coercion or voluntary participation (Newbury, 1980; Howe, 1984). The extent to which people were kidnapped, tricked into indenture or volunteered varied greatly over time and space. Exploitation was often hidden behind lurid and exciting tales of 'cannibal cargoes' while the labour trade 'fluctuated wildly between varying levels of coercion on the part of recruiters and varying levels of willingness on the part of first-time and returning recruits' (Banivanua-Mar, 2007, p. 28). The notion of indigenous islanders rudely snatched by slavers from an idyllic habitat gave way to a complex mix of pressures and influences. Not all labour migrants had to be coerced; many went willingly, lured by the attractions of wealth, new goods and adventure, if not exactly in conditions of their own choosing and to where their destinies were uncertain (Adams, 1990). The fewer women migrants were more likely to be fleeing difficult domestic circumstances, since men usually opposed their recruitment. Paamese did not allow women near the beaches when blackbirders were known to be in the area. Where women had somewhat more independence, as at Ambae (New Hebrides), they were more likely to be voluntarily involved (Shineberg, 1999). Based on a single ship's passengers, the heroic assumption of the Russian ethnologist, Nikolai Miklouho-Maclay, an opponent of blackbirding, was that in the 1880s a quarter of recruits were returnees, a further quarter were young people 'anxious to travel, or wishing to get arms', 15 per cent were 'obtained by means of different artifices and lying promises' and 15 per cent 'sold by relatives or chiefs', 10 per cent were escaping victorious enemies, 5 per cent escaping famines and drought and 5 per cent taken by force (Scarr, 1967, pp. 22–23). Calculation, internal coercion, attraction and 'the exigencies of a harsh environment' all played a part (Shineberg, 1999, p. 89), but few could possibly have known more than a fraction of what awaited them. The labour trade reflected a desire to travel but also a need to escape privations and limited opportunities at home; most workers had some degree of agency that reflected individual desires and local pressures alongside intense persuasion. They moved from largely subsistence agricultural systems, where income generation was trivial, in search of money, goods and experiences.

Migration combined some degree of economic necessity at home and coercion and incentives from recruiters. Any deception and duplicity could not have been sustained for long, as local ties and social networks were effective in transmitting knowledge about the distant circumstances, well evident in Tanna as on other colonial frontiers (Bonnemaison, 1994; Brass & Bernstein, 1992). Indeed on the larger islands the labour frontiers soon moved inland as coastal islanders became more demanding. Melanesians were not powerless and retaliation to illegal and deceptive recruiting was often considerable, especially as islander access to guns made their opposition to recruitment vessels more violent and effective. The 1871 murder of Bishop John Patteson in Solomon Islands was one extreme measure of the strength of local opposition to the overly coercive role of Europeans in blackbirding. Resistance to the labour trade became so strong in most parts of the New Hebrides and Solomon Islands that recruitment shifted northwards to the New Guinea islands. Even so, while conditions in Australia might not have been pleasant, nor recruitment gentle, Queensland was rather more favoured than some other destinations, from South America to Samoa, where treatment was routinely harsh and brutal (Firth & Munro, 1990).

Young unmarried men aged between 16 and 25 years, and initially even younger, made up the bulk of the migrants. In Solomon Islands a quarter of recruits were aged less than 16 (Corris, 1973). Mortality rates were high, and diets inadequate throughout the era, and punishment of illegal traders and violent plantation owners was non-existent. Over time, as labour migrants signed on for second periods, the age structure changed. Those who stayed for a second phase were able to bargain for higher pay and shorter contracts, and could move to more congenial plantations. As some stayed on, by 1891 the islander population of Australia had reached about 9500, less than 10 per cent of whom were women. Women who were recruited had to be accompanied by their husbands and generally undertook similar work in the cane fields to that of the men. Not all employment was in the fields and women were given jobs as 'housegirls' and servants but were also consorts for men (Shineberg, 1999; Banivanua-Mar, 2007). They were paid a lower rate than men, but were widely regarded as valuable recruits, and the constraints to more female recruitment were entirely on the supply side. Women were also more likely to wish to stay on in Australia after their contract ended rather than return to more unequal social circumstances.

At the same time that blackbirding boomed, greater competition for labour from planters in Fiji, Samoa and the New Hebrides made finding labour more difficult, especially since French and German governments imposed fewer restrictions on traders, so once again the labour frontier moved outwards. The Gilbert Islands (Kiribati) became a new labour reserve; with some 11,000 Gilbertese enlisted in the nineteenth century for places as far apart as Tahiti, Samoa, Fiji, Mexico and Guatemala, Gilbert Islanders were probably proportionately the most active labour migrants in the Pacific region, partly since the recruitment of couples was normal. Many enlisted voluntarily, especially in the drought-prone southern islands, although fewer than 200 came to Australia (Munro, 1992). Unlike in the Melanesian islands women made up a significant proportion—perhaps 40 per cent—of recruits. Ellice Islanders (Tuvalu) might have been similarly active but their populations had yet to revive after the depredations of the Peruvian slave trade. Polynesians did come to Australia from Tonga and Samoa but numbers were few.

Groups of men were often recruited from the same village and, in Australia, would work for the same master. Workers were highly organised, with collections of kinsmen and neighbours deciding themselves whether to re-enlist or return home at the end of their terms (Saunders, 1982). Recruitment had to become more subtle and persuasive, and individuals gradually became 'fairly free agents' (Howe, 1984, p. 342). Recruits from different islands were often kept apart for fear of violent confrontations. Tannese and Malaitans were much feared by other workers. On the plantations inter-island conflicts were not unusual and masters exploited such rivalries in order to divide and rule, and reduce concerted protests against the authoritarian structure of plantation work and life. But, over time, especially as the rigid plantation system weakened, and workers returned, labour migrants were able to engage in effective collective bargaining and negotiation.

Indentured labourers worked long hours (often ten-hour weekdays and five-hour Saturdays) usually lived in barracks and experienced monotonous and inadequate diets. Conditions were better where labourers were employed on smaller farms rather than plantations. Sick workers went to poorly equipped and overcrowded hospitals. Islanders complained about the monotony rather than the work itself, but as many as half of the workers—often those accompanied by wives—signed on for a second term (Corris, 1973). Workers learned the use of unfamiliar tools, the value and utility of money, new styles of dress and the pleasures and torments of liquor and gambling. However, 'the most determined assault upon the Melanesians' customs, beliefs and values was that made by Christian missionaries of various denominations' (Corris, 1973, p. 92) which contributed to literacy, Christian conversion and a means of acquiring prestige and status amongst others, to the exclusion of customs and 'sins' such as dancing.

The economic benefits from blackbirding and labour migration were almost exclusively located in Queensland, where the cane industry flourished. Melanesians were anxious to withdraw from plantation labour, as soon as some form of local development became feasible, but such opportunities were rare (Adams, 1990). In the New Hebrides especially, migrants returned to subsistence agricultural systems, albeit with new tools and clothes, but with no means of sustaining any real change in their economic status. That largely explains the ease of subsequent recruitment of previous workers for further employment in Australia. However, the flow of goods and ideas (despite the authority of older leaders) assisted in missionisation, and labour migration contributed to significant social change in much of island Melanesia. A number of prominent local men emerged as intermediary recruiters—in effect compradors—and gained local status and authority, perhaps none more so than Kwaisulia of Malaita who had learned English in Queensland (Corris, 1970). They too were not averse to fraud and trickery of other Melanesians and of the traders (Graves, 1986). Steel tools (including fishing hooks) and weapons, household implements, lamps and kerosene, clothes and tobacco were all valuable items to be purchased and brought back, benefiting household life and making agriculture and fishing easier and rather more productive. Those who brought them gained prestige and enjoyed superior consumption levels. However, various kinds of guns too were particularly popular trade goods (Scarr, 1967). A few returnees managed to become small entrepreneurs, benefiting from some literacy and language skills, so enabling a degree of leadership and authority as intermediaries. The knowledge that useful goods could be earned by labour migration had contributed to migration becoming increasingly voluntary. Even then goods were often purchased at 'company stores' at fraudulent rates. Most early accounts of the labour trade simply record a combination of trade store goods and Christianity, and the conflict between Christianity and traditional beliefs, that amounted to a 'power struggle', especially where younger men returned to oppose elders (Allen, 1968). Neither recruiters nor missionaries had any interest in what happened to returnees, other than their conversion to Christianity, so the actual impact of their return on themselves and local communities is poorly recorded. Men who returned from Queensland became 'modern' leaders, championing Christianity, opposing some aspects of traditional culture and planting new commercial crops. As early as 1876 a government agent in the New Hebrides observed that returned workers were responsible for cutting roads and building improved houses (Scarr, 1970). Rather later, in Solomon Islands, one returnee had purchased land with the money he had earned and began a coffee plantation (Shineberg, 1999). Similar processes occurred at least in Malaita, Tanna and Ambae (Allen, 1968) but invariably on a small scale and with no lasting guarantee of success.

Since migration was the only effective source of income generation a degree of continuity was inevitable. Labour migration influenced changes at the same time as missionisation and the first limited colonial administration. In different ways islands and islanders were coming under alien influence and control. What was not however changing was the achievement of any degree of autonomous economic development. Nonetheless, as the missionary, Maurice Frater, observed for the first decade of the twentieth century in the central islands of the New Hebrides: sometimes too many men had migrated—'emigration may be carried too far, and become detrimental to the well-being of the islands'. He, therefore, welcomed a trend where local people were cultivating their own cotton around the villages: 'they are finding it more profitable to work in their own plantations than to indenture themselves for a period of years to the white settlers' (1922, pp. 225, 216). Local competition and greater autonomy also made recruitment for Australia more difficult, and by the 1890s, few people from coastal villages were going to Australia as some local employment opportunities now existed (Bedford, 1973b), but by then, 'a large proportion' of New Hebrideans had already worked somewhere beyond the archipelago (Newbury, 1980).

Recruitment ended early in the twentieth century as the Australian colonies entered into federation, and the nation was born. The Pacific Island Labourers Act of 1901, part of a wider White Australia policy, alongside trade union opposition, effectively ended the labour trade. Pacific island labour had become more expensive, and labour demand had increased within Melanesia (from local traders, plantation owners and others) notably in the southern New Hebrides (Graves, 1993). Queensland cane growers were compensated for their enforced shift to employing more expensive European labour. Ironically, the southern Europeans who replaced them were long regarded derogatorily as 'black Italians' producing 'black sugar' (Moraes-Gorecki, 1994). It was not the end of indentured labour in Australia, which continued through both the employment of Asian workers in the pearling industry and the effective extension of the indentured system in Australia's colony of Papua to cover Melanesian domestic servants in Australian households (Martinez, 2005; Davies, 2019).

Most labour migrants returned home, but some stayed on and resisted deportation, with one outcome being a small descendant population of South Sea Islanders (Kanakas) living in Australia, centred on Mackay (Moore, 1985). A century after blackbirding and indentured labour had ended, some 6830 people in Queensland stated in the 2016 census that they were descendants of the Pacific Island workers. They constituted the first phase of a diasporic population of Pacific islanders in Australia. Around this time the links between Vanuatu and Queensland again increased but in quite different form. Vanuatu and Solomon Islands became destinations for roots tourism and diaspora travel as South Sea Islanders 'returned' to experience the homelands of their ancestors (Cheer & Reeves, 2013). No less a figure than Mal Meninga, the Australian rugby league coach, journeyed 'back' to Tanna to discover his heritage. In 2013, the 150th anniversary of the start of the labour trade, the Vanuatu government was still pressing for compensation from Australia (who carefully avoided the issue) and in Tanna, for example, inherited memories of the trade are almost exclusively negative.

Overall as many as perhaps 300,000 Pacific islanders were engaged in contract labour between 1840 and 1915, some more than once, suggesting that the Western Pacific was 'the most exploited labour reserve anywhere during the nineteenth century' (Graves, 1986, p. 237). The New Hebrides contributed many of these, perhaps 40,000 in Australia, along with 14,000 in Fiji and 10–13,000 in New Caledonia (Adams, 1990) while the Gilbert Islands was also disproportionately involved in the labour trade but mainly to destinations other than Australia. Larger islands like those of New Guinea, Fiji and Samoa were more self-contained and required labour imports; by the 1870s, colonial governments of Fiji, Samoa, Tahiti and Hawai'i had turned to recruiting labour from Asian sources, and Melanesian labourers worked alongside Chinese in Samoa and Indians in Fiji.

Early in the twenty-first century, long after blackbirding had ended, a new phase of labour migration from Vanuatu and elsewhere in the Pacific began, with labour migrants being recruited to work in the agricultural sectors of New Zealand and Australia. Until the belated inclusion of Solomon Islands in this new programme, Vanuatu was the only country involved in the two phases of migration. What is remarkable is how similar the two phases proved to be (Chap. 11).

The Diversification of Migration

The end of blackbirding early in the twentieth century brought something of a century-long hiatus in long-distance agricultural labour migration, although in Micronesia, during the inter-war years, the Japanese established small plantations for sugar cane and pineapples from Palau to the Marshall Islands, and agricultural workers came from nearby islands. In the New Hebrides, Papua, New Guinea and Solomon Islands, the plantation economies that had been established in the late nineteenth century used only local migrant workers and survived until after independence. Only then did they begin to fade in a few coastal regions, such as East New Britain and Bougainville, where other income-generation opportunities became generally available. In the New Hebrides, islanders were willing to work on local plantations when copra prices were low but when prices were high, as in the 1920s, they preferred to stay at home and sell their copra to traders. The German ethnologist, Felix Speiser, observed of the northern New Hebrides in 1910:

> In years past the natives crowded the recruiting schooners by hundreds, driven by the greed for European luxuries, by desire for change, and inexperience; today this is the case in but very few and savage districts. Generally the natives have some idea of what they may expect; moreover by trading with coprah they can buy all they need and want. They enlist nowadays from quite different motives. With young people it is the desire to travel and to 'see the world' and to escape the strict village laws that

> govern them, especially in sexual matters, and to get rid of the supervision of the whole tribe. Sometimes, but only in islands poor in cocoanut trees, it is the desire to earn money to buy [in brideprice] a woman, a very expensive article at present. They may seek refuge in the plantations from persecution of all sorts, from revenge, or punishment for some misdeed at home'. (1913, pp. 54–55)

In the absence of other alternatives, the plantation economy survived throughout Melanesia, despite being so exploitative that even the Lieutenant-Governor of Papua, Hubert Murray, had described it in the 1930s as 'really rather like slavery' (quoted by Fitzpatrick, 1980, p. 57). Nonetheless, as some coastal plantations were experiencing economic problems, contract labour migration was still contributing to 'opening up' some of the more remote part of the highlands, very much the last labour frontier in the Pacific (Ward, 1990). But, quite quickly, Papua New Guineans, like Melanesians elsewhere, preferred to remain on their land, pursuing cash cropping to earn 'the money that grows on trees' (Brookfield, 1968) or move into urban areas, in both cases to escape the powerlessness that existed on plantations (Curtain, 1980).

Throughout the colonial era, in the Gilbert and Ellice Islands Colony, migrants were employed in Line Islands coconut plantations, often several hundred men at a time, a process that continued until independence. In Vanuatu, PNG and Solomon Islands the 'last' plantation workers came mainly from the most remote regions where few alternative income-generating opportunities existed, notably the remote highlands and outer islands (e.g. Larson, 1981; Wilson, 1966; Bedford, 1973; Adams, 1990). In some respects, little had changed. As Peter Corris suggested in 1973, when both PNG and Solomon Islands remained colonies:

> The motives of Melanesians who engage for plantation work at the present time do not seem to differ greatly from those of their nineteenth-century counterparts. The wish to escape from the dullness of village life, the desirability of European goods and the attractiveness of travel are still powerful forces; the most important difference is that today New Guineans

and Solomon Islanders require cash to pay taxes and many seek employment for that reason. (1973, p. 164)

Employment on plantations, typified as circular migration, faded after independence as alternatives diversified, and rural-urban migration became an option for some. The plantation system entered into a steady decline throughout Melanesia, but the combination of oil palm development and resettlement brought one more form of migration and agricultural employment, as people moved from some densely populated areas to coastal oil palm estates, mainly in PNG (Koczberski & Curry, 2005). Where plantation migration survived it was usually seen as a short-term means of acquiring enough capital to establish commercial agriculture at home (Bastin, 1985 Connell, 1985). Throughout the Pacific there remained places where local income earning opportunities were few and migration, whether to the declining number of agricultural schemes or to urban areas, meant some possibility of release from poverty of opportunity.

Other migration flows continued to characterise Pacific islands. By the late nineteenth century New Guinea had provided gold from various sites. The gold rushes of the 1850s in Australia created a mobile population of miners and, by the 1870s, many had moved on to small mines in the Papuan islands before shifting inland into New Guinea. Local Melanesians worked for the miners, acquiring the same trade goods as elsewhere, and valuing weapons particularly as a means of securing local power. Much the same was true of New Caledonia, with the first nickel boom of the 1870s bringing more workers from the New Hebrides (Nelson, 1976; Shineberg, 1999). Indigenous Melanesian (Kanak) workers were rare and convicts proved inadequate. Death rates among the New Hebridean Melanesians were particularly high so that, in the twentieth century, Asian labour was also introduced.

Throughout the twentieth century hundreds of workers from the Gilbert and Ellice Islands also worked under temporary labour migration contracts both in the phosphate mining industry on Banaba (Kiribati) and Nauru. They were not the only mines. From 1909 until 1954 phosphate mining took place on Angaur (Palau), originally by the Germans, then the Japanese. It provided jobs for Micronesians, but there was at least circumstantial evidence that, in the Japanese era, many were 'transported far from home for a year of exhausting work totally unsuited to their temperament, [being] recruited for mine work against their will' (Peattie, 1988, p. 83). Makatea in French Polynesia, mined from 1917 to 1964, was the third largest phosphate mine in the Pacific, with workers coming from as far as the Cook Islands (leaving a twenty-first-century legacy of a ghost town, Vaitepaua, and disputes over the possibility of restarting the mine).

Urbanisation

For most of the twentieth century, with the exception of Kiribati and Tuvalu, migration was mainly local and usually to plantations. Colonial trading posts and administrative needs combined in urban centres, first in Polynesia and much later in Melanesia. Amidst the flows of Pacific islanders, European and Chinese traders had established trade stores: Germans in Micronesia and Samoa, French in the French territories and Chinese merchants almost everywhere, even on the smallest atolls. Urbanisation in Polynesia was faster than elsewhere. Urban centres developed rapidly in Fiji in the inter-war years, but it was not until after the Second World War that towns grew elsewhere in Melanesia. In Kiribati and Tuvalu they came even later. Throughout Melanesia, islanders were welcome in towns as workers but unwelcome as urban residents. In Vanuatu, neither Port Vila nor Luganville had more than 2000 inhabitants by 1955 and less than 10 per cent of these were ni-Vanuatu. Rapid growth eventually brought their populations to 15,000 and 7350, respectively, by 1970. The same phases of growth were evident in Solomon Islands and PNG, until a much more rapid period of growth in the post-independence years when the Melanesian capitals grew faster than anywhere else in the region. The two towns of Vanuatu now host about a third of the country's population, spilling over from their official boundaries (Connell & Lea, 2002; Trundle et al., 2018; Petrou, 2020), creating doubt over the actual numerical extent of urbanisation (Table 2.1). Virtually since independence, urban populations have been increasing faster than those in rural areas, through both high levels of urban fertility and migration. In this century urbanisation is ubiquitous, with Melanesian towns and cities, in particular, expanding fast. By contrast in Tonga and Samoa the towns are being bypassed by longstanding flows of migrants that have resulted in a growing diaspora, especially in New Zealand.

Urbanisation in Melanesia and Micronesia has posed challenges for effective planning, management and employment policies, contributing to an 'over-urbanisation' characterised by high unemployment rates (typified in the 'youth bulge'), inadequate housing and services, alongside intensified environmental problems. Such issues have reached crisis proportions in South Tarawa, the capital of Kiribati resulting in a 'perfect storm' of economic problems intensified by environmental degradation, mismanagement and extreme pressure on scarce resources (Storey & Hunter, 2010). Both there and in Funafuti, the capital of Tuvalu, land has been reclaimed from lagoons for residential use.

Contributing to urbanisation has been depopulation of remote places and 'outer islands' (a relatively new concept in the Pacific), as formal employment is scarce, education and health services are limited, and skilled workers are reluctant to be posted there. Almost every Pacific atoll, other than the urban centres, has lost population in this century: South Tarawa and Funafuti now hold more than half their nations' populations. Less extreme patterns of migration characterise all PICs, although migration from small Vanuatu islands such as Paama has been long-established (Petrou, 2020). Despite the active discouragement of urban migration and residence, a growing number of urban residents are now of a second or higher generation with few ties to 'home' areas and limited ability and will to return (Kraemer, 2013; Petrou, 2020; Petrou & Connell, 2018). In the Melanesian and atoll PICs, where significant informal economies have not made up for the lack of formal employment, over-urbanisation has resulted in greater urban interest in international migration, rather than return migration to small island homes. Rural-urban migration is now mainly a Melanesian and atoll state phenomenon and is something of an alternative to international migration, where that is not feasible. The Melanesian towns and cities are becoming a little more like the Polynesian capitals that are stepping stones (or simply bypassed) as people seek out international migration opportunities.

The Global Long Boom: International Migration

International migration from the Pacific islands effectively restarted in the 1960s, particularly from Samoa after its early independence in 1962, and was spurred by the global long boom, demand for workers in more developed countries and the arrival of jet transport. This migration was mostly to metropolitan states: New Zealand and the United States. Tonga soon followed, along with the Cook Islands and Niue. Migration relieved growing population pressure and initiated a flow of remittances to the islands. Often intended as temporary, migration gradually became more permanent, and the demographic balance of many states shifted outwards. By the end of the century the Polynesian PICs had become unusually dependent on remittances from overseas migrants in the Pacific Rim. With rare exceptions, such as the migration that followed the late 1960s nickel boom in New Caledonia (mainly from other French territories and the New Hebrides), the lack of intra-regional migration was indicative of limited island economic development.

Uniquely, New Zealand instituted several immigration schemes to encourage workers from the Pacific Islands to migrate. A quota system introduced in 1962 admitted 1000 immigrants per year from the then Western Samoa, and a Tokelauan resettlement programme was initiated in response to cyclone devastation, overcrowding and limited opportunities in Tokelau (Wessen et al., 1992). In 1967, the then New Zealand Prime Minister, Sir Keith Holyoake, announced a plan to recruit 100 Tongan workers every 6 months for industrial firms located in the Hutt Valley. The workers were issued with temporary work permits and allowed to stay in the country for six months. Such migration incentives resulted in a chain migration of Polynesians to New Zealand, encouraged by improved transport links, dissatisfaction with island conditions, and job opportunities in New Zealand. Because the 1960s and early 1970s were years of economic expansion and labour shortages, temporary visas and quotas were not strictly enforced, and many migrants overstayed.

For over half a century, uneven development between the PICs and metropolitan countries resulted in high levels of migration, especially in the Polynesian PICs and spilling over from New Zealand to Australia. So significant has international migration become that families created small-scale metaphorical 'transnational corporations of kin', seemingly allocating family members to different countries, to diversify livelihood opportunities. Simultaneously, a 'culture of migration' emerged in several countries, where migration was normative, anticipated and unexceptional (Connell, 2008). Smaller dependent territories in Micronesia and Polynesia, including the Cook Islands, resisted a political independence that might deny them migration opportunities. The outcome has been a now falling population in at least the smallest Polynesian state, Tuvalu.

Over time acquiring marketable skills played a valuable role in enabling migrants to gain better jobs, and migration has been more evidently linked to skills, such as teachers, health workers and sportsmen—gender bias intended (Connell, 2009). Indeed, the dominance of skills in migration streams moved the Speaker of the Tuvalu parliament to suggest: 'we should focus on rugby, not only for the rewards and fame, but [since playing rugby] can also pave the way for permanent residence in New Zealand' (quoted in Shen & Binns, 2012, p. 77). Thousands of Fijians have worked for special United Nations forces, mostly in the Middle East and Africa. Others are employed as teachers and nurses in Tuvalu and the Marshall Islands. Samoans have long benefited from guestworker permits from neighbouring American Samoa, mainly to work in the Starkist tuna cannery. When COVID-19 trapped temporary workers overseas in 2020, as many as 400 Samoans—all anxious to return home—were working on 17 different cruise lines in the Caribbean (RNZ, 26 June 2020). Others, similarly delayed, included members of the Samoan Police Force, serving with the UN in South Sudan. Losses of skilled labour have become serious in several PICs, with small states losing disproportionately more skills than larger states. Limited return migration has not remedied this problem and remittances may not always compensate for skill drains.

Migration increasingly involved households rather than individuals, moving to metropolitan states, and gradually becoming permanent migrants with several Polynesian migrant generations now resident overseas. By contrast permanent international migration from PNG, Solomon Islands and Vanuatu is rare. That has partly resulted from different colonial relationships, as migration typically followed 'colonial' lines, notably from Samoa (and the Cook Islands, Niue and Tokelau) to New Zealand. Conversely, Kiribati and Tuvalu, and the Melanesian states, were latecomers to international migration to metropolitan states, without migration corridors to the United Kingdom, Australia or elsewhere. In that respect, Kiribati, Tuvalu and also Nauru have even been referred to as 'orphan states' that would benefit from preferential access to international labour markets (Kagan, 2015). Nonetheless, migration from Pacific Island Countries (PICs) has gradually increased, extended, diversified and become more skilled.

Structural circumstances have changed since the independence of the PICs. Many now benefit from some preferential arrangements, notably in New Zealand, whether permanent migration opportunities (the lotteries of New Zealand, and also the United States) or the Recognised Seasonal Employer (RSE) Scheme, Seasonal Worker Programme (SWP) and Pacific Labour Scheme (PLS) for temporary employment. Islanders, particularly Tongans and Samoans, have migrated to the United States by various means, some facilitated through membership of the Mormon Church (Church of Jesus Christ of Latter-Day Saints), while others have been able to migrate through the Green Card lottery system. Since 2002, New Zealand has offered a Pacific Access Category, which allocates places to several PICs. It offers 1100 places a year to Samoa (its former colony), 250 places a year to Tonga and Fiji and 75 for Kiribati and Tuvalu, but applicants are only granted a visa if they receive a job offer from New Zealand, a significant constraint for the atoll states where internationally marketable skills are scarce (Kagan, 2016b). Demand exceeds supply. No other Pacific Rim countries have similar programmes (although the Compacts of Free Association between the United States and the three Micronesian states of Palau, Federated States of Micronesia and Marshall Islands enable unlimited migration to the United States, circumstances that account for the lack of interest in the SWP in these PICs). In a multitude of ways, migration streams are focused on the Pacific Rim.

The widespread demand for migration opportunities is reflected in one other migration strategy: overstaying on tourist (or other) visas. After 1986 New Zealand provided visa waiver migration to citizens of several PICs, but the scheme was cancelled in 2002 mainly because many migrants overstayed rather than returning home. At the start of the century New Zealand had some 7000 overstayers, half of whom were from Tonga and Samoa. By the end of 2018 that number had doubled to almost 14,000, again including just under 5000 from PICs, especially Samoa and Tonga (the two largest sources), but with several hundred from each of Fiji and Tuvalu. In Australia too, Tonga and Samoa have some of the highest rates of overstaying, but numbers are much smaller (Chand, 2006). Finding and deporting illegal migrants proved too expensive to be a valuable administrative strategy, and many had found worthwhile employment (Nishitani & Lee, 2019). While overall numbers are small it has become a common phenomenon in New Zealand and Australia, and in the United States, a continued indication of the desire and determination to migrate.

Hazards and Resettlement

While most migration has involved some degree of choice about timing, destination, social composition or objectives, in some exceptional circumstances migration has been largely involuntary. Since the 1930s in Kiribati people from some southern atolls, affected by drought and excessive population pressure on resources, were resettled in the Phoenix Islands and, after they too proved drought-prone, later went to Solomon Islands. A decade later, a group of people from the island of Vaitupu (Tuvalu), concerned that their home island was in danger of overpopulation, moved to the island of Kioa (Fiji). A similar move took people from the overcrowded atoll of Niutao (Tuvalu) to unpopulated Niulakita. Most resettlement has followed more dramatic environmental hazards, mainly volcanic eruptions and tsunamis, and been locally unsuccessful with migrants seeking to return home, land disputes intensifying, friction occurring with nearby people and challenges in adapting to different ecological zones and cultures (Connell, 2012; Connell & Lutkehaus, 2017a, 2017b; McAdam, 2014). While the resettlement of Banabans in Fiji in the 1940s was relatively successful, since an 'empty' island without local claimants was available and at the time involved no more than about 1000 people, even some 70 years later uncertainty still reigns over land tenure, raising questions about exploitation, citizenship and sustainability (Connell & Tabucanon,

2016). Resettlement of Manam and Carteret Islanders in PNG has been typically thwarted by 'host' landowners, the impossibility of gaining adequate access to land or land rights and government inactivity. The problems experienced by quite small populations moving short distances in similar cultural contexts are indicative of the potential future problems facing environmental migrants. Settlers have been perceived as outsiders and rival claimants to valuable coastal resources. Although Kiribati has purchased a small island in Fiji, possibly for resettlement, the 'welcome' that might be attached to migration of i-Kiribati is likely to be ambiguous (Hermann & Kempf, 2017). Nevertheless, Fiji has expressed a willingness to accept citizens from both Kiribati and Tuvalu should they be forced to leave. In the 1950s after phosphate mining had degraded their island, Australia offered Nauru the possibility of settling on an island in Queensland but the offer was rejected (Tabucanon & Opeskin, 2011). Land issues have constrained internal migration and resettlement throughout the Pacific region, access to land is more difficult now than at any time in the past and migratory responses to hazards are more likely to demand distant international solutions. In the context of climate change that was put to a particular legal test in New Zealand in 2014 after a temporary resident family from Kiribati sought to remain there on the grounds that they were environmental refugees since climate change would have adverse effects on their livelihoods were they to return home. The New Zealand Immigration and Protection Tribunal rejected their claim but permitted them to remain because of their strong family ties within New Zealand (McAdam, 2015). The migration of Tokelauans to New Zealand in the 1960s after cyclone damage to their home atolls may yet prove a precedent for international migration.

Remittances

Other than migration following hazards, almost all migration has been underpinned by a loosely economic rationale; whether through direct income earning opportunities or indirectly through developing the capacity for higher status and salaried employment. Consequently, migration has been associated with often significant flows of remittances, mainly to the homelands, villages and households of migrants. The earliest migrants brought money and goods from abroad, at the end of their employment. More recent migrants have been much more likely to send money home at regular intervals or in response to requests and crises.

The flow of remittances back home, as cash or goods, eventually resulted in both Tonga and Samoa being two of the top 20 countries in the world in terms of the contribution of remittances to GDP. Migrants contribute to development at home and are well placed to do so through their familiarity with local needs. Their remittances contribute both directly and indirectly to welfare and human capital formation. Remittances constitute an effective, informal and family-based system of social protection for migrants' kin. At the same time, remittances may help to sustain a 'migration syndrome', while imposing sacrifices on migrants, and social costs at home and in destinations. Remittances, whether from more or less permanent migrants, and now from seasonal workers too, are of growing importance, notably in the smaller islands of Melanesia, where they have contributed to the emergence of a 'singaut' economy' as those at home request support (Rasmussen, 2015; Wilson, 2013; Petrou & Connell, 2018). Both unskilled and skilled migrants sustain remittance levels at high levels, over long time periods, often more than 20 years, while skilled migrants and returnees may also provide advice and assistance ('social remittances') on education, health, business and other matters (Connell & Brown, 2004; Brown & Connell, 2015). Remittances have made a significant contribution to improved housing and water supplies, better access to health and education, more modern forms of consumption, including solar panels, smartphones and generators, and small-scale services such as cafes or petrol stations, small stores and taxis. Over time they have enabled some diversification out of agriculture, an increased flexibility of livelihood choices and a reduction in both poverty and inequality (Brown et al., 2014). Remittances also support local organisations (especially churches) and cater to immediate needs.

Increased international migration has brought larger remittance flows, especially in comparison with aid flows, and are more stable than other sources of national and household finance. Remittances compensate in the aftermath of crises and disasters, whether physical (especially cyclones) or economic, being counter-cyclical, in response to economic and environmental shocks. That was evident in Samoa, for example, following both the 2009 tsunami and Cyclone Evan in 2012 (Le De et al., 2015) and repeated throughout the Pacific in the wake of COVID-19. Growing recognition of the value of remittances has brought some institutional interest in managing flows and 'harnessing' remittances to finance investment and economic growth. Remittances have become a global 'development mantra' (Chap. 4), in a climate of aid fatigue, uncertainty about private investment, the extent of international migration and the propensity in some countries to train workers for migration.

Remittances are so substantial in most small PICs, and many small islands in Melanesia, that both PICs and islands have been described as having Migration, Remittances, Aid and Bureaucracy (MIRAB) economies, an acronym first devised for Kiribati and Tuvalu, alongside Tokelau and the Cook Islands (Bertram & Watters, 1985), and now more broadly relevant. Sustained remittances are crucial for national development and are likely to become yet more important in the future, even in a global context of widespread international opposition to migration, should both domestic expectations and aid fatigue increase.

National Strategies and Migration Schemes

If most migration from PICs has been independent of government involvement and intervention (or even facilitation), as migration became more obviously an income-generating development strategy, governments began to intervene to develop and regulate distant employment opportunities. A consequence of limited economic growth and a growing dependence on remittances was that various countries at different times sought new formal migration and employment opportunities within and outside the Pacific region. That was spearheaded by Kiribati and Tuvalu, first in responding to mining opportunities in Nauru, for most of the twentieth century and subsequently, beginning in the 1960s, with the establishment of national Marine Training Schools to enable workers to acquire the skills needed to obtain employment on merchant shipping lines.

The Precedence of Kiribati and Tuvalu

Migration has always been a function of development, disadvantage and opportunity. Within the Pacific region enormous differences exist between PICs with very high population densities—on less fertile land—as in the atoll states (notably 400 people per km^2 in Tuvalu) compared with the much lower densities in the larger, high island, Melanesian states of PNG (with 20 people per km^2), Solomon Islands and Vanuatu. Weak and narrow domestic economies, scarce employment opportunities and intermittent hazards have made the two states of Kiribati and Tuvalu more likely to be involved in migration in some regulated form since the early nineteenth century. In the second half of the twentieth century, migration to Nauru, Banaba and then employment on merchant ships was economically important. Distinctively, these migration flows were encouraged by both the colonial power and just as strongly by the independent states that followed. For seafaring employment, workers were locally trained. Until the emergence of guestworker schemes, PICs had rarely engaged in comparable contract labour schemes.

Migration to both Banaba and Nauru was of particular importance from 1955 to the end of the century. The Australian administration in Nauru gave the Gilbert and Ellice Islands precedence since the colony 'had widespread unemployment and little hope of economic prosperity, so that money brought back to the colony by indentured labourers was very important' (Viviani, 1970, p. 127). Workers were contracted for two years but contracts could be extended for up to six years. Residence was only permissible while working, but extended over several decades for some workers, despite the Kiribati government actively discouraging renewal of contracts to achieve greater equity in access to scarce employment opportunities.

At least in the 1970s migration from the colony was carefully regulated. For many years recruitment excluded the two northernmost islands since it was felt that people from there had less need of wages than those in the drier southern islands. The colonial administration also placed limits on numbers from any one island. Actual decisions on who might go to Nauru were made at the atoll and village level based on the perceptions of local leaders over who had the greatest need (Connell, 1983: 33–34). By then, remittances were the most important source of income on almost every atoll. Likewise in the early 1980s government and quasi-government jobs were also allocated through three-year contracts and those in Tarawa allocated to islands according to population size and density, and their ability to produce copra, again intended to favour islands where development opportunities appeared least.

After the Second World War, Kiribati and Tuvalu pioneered another form of international migration in the region. A Marine Training School was set up in 1968 in Tarawa to train both i-Kiribati and Tuvaluans as merchant seafarers for overseas shipping lines, mostly operating from Europe. A separate School was set up in Tuvalu after independence in 1979, training 60 seafarers a year. Initially, the programmes catered only for men but were later extended to women. Seafarers necessarily returned to their home countries at regular intervals and at the end of extended working periods. While private remittances exceeded the cost of public training, there were social costs; as one i-Kiribati commentator suggested, the participants 'dehumanise themselves, destroy local customs and cause envy and resentment in the rest of the local population' (Toatu, 1980, p. 31). A rising incidence of STDs and HIV/AIDS was a further consequence (Dennis, 2003). Finding employment for graduates from both Schools eventually became more difficult, with challenges from European unions, and numbers from both countries were stabilising by the 1980s, after the two countries achieved independence. By the late 1990s, Kiribati had around 1400 workers employed on overseas ships. Opportunities for both mining and seafaring were, however, becoming scarce and earnings were relatively poor, but vastly superior to what could be obtained at home, even when paid employment was available. However, both the Banaba and Nauru mines had closed by the end of the century, and demand for Pacific islanders as shipping crew has continued to fall with greater global competition. In this century opportunities briefly extended to cruise ships, providing employment for about 120 i-Kiribati women, between 2004 and 2012. Numbers were small, management was challenging and many women became pregnant and unable to complete their contracts (Kagan, 2016a) so that the scheme eventually ended.

In both countries migration and employment policies sought to relate migration to equity. For government employment recruiting would only take place on home islands to discourage adventitious migration. Supported by the colonial administration, migration was intendedly 'part of a pattern of circular migration from the village to the phosphate islands or other places of employment and then back to the villages' (Watters & Banibati, 1977/1984, p. 127). Thus:

> Gilbertese did not turn to being sailors or phosphate workers as life endeavors. Instead they became activities primarily for young men who would spend a few years at one or another (or perhaps both) and then return to the Gilberts. Viewed from another perspective, families sent out their young men to take advantage of these impermanent resources and to bring and send home their proceeds. (Knudson, 1981, p. 94)

Writing of Kiribati, Bill Geddes concluded in the 1980s that:

> Kiribati communities have institutionalised migrant labour. Households readily compensate for the absence of those involved, with remaining household members broadening their roles to include the necessary contributions of absentees. In turn, migrant labourers accept their responsibility to remit money and goods to their home. (1988, p. 43)

In Tuvalu a similar perspective prevailed:

> Nanumeans tend to view periodic off-the-island employment as an enjoyable interlude in a basically subsistence life…for most Tuvalu people overseas employment is a temporary experience to be enjoyed and valued for itself, as well as for the money it produces. (Chambers, 1975, p. 29)

Anne Chambers thus concluded:

> The present system of temporary, rotating employment off the island by a wide sector of the community has enabled Tuvalu to reap economic benefits from its export of labour with minimal social sacrifice. Off the island workers tend to retain a commitment to their home island … In developing any future employment opportunities Tuvalu would be wise to perpetuate the present pattern of temporary and rotating employment. (1975, p. 103)

However, some migrants were reluctant to return home because of the absence of formal sector employment, distaste for traditional foods, dislike of hard work in the taro pits and a rigidity and monotony of social life. Subsistence work was to be disdained (Connell, 1983). Some returned only as far as the main town and urbanisation became one concomitant of migration.

Both Tuvalu and Kiribati were relatively poor, had participated in and instigated resettlement schemes in colonial times to relieve very high population densities in drought-prone islands, had few development options and had no politico-migration ties to the Pacific Rim states. Beyond mining and seafaring, some workers had found employment in Vanuatu and the Marshall Islands, under various arrangements, and Kiribati had also sought employment opportunities in a range of countries, including PNG, Saudi Arabia, New Caledonia and in the British army, without any success. In mid-1977 a group of 12 Tuvaluans had managed to obtain employment in New Zealand under a similar work scheme to that operating between New Zealand and Tonga and Samoa, but the New Zealand economy was experiencing stagnation and the Tuvalu work scheme collapsed following trade union pressures in New Zealand (Chap. 4).

By the early 1980s, Tuvaluans at least felt that much wealthier Australia and New Zealand should offer quotas of jobs for Tuvaluan workers. In both Tuvalu and Kiribati all kinds of labour migration had been sought out by governments, households and individuals and used as a means of acquiring capital and contributing to household and national development. More than other PICs they therefore seemed particularly likely to be enthusiastic about and benefit from the future RSE and SWP schemes.

Policy Detours

Beyond Tuvalu and Kiribati, little significant short-term migration away from the PICs occurred until the 1960s, nor was there long-term migration or great interest in supporting and organising migration. It was not until 1971 that any government in the region expressed interest in increased migration to Australia, when the Prime Minister of Fiji, Ratu Sir Kamisese Mara, attacked Australia's 'racist' migration policies and recommended the establishment of a guestworker scheme. One academic commentator then concluded that both New Zealand and Australia might do more in this regard to assist with Fiji's unemployment problem, and that 'Australia in particular could probably afford a more liberal policy and allow in unskilled workers from Fiji' (Ward, 1971, p. 223). In 1974 Australian Catholic Relief made a study of the possibility of a Tongan guestworker scheme, and at several annual conferences of South Pacific Labour Ministers in the 1970s and early 1980s similar proposals and requests were reiterated, and just as regularly rejected on the grounds that, unlike New Zealand, Australia required skilled labour, which would result in a brain drain, and preferred a non-discriminatory migration policy. Such responses provoked resentment in the PICs, and pleas that Australia be more like New Zealand, which had a limited temporary labour migration scheme. However, the Australian Standing Committee on Foreign Affairs and Defence, charged with reviewing the relationship between Australia and the South Pacific, regarded any introduction of a guestworker scheme as 'impractical' at a time of high unemployment (Australia., 1978, pp. 15–16). A year later the Harries Report on Australia's relations with the 'Third World' again rejected the idea of guestworker schemes for Pacific islanders on the grounds that this would cause dissatisfaction in the PICs if it became necessary to wind up the schemes because of changed economic conditions, and that guestworkers would be regarded as second-rate citizens (Australia, 1979).

The issue of migration from the PICs to Australia was further considered by the Committee to Review the Australian Overseas Aid Program (Jackson Committee) in 1984. For the first time that gave detailed consideration to the needs of PICs and singled out Kiribati and Tuvalu as having greater need of assistance than other PICs, with development unlikely to be satisfied by conventional aid policies, training programmes or trade policies. It gave them a special significance denied to all other states except the former Australian colony of PNG:

> Kiribati with a population of 60,000 and Tuvalu, with a population of 8000, have special problems. Their remote and minute land areas are

> heavily populated. They depend very much on remittances from their emigrants and from foreign aid. Their long-term development prospects are discouraging. In view of structural problems which are beyond their control and beyond the reach of aid, Australia should make available limited opportunities for immigration from Kiribati and Tuvalu … Australia should establish a special immigration quota to deal with the unique problem … Potential migrants should be trained in skills which would help them find employment in Australia. (Jackson, 1984, pp. 8, 181)

The then Australian High Commissioner to Fiji and Tuvalu stated: 'to put it bluntly we will take the young men who have been through the marine training school there. They can work in the mines and in the outback and can take their young ladies with them' (quoted in Keith-Reid, 1984). That unsubtle response was ignored. Equally, such a radical suggestion concerning countries with which Australia had limited contact was ignored, despite some discussion of how such a policy might be put into practice, elaborated on in an unpublished paper by the geographer, Diana Howlett (1985), on potential Pacific 'boat people' that proposed a humanitarian programme for migration from both states.

While these debates and deliberations meandered on without resolution, various PICs had sought other options and possible long-term migration channels. Several countries including Kiribati and Tuvalu, Tonga and Fiji, all sought employment opportunities in New Caledonia's nickel mines in the nickel boom of the 1970s, but were rejected in favour of more 'docile' workers from other French Pacific territories. (In this century, as new mines opened, New Caledonia again rejected regional interests in favour of Filipino workers). Nonetheless, many workers went from Vanuatu to New Caledonia; more than 3000 were reportedly there in the early 1970s, on three- and six-month contracts. Remittances were considerable and wages were 'so much higher than at home that the road to New Caledonia has become the brightest prospect to many a New Hebridean: leaders and employers on the other hand view this exodus as a severe drain on the limited labour available for development of the New Hebrides themselves' (Brookfield, 1972, p. 121). As many as 10,000 ni-Vanuatu worked in New Caledonia at the peak but by the mid-1970s the boom was over and most had returned home.

In 1987 yet another review of Australian's relations with the region recognised the recurrent pressures on Australia for migration opportunities. This review expressed greater sympathy with migration as a means of reducing population pressures and providing valuable remittance flows but felt that the benefits for individuals and families were at a cost to society as a whole. It cautiously concluded that 'especially for those countries with very poor prospects for self-sustaining development and poor standards of living, opening up of migration policy may be an essential adjunct to aid' (AIDAB, 1987, p. 390). The wide-ranging 1988 *Immigration to Australia* Report focused on permanent migration, although a commissioned report from Stephen Castles (1988) discussed temporary migration schemes but argued against them, since they were likely to involve nearby low-income Pacific islands and would probably lead to overstaying and permanent settlement so that social, economic and political costs would outweigh economic benefits.

A subsequent parliamentary review of Australia's Relations with the South Pacific (Parliament of the Commonwealth of Australia, 1989, p. 134) again rejected any guestworker scheme 'primarily because of the inability of such a scheme to provide a constant source of employment and hence reliability and security for the participating countries' and because the unskilled work would not improve the 'skill base of the donor countries'. Once again the review considered the particular problems of Kiribati and Tuvalu, recognised the decline of overseas employment opportunities and noted a new threat from climate change 'and the widespread perception of eventual inundation' (1989, p. 134). It thus recommended that migration should be considered as an adjunct to the aid programme for both countries and that Australia's immigration policy should have the flexibility to allow for such special relationships, but ultimately it only supported a work experience programme. Once again nothing came of the recommendation. Despite growing sensitivity to the needs of at least some PICs, the possible advantages of any schemes, whether as economic strategies or as soft power, were not reflected in policy shifts. That position was retained well into the twentieth century, when the seasonal worker scheme finally came to fruition.

Through the twentieth century, Kiribati and Tuvalu remained unusual in actually training workers for migration as 'seafarers'. That instituted the start of greater interest in, and a wider process of, training people for migration to achieve what the former Kiribati President, Anote Tong, much later called 'migration with dignity', that is employment in occupations which enabled some degree of self-esteem and higher incomes, rather than low-status, low-income, entry-level positions, and which were also likely to offer some permanency and prospects and enable a more substantial flow of remittances. The then Prime Minister of Tonga, Feleti Sevele, had similarly remarked: 'Skilled workers are our best export. I'd rather people go overseas with some skills. At least they don't just sweep the floors' (quoted in Connell, 2013, p. 183). By the 1980s the migration of many skilled islanders had begun to create a significant skill drain in several social and economic sectors.

APTC and KANI

Despite some concern over skill losses to metropolitan countries, other PICs have given increased attention to training workers for employment either overseas or at home. Vocational education has been upgraded in several countries. Two new initiatives in this century include the multinational Australia-Pacific Technical College (APTC) and the Kiribati Australia Nursing Initiative (KANI), both funded by the Australian Department of Foreign Affairs and Trade (DFAT). The APTC enables Pacific islanders to gain Australian skills and qualifications in a variety of trade areas that would enable better access to overseas labour markets, although in the first few years most graduates remained in their home countries.

Since 2006 APTC has provided vocational training on its five campuses in Fiji, Samoa, Vanuatu, PNG and Solomon Islands, covering a range of areas, including painting and decorating, automotive repair, plumbing, tourism and health services. After a decade, APTC had proved successful in skills development but unsuccessful in fostering labour mobility; less than 2 per cent of the graduates migrated to Australia or New Zealand after completing the courses. By 2016 only 276 out of 10,087 graduates had found residence and employment overseas (Doyle & Howes, 2015) and by 2019 only 3.5 per cent of 895 graduates surveyed by Richard

Curtain and Stephen Howes (2021) had migrated overseas. Migration was hampered by administrative issues relating to skills assessment, the task of matching graduates with potential employers in Australia and few job vacancies, despite roughly half the graduates wishing to move overseas (Clemens et al., 2015). Most APTC students already had a job and those seeking to migrate were less likely to be released by their employers to take part in the programme. The scheme did, however, meet PIC needs in some areas, such as plumbing, where local shortages were considerable.

The third stage of the APTC programme (subtly renamed Training Coalition rather than Technical College) from 2018 to 2022 was intended to be better integrated both with Pacific Technical and Vocational Education and Training (TVET) schemes and with potential employers. It aimed to support the emerging Pacific Labour Scheme (Chap. 10) with renewed emphasis on labour mobility, and divide students in the programmes into either a domestic or a labour mobility track; those in the labour mobility track would receive extra training from the Pacific Labour Facility to meet migration requirements. The objective was to ensure that graduates would have an internationally recognised, accredited Australian qualification. APTC was intended to work with national governments, the private sector and civil society organisations across the same nine PICs that participate in the SWP: Fiji, Samoa, Vanuatu, PNG, Solomon Islands, Nauru, Tuvalu, Tonga and Kiribati. The aim was to make APTC training programmes relevant and aligned with current and future career opportunities where skilled employees were in high demand. Nonetheless, the focus of APTC was rather more on the benefits to Australia than to the PICs, with no detailed consideration of the short- or long-term consequences of what might become a substantial drain of skilled workers (unlike the return migration implicit in the SWP program). Its effectiveness and impact are yet to be established.

KANI was a short-lived sectoral training scheme, where i-Kiribati nurses were trained in Australia, principally to remain there and generate remittances rather than transfer superior skills to Kiribati. In a sense it was a subset and microcosm of APTC: a specific programme in one skilled sector and for one country (but which also addressed a nursing shortage in Australia). The scheme was intended to run between 2007 and 2013, with at least two years of training provided by Griffith

University in Brisbane. Once again Kiribati was singled out because of its particular needs. The scheme took in an annual cohort of about 30 potential nurses, whose rationale for entering the programme was to improve the livelihoods of their families. All wanted to stay in Australia, in the absence of good employment opportunities in Kiribati, though the early graduates found it difficult to find such jobs; the best graduates tended to remain and weaker students to return (O'Brien, 2013). It proved an extremely expensive programme, with doubtful benefits other than for the graduates, and quite quickly ended.

The Diversity of Migration

Migration has long been a feature of Pacific existence, even before nineteenth-century 'blackbirding', with islanders being seen as particularly mobile people. By the end of the nineteenth century, in remote atolls like those of Tokelau, 'the idea of permanent migration, involving a severance of many ties with the home island and of seeking one's fortune elsewhere, is well established in … life and thought. For the past 70 years or so it appears to have been accepted that some of nearly every group of siblings must *tahe* ('emigrate') simply because the local resources are seen as insufficient' (Hooper & Huntsman, 1973, pp. 403–404). An already-institutionalised culture of migration was emerging: a long history of migration, characteristic of the atoll states, with their early government involvement, but not of the larger Melanesian states. Although most migration was temporary, it was already underpinned by the desire to benefit both the welfare of the migrants and of family members who remained at home.

In the last half century, after a hiatus following the abrupt end of the problematic blackbirding era, there has been accelerated international migration, especially from the smaller Polynesian and Micronesian states, and ubiquitous rural-urban migration, with declining populations in national peripheries and growing high density urban concentrations. Migration has taken increasingly diverse forms, as it has gone from temporary to more permanent, unskilled to skilled and become more likely to be organised by households and later by governments: a measure of the widespread and growing significance attached to its value.

Migration has usually been a response to perceived socio-economic inequalities that in recent years have been accentuated by both a rise in expectations of incomes and living conditions and greater familiarity with circumstances in overseas countries. The PICs responded to economic booms elsewhere and resultant demands for workers. Changing aspirations, the increased necessity and desire to earn cash, a preference for regular employment, and—ironically—distaste for agricultural work at home gradually gave migration a more obvious economic rationale: a movement in search of wage and salary employment, scarce at home, and some benefit from income generation and consequent social mobility. Even in the more marginal environments of the atoll states migration has remained primarily an economic phenomenon (Shen & Gemenne, 2011; Shen & Binns, 2012; Connell, 2015). Environmental factors may gradually become more significant, as climate change, by stimulating more frequent and more intense cyclones, is likely to contribute further to 'slow onset' migration and quests for resettlement.

One outcome of these international migration flows has been a substantial population of Pacific islanders in the metropolitan states bordering the Pacific. By 2020, about a million people of Pacific ancestry lived outside the island PICs and territories, compared with a resident population of about 11 million (three-quarters of whom are in PNG). Many of these, especially in the United States and New Zealand, are from dependent territories, like the Marshall Islands and the Cook Islands, where there are no legal restrictions on access. Ultimately migration in search of superior livelihoods is a continuation of past practices, but where jet planes are faster, more capacious and travel greater distances than canoes. That is abundantly true of atolls and the atoll states. The growing numbers in the diaspora, the multiple migration strategies and schemes, the crucial role of remittances (hence the MIRAB states), training and education for migration, and even the extent of overstaying, all point to the continued unmet demand for migration. This desire for migration also contributed to, perhaps surprisingly, one final twentieth-century flourish of seasonal agricultural schemes in New Zealand that proved the unexpected precursor of something much bigger as the PICs entered into a new era of simultaneous temporary and permanent migration. That is the subject of the next chapter.

Bibliography

Adams, R. (1990). Plantation Labour in Vanuatu. In C. Moore, J. Leckie, & D. Munro (Eds.), *Labour in the South Pacific* (pp. 140–143). James Cook University.

AIDAB. (1987). *Australia's relations with the South Pacific*. International Development Issues No. 2.

Allen, M. (1968). The establishment of Christianity and cash cropping in a New Hebridean community. *Journal of Pacific History, 3*, 25–46.

Australia. (1978). *Australia and the South Pacific*. Senate Standing Committee on Foreign Affairs and Defence.

Australia. (1979). *Australia and the Third World*. Senate Standing Committee on Foreign Affairs and Defence.

Banivanua Mar, T. (2019). 'Boyd's Blacks': Labour and the making of settler lands in Australia and the Pacific. In V. Stead & J. Altman (Eds.), *Labour lines and colonial power: Indigenous and Pacific Islander labour mobility in Australia* (pp. 57–73). ANU Press.

Banivanua-Mar, T. (2007). *Violence and colonial dialogue. The Australian-Pacific Indentured Labour Trade*. University of Hawai'i Press.

Barlass, T. (2021, November 14). *National park to be renamed over slave link* (p. 2). Sun-Herald.

Bastin, R. (1985). Weasisi mobility: A committed rural proletariat? In M. Chapman & R. Prothero (Eds.), *Circulation in population movement. Substance and concepts from the Melanesian Case* (pp. 175–190). Routledge and Kegan Paul.

Bedford, R. (1973b). *New Hebridean mobility: A study of circular migration*. ANU Department of Human Geography Publication HG/9.

Bertram, G., & Watters, R. (1985). The MIRAB economy and South Pacific microstates. *Pacific Viewpoint, 26*(3), 497–519.

Bonnemaison, J. (1994). *The tree and the Canoe. History and ethnography of Tanna*. University of Hawaii Press.

Brass, T., & Bernstein, H. (1992). Introduction: Proletarianisation and deproletarianisation on the colonial frontier. *Journal of Peasant Studies, 19*(3), 1–40.

Breman, J. (1990). *Taming the coolie beast. Plantation society and the colonial order in Southeast Asia*. Oxford University Press.

Brookfield, H. (1968). 'The money that grows on trees': The consequences of an innovation within a man-environment system. *Australian Geographical Studies, 6*(2), 97–119.
Brookfield, H. (1972). *Colonialism, development and independence. The case of the Melanesian Islands in the South Pacific*. Cambridge University Press.
Brown, R., & Connell, J. (2015). Migration and remittances: A multidisciplinary synthesis. In J. Connell & R. Brown (Eds.), *Migration and remittances* (pp. xiiv–lxviii). Edward Elgar.
Brown, R., Connell, J., & Jimenez-Soto, E. (2014). Migrants' remittances, poverty and social protection in the South Pacific: Fiji and Tonga. *Population, Space and Place, 20*(5), 434–454.
Castles, S. (1988). Temporary migrant workers—Economic and social aspects, in committee to advise on Australia's Immigration Policies. In *Immigration. A commitment to Australia* (pp. 1–10). : Australian Government Publishing Service.
Chambers, A. (1975/1984). *Nanumea* : ANU Development Studies Centre.
Chand, S. (2006). Labour mobility for sustainable livelihoods in Pacific Island States. In M. Powles (Ed.), *Pacific futures* (pp. 127–141). Pandanus Books.
Cheer, J., & Reeves, K. (2013). Roots tourism: Blackbirding and the South Sea Islander diaspora. *Tourism Analysis, 18*, 245–257.
Clemens, M., Graham, C., & Howes, S. (2015). Skill development and regional mobility: Lessons from the Australia-Pacific Technical College. *Journal of Development Studies, 51*(11), 1502–1517.
Connell, J. (1983). *Migration, employment and development in the South Pacific. Country Report No. 7. Kiribati*. SPC and ILO.
Connell, J. (1985). Cocoa, copper and cash: Terminal, temporary and circular mobility in Siwai, North Solomons. In M. Chapman & R. Prothero (Eds.), *Circulation in population movement. Substance and concepts from the Melanesian Case* (pp. 119–148). Routledge and Kegan Paul.
Connell, J. (2008). Niue: Embracing a culture of migration. *Journal of Ethnic and Migration Studies, 34*, 1021–1040.
Connell, J. (2009). *The global care chain. From the Pacific to the World*. Routledge.
Connell, J. (2012). Population resettlement in the Pacific: Lessons from a Hazardous History. *Australian Geographer, 43*, 127–142.
Connell, J. (2013). *Islands at risk*. Edward Elgar.
Connell, J. (2015). Vulnerable Islands: Climate change, tectonic change, and changing livelihoods in the Western Pacific. *The Contemporary Pacific, 27*(1), 1–36.

Connell, J., & Brown, R. (2004). The remittances of migrant Tongan and Samoan Nurses in Australia. *Human Resources for Health, 2*(2), 2.

Connell, J., & Lea, J. (2002). *Urbanisation in the Island Pacific. Towards sustainable development*. Routledge.

Connell, J., & Lutkehaus, N. (2017a). Escaping Zaria's fire? The volcano resettlement problem of Manam Island, Papua New Guinea. *Asia Pacific Viewpoint, 58*(1), 14–26.

Connell, J., & Lutkehaus, N. (2017b). Environmental refugees? A tale of two resettlement projects in Papua New Guinea. *Australian Geographer, 48*(1), 79–95.

Connell, J., & Tabucanon, G. (2016). From Banaba to Rabi: A Pacific model for resettlement? In S. Price & J. Singer (Eds.), *Global implications of development, disasters and climate change* (pp. 91–107). Routledge.

Corris, P. (1970). Kaisulia of Ada Gege: A strongman in the Solomon Islands. In J. Davidson & D. Scarr (Eds.), *Pacific Islands portraits* (pp. 253–265). ANU Press.

Corris, P. (1973). *Passage, port and plantation. A history of Solomon Islands Labour Migration 1870–1914*. Melbourne University Press.

Cullen, R. (2012). Empire, Indian indentured labor and the colony: The Debate over 'coolie' Labor in New South Wales, 1836–1838. *History Australia, 9*, 84–109.

Curtain, R. (1980). The structure of internal migration in Papua New Guinea. *Pacific Viewpoint, 21*(1), 42–61.

Davies, L. (2019). A regulated labour trade across the Torres Strait: Papuan and New Guinean Domestic Labourers in Australia. In V. Stead & J. Altman (Eds.), *Labour lines and colonial power: Indigenous and Pacific Islander Labour mobility in Australia* (pp. 75–102). ANU Press.

Dennis, J. (2003). *Pacific island seafarers: A study of the economic and social implications of seafaring on dependents and communities*. South Pacific Commission.

Docker, E. (1970). *The Blackbirders. The recruiting of South Seas Labour for Queensland* (pp. 1863–1907). Angus and Robertson.

Doyle, J., & Howes, S. (2015). *Australia's Seasonal Worker Program: Demand-side constraints and suggested reforms*. World Bank.

Firth, S., & Munro, D. (1990). Compagnie et Consulat: Loies germaniques et emploi des travailleurs sur les plantations de Samoa, 1864–1914. *Journal de la Société des Océanistes, 91*, 115–134.

Fitzpatrick, P. (1980). *Law and State in Papua New Guinea*. Academic Press.

Frater, M. (1922). *Midst Volcanic Fires*. James Clarke.

Geddes, W. (1988). *South Pacific Emigration Study*. Deakin University.
Gleeson, P. (2020a, June 18). From Louisiana to Queensland: How American slave owners started again in Australia, *The Conversation*.
Graves, A. (1986). Colonialism and indentured labour migration in the Western Pacific, 1840–1915. In P. Emmer (Ed.), *Colonialism and migration: Indentured labour before and after slavery* (pp. 237–259). Martinus Nijhoff.
Graves, A. (1993). *Cane and labour. The political economy of the Queensland Sugar Industry 1862–1906*. Edinburgh University Press.
Hermann, E., & Kempf, W. (2017). Climate change and the imagining of migration: Emerging discourses on Kiribati's Land Purchase in Fiji. *The Contemporary Pacific, 29*(2), 231–263.
Hooper, A., & Huntsman, J. (1973). A demographic history of the Tokelau Islands. *Journal of the Polynesian Society, 82*(4), 366–411.
Howe, K. (1984). *Where the waves fall. A New South Sea Islands History from First Settlement to Colonial Rule*. Allen and Unwin.
Howes, S. (2021, July 23). *Foreign workers on Australian farms: Not if but how.* ANU DevPolicyBlog.
Howlett, D. (1985). Australia's next boat people? The need for a concessionary migration policy for Kiribati and Tuvalu, unpublished mimeo.
Jackson Report (Committee to Review the Australian Overseas Aid Program). (1984). *Report of the committee to review the Australian Overseas Aid Program.* Australian Government Publishing Service.
Kagan, S. (2015, January 20). Making the case for preferential access to labour markets for Kiribati and Tuvalu migrants. ANU DevPolicyBlog.
Kagan, S. (2016a). 'On the ship, you can do anything': The impact of international cruiseship employment for I-Kiribati women. *Pacific Studies, 36*(1), 35–51.
Kagan, S. (2016b). The role of remittances in risk management and resilience in Tuvalu: Evidence and potential policy responses. In A. Milan, B. Schraven, K. Warner, & N. Cascone (Eds.), *Migration, risk management and climate change: Evidence and policy responses* (pp. 175–191). Springer.
Keith-Reid, R. (1984, July 10). Australia's new vision of islands. *Islands Business*, 66.
Knudson, K. (1981). Adaptational persistence in the Gilbert Islands. In R. Force & B. Bishop (Eds.), *Persistence and exchange* (pp. 91–99). Honolulu.
Koczberski, G., & Curry, G. (2005). Making a living: Land pressures and changing livelihood strategies among oil palm settlers in Papua New Guinea. *Agricultural Systems, 85*, 324–339.

Kraemer, D. (2013). *Planting roots, making place: An ethnography of young men in port Vila, Vanuatu*. PhD thesis, London School of Economics.

Lal, B. (2004). *Girmitiyas. The Origins of the Fiji Indians*. Fiji Institute of Applied Studies.

Larson, E. (1981). The effects of a plantation economy on a Polynesian population. In K. Michaelson (Ed.), *And the poor get children. Radical perspectives on population dynamics* (pp. 180–193). Monthly Review Press.

Le De, L., Gaillard, J.-C., Friesen, W., & Matautia-Smith, F. (2015). Remittances in the face of disasters: A case study of rural Samoa. *Environment, Development and Sustainability, 17*, 653–672.

Loy-Wilson, S. (2017). Coolie Alibis: Seizing gold from Chinese Miners in New South Wales. *International Labor and Working Class History, 91*, 28–45.

Martinez, J. (2005). The end of indenture? Asian workers in the Australian Pearling Industry, 1901–1972. *International Labor and Working Class History, 67*, 125–147.

Maude, H. E. (1981). Slavers in paradise. The Peruvian Labour Trade in Polynesia, 1862–1864. : ANU Press.

McAdam, J. (2014). Historical cross-border relocations in the Pacific: Lessons for planned relocations in the context of climate change. *Journal of Pacific History, 49*, 301–332.

McAdam, J. (2015). The emerging New Zealand jurisprudence on climate change, disasters and displacement. *Migration Studies, 3*(1), 131–142.

McCall, G. (1976). European impact on Easter Island: Response, recruitment and the Polynesian experience in Peru. *Journal of Pacific History, 11*, 90–105.

McKinnon, A. (2019, June). Blackbirds. Australia had a slave trade?, *The Monthly*, 44–55.

Moore, C. (1985). Kanaka. *A History of Melanesian Mackay*. Institute of Papua New Guinea Studies, Port Moresby.

Moraes-Gorecki, V. (1994). 'Black Italians' in the sugar fields of North Queensland: A reflection on labour inclusion and cultural exclusion in tropical Australia. *Australian Journal of Anthropology, 5*(3), 306–319.

Munro, D. (1992). Gilbert and Ellice Islanders on Queensland Canefields, 1894–1899. *Journal of the Royal Historical Society of Queensland, 14*(11), 449–465.

Nelson, H. (1976). *Black, white and gold. Goldmining in Papua New Guinea 1978–1930*. ANU Press.

Newbury, C. (1980). The Melanesian Labor Reserve. Some reflections on Pacific labor markets in the nineteenth century. *Pacific Studies, 4*(1), 1–25.

Nishitani, M., & Lee, H. (2019). Becoming 'Overstayers': The coloniality of citizenship and the resilience of Pacific Farm Workers. In V. Stead & J. Altman (Eds.), *Labour lines and colonial power: Indigenous and Pacific Islander Labour Mobility in Australia* (pp. 159–184). ANU Press.

O'Brien, L. (2013). *'Migrating with dignity'. A study of the Kiribati-Australia Nursing Initiative (KANI)*. MA thesis, University of Kansas.

Parliament of the Commonwealth of Australia. Joint Committee on Foreign Affairs Defence and Trade. (1989). *Australia's relations with the South Pacific*. Australian Government Publishing Service.

Paton, F. (1903). *Lomai of Lenakel. A hero of the New Hebrides*. Hodder and Stoughton.

Paton, J. (1891). In G. John & D. D. Paton (Eds.), *Missionary to the new Hebrides. An autobiography*. Hodder and Stoughton.

Peattie, M. (1988). *Nan'yo. The rise and fall of the Japanese in Micronesia, 1885–1945*. University of Hawaii Press.

Petrou, K. (2020). *If everyone returned the Island Would Sink: Urbanisation and migration in Vanuatu*. Berghahn.

Petrou, K., & Connell, J. (2018). 'We don't feel free at all': Temporary ni-Vanuatu workers in the Riverina, Australia. *Rural Society, 27*(1), 66–79.

Rasmussen, A. (2015). *In the absence of the Gift. New forms of value and Personhood in a Papua New Guinea Community*. Berghahn.

Robertson, H. (1902). *Erromanga. The Martyr Isle*. Hodder and Stoughton.

Saunders, K. (1982). *Workers in bondage. The origins and bases of unfree labour in Queensland 1824–1916*. University of Queensland Press.

Scarr, D. (1967). Introduction. In W. E. Giles (Ed.), *A Cruise in a Queensland labour vessel to the South seas* (pp. 1–32). ANU Press.

Scarr, D. (1970). Recruits and recruiters: A portrait of the labour trade. In J. Davidson & D. Scarr (Eds.), *Pacific Islands portraits* (pp. 225–251). ANU Press.

Shen, S., & Binns, T. (2012). Pathways, motivations and challenges: Contemporary Tuvaluan migration to New Zealand. *GeoJournal, 77*, 63–82.

Shen, S., & Gemenne, F. (2011). Contrasted views on environmental change and migration: The case of Tuvaluan migration to New Zealand. *International Migration, 49*, 224–242.

Shineberg, D. (1967). *They came for sandalwood: A study of the sandalwood trade in the South-West Pacific 1830–1865*. Melbourne University Press.

Shineberg, D. (1999). *The people trade. Pacific Island Laborers and New Caledonia, 1865–1930*. University of Hawai'i Press.

Speiser, F. (1913). *Two years with the natives of the Western Pacific*. Mills and Boon.
Storey, D., & Hunter, S. (2010). Kiribati: An environmental 'perfect storm'. *Australian Geographer, 47*, 161–181.
Tabucanon, G., & Opeskin, B. (2011). The resettlement of the Nauruans in Australia: An early case of failed environmental migration. *Journal of Pacific History, 46*, 337–356.
Thurnwald, R. (1932). The psychology of acculturation. *American Anthropologist, 34*(4), 557–569.
Toatu, T. (1980). Seamen and cultural change in Kiribati. *Pacific Perspectives, 8*, 31–32.
Trundle, A., Barth, B., & McEvoy, D. (2018). Leveraging endogenous climate resilience: Urban adaptation in Pacific Small Island developing states. *Environment and Urbanization, 31*(1), 53–74.
Viviani, N. (1970). *Nauru. Phosphate and Political Progress*. ANU Press.
Ward, M. (1971). *The role of investment in the development of Fiji*. Cambridge University Press.
Ward, R. (1990). Contract labor recruitment from the highlands of Papua New Guinea, 1950–1974. *International Migration Review, 24*, 273–296.
Watters, R. with Banibati, K. (1977/1984). Abemama. ANU Development Studies Centre.
Wessen, A., Hooper, A., Huntsman, J., Prior, I., & Salmond, C. (1992). *Migration and health in a small society. The case of Tokelau*. Clarendon Press.
Wilson, J. (1966). *Economic survey of the new hebrides*. HMSO.
Wilson, K. (2013). Wan Laki Aelan? Diverse development strategies on Aniwa, Vanuatu. *Asia Pacific Viewpoint, 54*(2), 246–263.

4

The Revival of Guestwork

Pacific migration diversified over time, from different places, to different destinations and for different reasons. However, loosely economic motives were never far away—whether by those who sought out migrant workers or by the migrants themselves, or indeed by their families who sponsored them, went with them or waited for their remittances and their return. During the twentieth century migration had shifted from being dominated by short-term employment in agriculture, through more diverse employment and longer-term work in mines, on ships and in growing urban centres, eventually to the more skilled migration of health workers, sportsmen, bureaucrats and others. Migration became a family, rather than a male preoccupation, and was increasingly more likely to be permanent, as diaspora populations grew, in much of Polynesia at least, to become more numerous than the populations that remained at home. Amidst this growing diversity of migration, with Pacific islander households so often taking on new 'modern' livelihoods, and many living in metropolitan cities, demand for agricultural labour grew in New Zealand and Australia, as it was also doing in western Europe and North America. New Zealand and Australia turned to their nearest neighbours. Once again, Pacific islanders sought out, and were organised and encouraged to take up, agricultural work, this time for much shorter time periods.

K. Petrou, J. Connell, *Pacific Islands Guestworkers in Australia*,
https://doi.org/10.1007/978-981-19-5387-3_4

This chapter examines the rationale for the re-establishment of a new 'labour trade', first in New Zealand and then in Australia, as Stephen Castles' 'new age of migration' reached Australasia in unexpected form. The 'new age' was initiated through a small-scale New Zealand trial that became a prelude to the substantial Recognised Seasonal Employer (RSE) and Seasonal Worker Programme (SWP) schemes. We examine here how, why and where the new era came about and then assess the earliest schemes in New Zealand and Australia. We consider how the new RSE and SWP schemes were welcomed by all the Pacific Island Countries (PICs), despite significant variations in participation rates, and how they gradually expanded, suggesting an initial 'triple win'.

A Twentieth-Century Twist—Back to Agriculture

While twentieth-century migration was generally becoming focused on skills, one more twist occurred. A few guestworker schemes had belatedly sprung up late in the century, but escaped analytical interest. Between 1978 and 1986, New Zealand developed a South Pacific Work Permit Scheme, after being unable to obtain local labour for unskilled rural tasks such as scrub cutting, forestry, tobacco picking and market garden work. The scheme employed a few hundred workers each year, but Fiji and Tonga were the only two countries to be significantly involved. Fiji sent around 500 workers a year in the 1970s. Employment was restricted to rural areas, visas were only issued for four months and workers were not allowed to return to New Zealand for another twelve months. The four-month period was set by Fiji to enable the scheme to reach as many people and regions as possible and ensure that any disruption caused by workers being in New Zealand was minimised. There was some evidence that temporary migrants were chosen at village level according to who might benefit most. Employers sought longer periods of employment, and more return migration, arguing that they would have to engage in too much training with only short-term workers. The scheme proved particularly valuable by employing workers in activities disdained by

New Zealanders, with one distinctive group being Muslim Indo-Fijians employed as halal slaughtermen of lamb for Middle Eastern export markets. The scheme only ended following a downturn in the New Zealand economy.

Workers were mainly young men in their 20s, and despite Fiji's efforts, many became repeat migrants. Workers were expected to use their savings, from incomes roughly five times what they might have earned in Fiji, for investments in housing, education and business development. That appears to have been what transpired, although significant proportions were also used for collective village projects. The scheme proved particularly valuable for parts of Fiji badly affected by droughts and floods where savings were used for reconstruction. But it was not an unalloyed success. Fijian scrubcutters in the Napier area in 1971 complained of 'slavery' and earned less than NZ$300 (A$268) over four months, through unfair pay rates, while working ten-hour days. Conversely tobacco famers complained of unreliable workers (de Bres et al., 1974, pp. 12–15). Although the Fijian government emphasised the need for skill acquisition and training rather than the potential for remittances, skills transfer was non-existent and that policy focus gradually disappeared. Nonetheless the Fijian Minister of Labour was enthusiastic enough to state that this particular provision of employment opportunities was 'the greatest form of overseas aid any overseas government can offer' (quoted in Levick & Bedford, 1988, p. 16) while all the workers were interested in returning.

Several hundred workers from Tonga worked in New Zealand on the same scheme. They were selected by ballot, and were not allowed to return for a second year, to balance opportunities between regions and individuals (Kite, 1975). Numbers declined in the 1980s as local workers became available. While no more than around 380 Tongans participated, even in the best year the demand was much greater; in one year, almost 9000 applications were received (Connell, 1983). A much smaller number of workers travelled from Samoa, partly because there were long-term migration possibilities for Samoans in New Zealand. In distinct contradiction to Fiji, the scheme was regarded as a failure in Samoa. However, workers were able to save between NZ$4700 (A$4204) and NZ$7600 (A$6799) per year, and were chosen according to their personal circumstances, their plans for using the income, their alternative sources of income and degree of poverty, and the consequences of their absence for their dependents (Macpherson, 1981). Once again workers were being selected and sponsored according to their perceived needs and some regard for equity. An even smaller group of workers had arrived in New Zealand from Tuvalu in the mid-1970s, but that source too disappeared as opportunities declined.

While the South Pacific Work Permit Scheme officially ended in the 1980s, it had become increasingly viewed negatively in New Zealand with changes to formal migration policy that favoured more long-term skilled migration; new perceptions over the relationship between employment, migration and foreign policy (Levick & Bedford, 1988); and the greater ease of finding workers in New Zealand as unemployment increased. Both politics and economics contributed to its demise. However, the scheme was viewed positively by the workers and by both the Fijian and New Zealand governments—an early version of the 'triple win'. It proved remarkably similar to the schemes again adopted by New Zealand a quarter of a century later: an accidentally and largely forgotten precursor to twenty-first-century developments.

By contrast, while there was never a period in post-war Australian development when temporary migration was not seen by some as a solution for labour shortages, in the twentieth century Australia was never tempted to seriously consider such schemes. In 1973 the topic of guestworkers entered parliamentary debates, with Al Grassby, the flamboyant Minister for Immigration in the Whitlam Labor government, declaring: 'There have been suggestions that we might have guestworkers from the South Seas for harvest purposes; guestworkers from Norway to become fishermen temporarily; guestworkers from China to build roads in Victoria; and guestworkers from Europe to fill some of the gaps in the building trade' (quoted in Hansard, 25 May 1973). Such suggestions, then as subsequently, came from employer organisations and were opposed by the trade union movement. It was more than three decades before Australia finally gave way.

A Theoretical Detour … and a Context

New Zealand's brief phase of guestwork during the 1970s–1980s was situated within changing perspectives on the relationship between migration and development. For most of the twentieth century, a widespread perception persisted that migration and development were somewhat incompatible, with migration antithetical to notions of sustainable development and self-reliance. A dependence on remittances was viewed as somehow sneaky, uncertain and inconsistent with national development. Thus, countries like Ireland, where a century of migration had brought no fundamental underlying changes, were seen as failing and incapable of developing their own national resources. While J. K. Galbraith had forcibly argued that migration actually brought development, this was too often denied in the countries that absorbed the migrants:

> As a remedy for poverty, [migration] focuses with precision on those for whom such a policy is alone workable and for whom it must be designed—those who are motivated to improve their economic position. No effort or money is wasted on those not yet so motivated. … Migration … is the oldest action against poverty. It selects those who most want help. It is good for the country to which they go; it helps to break the equilibrium of poverty in the country from which they come. What is the perversity in the human soul that causes people to resist so obvious a good? (1979, p. 109)

It might have been a manifesto for seasonal worker programmes, but few were listening.

However, as the world entered an 'age of migration', increased migration became inevitable. Multiple studies of migration ensued and have been variously summarised (e.g. de Haas, 2010, 2012; Massey et al., 1993; Castles, 1998; de Haas et al., 2020). Most focused on long-term, or seemingly permanent, migration and, with exceptions (e.g. Hugo, 1982), relatively few examined temporary and especially seasonal migration. It was considered a lesser phenomenon of no great consequence or endurance.

Gradually even neo-classical perspectives, based on supposedly rational utility-maximising decision-making that centred on increased earnings and the flow of remittances, with a potential bonus in the acquisition of skills and social capital, brought more support for pro-migration policies, and even for temporary migration (Wickramasekera, 2011). That was summarised in the new economics of labour migration (NELM) perspective where migration was a household strategy to diversify incomes to reduce risk and overcome relative deprivation within a particular community, and where remittances might lessen production and market constraints faced by households in the Global South. In this context migration decisions were driven not only by migrants' desires to improve their incomes but also by their interest in reducing their families' deprivation relative to other local families. Moreover, the greater probability of return, virtually guaranteed within short-term labour migration, induced migrants to focus on their savings.

Such narrow economic perspectives could easily be criticised by those who perceived social costs and also suggested that temporary migration schemes particularly could all too easily be 'where the South forgoes self-development in favour of being a long-range bedroom community to supply the labour for dirty, dangerous and difficult jobs in the north' (Ellerman, 2005, p. 617). In this perspective temporary migrants could be seen as a 'reserve army' of workers coerced into performing what could be the underpaid and precarious employment that local workers avoided and through which an underclass emerged. That critique was particularly directed at seasonal employment where socio-economic mobility was implausible, and extended into wider notions of core-periphery relations, limited agency and even the imposition of capitalist systems on 'traditional societies' so undermining their integrity, all of which suggested dependency and structural constraints to development in the periphery. Although long-term migration could reduce local unemployment and contribute to a shortage of workers (especially in certain skilled categories) and an ageing local population structure, seasonal migration largely avoided these problems.

Neo-classical migration models, with their emphasis on 'trickle-down' benefits via remittances (and their investment and development potential), improved consumption and increased human capital, were thus regarded as somewhat optimistic and embedded in a crude version of modernisation, that neglected the possibility of skill-drain, increased inequality and lack of agency, and ignored the structural constraints to development (de Haas, 2012). Minimally it was readily evident both that migration benefited some communities, places and people, at particular times and according to the structure of migration, but had quite different impacts elsewhere, and that much depended on what facets of development (and its definition) were given priority. As the focus of development shifted through the Millennium Development Goals (MDGs) to the greater subtlety and complexity of the Sustainable Development Goals (SDGs), the number of potential development variables increased considerably, amid greater uncertainty about agreed-upon endpoints, while environmental change emerged as an additional factor in migration.

Ever more complex and subtle arguments about which places, why them and who within them marked new considerations of gender, temporal shifts, the role and use of remittances, social costs and freedoms, all largely still embedded in contexts where development interventions retained sedentary roots which were focused on the control of mobility and tended to cast migration as a symptom of development failure (Bakewell, 2008). To a substantial extent, conceiving migration more positively necessitated reconceptualising development within a mobile world, although that somewhat exonerated seasonal migration, where assumptions of return were central. Further arguments shifted the balance of opinion towards the positive role of migration, a shift that paralleled the global rise of migration, frustrations of rural development and the more widespread championing of the role of remittances in development, even if partly by default.

In this century migration was consequently rediscovered as a means of facilitating development, offering a route to mitigating deepening inequalities. In lieu of aid, national governments, international funding organisations and diasporic organisations have all mobilised migrants to fund development initiatives in their home countries, sometimes pursuing the argument that migrants were more capable of remedying 'grassroots' problems than alien and bureaucratic aid agencies. But the jury was constantly out as global and local economies and societies evolved and what appeared successful in some places was not necessarily so in others. In the Pacific it was apparent that whatever the costs of development, more and more people were voting with their feet, and choosing or seeking to choose one or other form of migration. Even the seemingly final fling of guestwork had generated excessive interest far beyond the capacity of the programmes.

Hard Realities

Beyond often detached scholarly theory and debate was practice. Both globally and regionally seasonal migration schemes were becoming more common around the turn of the century. In the Pacific the practical legacy of the late twentieth century suggested there was some merit to early schemes that had brought Fijians and Tongans to New Zealand, alongside recognition that islanders were becoming established migrants, and that after around a quarter of a century of independence PIC economies were failing to grow. Broader debate over migration and development was embedded in a context where, by the 1980s, the MIRAB (Migration, Remittances, Aid and Bureaucracy) model (Chap. 3) had suggested the utility of migration for development, initially especially in the smaller PICs, where migration was normative, and domestic development opportunities scarce. While global theoreticians took little note of the Pacific, regional interests increasingly pointed to benefits from migration. Geoff Bertram (1986) argued that migration was a perfectly 'sustainable' development strategy as long as the remittances could be maintained over time; others however saw it as inadequate and unsustainable, merely fostering dependency and excessive consumption (Hayes, 1982). Bernard Poirine (1998) suggested that negative views failed to recognise that aid and remittances were not just handouts, but represented exchanges with the rest of the world, that made the most of PICs' scarce comparative advantages. Much of this debate, significant in the twentieth century, concerned effectively permanent migration, so that when short-term seasonal migration became important in this century, the validity and effectiveness of migration for development arguments were enhanced.

In the twenty-first century remittances have globally outstripped overseas aid, not least in some of the PICs, and especially Tonga and Samoa where this had occurred by the mid-1980s. That transition strengthened the arguments of Bertram, Poirine and others, for the Pacific, and had prompted a wider institutional recognition of the utility of migration and the virtues of self-help. Devesh Kapur (2004) began to emphasise what he called a 'remittances mantra', arguing that remittances were the most stable source of external finance, played a critical social insurance role in many countries, were generally pro-poor and were unlikely to experience a version of 'aid fatigue'. Kapur's study was written with some collaboration from Dilip Ratha of the World Bank, who enthusiastically embraced the notion. Indeed, unlike formal development aid, remittances avoided urban bias, mitigated bad governance, and were sent and received by those who knew the nature of local development problems (or at least knew what they wanted). That enabled an escape from extremes of uneven development, and supported welfare provision and some degree of human capital formation. While such global arguments centred on longer-term migration—where remittances were often substantial—all the then available evidence from the Pacific (Connell & Brown, 2005) indicated that similar circumstances existed in the region and the same arguments were relevant to short-term seasonal migration, if on a smaller scale. While Kapur had ignored the Pacific islands, the overall tide in favour of migration was turning and the World Bank was to play a further significant regional role.

Enter the World Bank

For more than a decade the Australian government had ignored the possible benefits of a labour migration scheme either for Australia or for the PICs. However, such schemes had begun in Canada, New Zealand was again moving towards its own scheme and prominent voices in Australia were calling for change. More practically, Australian farmers were having greater difficulty in securing workers. In 2003 a bipartisan Australian Senate Committee into Australia's relations with the Pacific region considered various submissions from agricultural groups who pointed to labour shortages that limited production and harvesting and suggested the possibility of six-month long working visas for people from 'rural regions which most need economic development' (Australian Government, 2003, p. 73). This was again opposed by the Department of Immigration, defending a non-discriminatory migration policy and challenging the need for unskilled migrants when unemployment rates were significant. They also pointed to the inability of guestworker schemes elsewhere to provide long-term benefits for either sending or receiving countries, and the potential for illegal overstaying. The former World Bank economist, Helen Hughes (2003), argued that it was paternalistic for Pacific islanders to be accorded special immigration conditions. Nonetheless, the Committee supported a pilot scheme that would address the needs of agricultural industry groups and would 'make available labour for seasonal work, the provision of training and knowledge to the labour employed through the scheme to ensure a transfer of skills and arrangements of remuneration … to ensure that these workers are not the subject of systematic exploitation' (Australian Government, 2003, pp. 75–76). It made no mention of particular PICs but suggested that the pilot be carried out in the 'smaller island countries' that would benefit most.

Despite the Senate Committee's Inquiry, it remained true, as Radio Australia pointed out, that

> the idea of Pacific 'guestworkers' is the great taboo of Australia's Pacific policy. It is raised by the region every time Australia asks, and often behind closed doors whenever political leaders meet. But Canberra has never embraced it, despite a series of recommendations over the past two decades from Australian inquiries. (Radio Australia, 19 May 2003)

Concern over seasonal labour schemes derived from longstanding fears that 'cheap foreign labour' would 'steal' Australian jobs (and indigenous workers should have priority) and undermine local wages and conditions, workers would overstay, and benefits to Australia and to the source countries would be trivial. In short, Australia preferred skilled migrants and a non-discriminatory migration policy (despite the existence of a Working Holiday Maker (WHM) scheme, begun in 1975, and special preferences for migration from New Zealand). PICs increased pressure on Australia, through the 2005 Pacific Plan and at the same year's Pacific Islands Forum, but John Howard's Liberal coalition government was not for turning (and nor was the New Zealand government), although the Labor Party was willing to consider the possibility. Meanwhile, supply and demand were evident as a small but significant number of Pacific islanders were entering Australia on three-month tourist visas and taking work in hospitals, shops and factories, and as fruit pickers. The Foreign Affairs Minister Alexander Downer was unconvinced and argued that Fijians did not have the skills to pick fruit (*Fiji Times*, 3 October 2005). Objections were even embedded in memories of blackbirding, apparent in the then Treasurer Peter Costello's opposition, and in the objections of Alexander Downer, in the face of increased pressure from PICs:

> I don't like the idea of guest workers I don't think we should go back to the schemes that existed back in the nineteenth century where cheap labour was brought in from the South Pacific, particularly into Queensland … Just to take unskilled people for the citrus blocks of the Riverina or South Australia or somewhere for a few weeks every year—I don't think the idea is going to work… In the end Fiji would get very little out of it. I think there'd be a certain amount of resentment in certain parts of Australia. (Quoted in Mares, 2007, p. 76)

It seemed curious to imply that any new scheme would be much the same as in the nineteenth century. Downer went on to say: 'The answer to the Pacific's large and growing unemployment problems does not lie in a few hundred unskilled young people coming to Australia to pick fruit for a few months of the year' (quoted in Maclellan & Mares, 2006, p. 32). It was not quite the last flourish of opposition.

Willingness to contemplate guestworkers was in large part the consequence of World Bank intervention. The Bank organised and introduced more sophisticated analysis of policy and practice and laid much of the groundwork for the eventual policy shift. Curiously, evidence suggests that funding for the study came via the Australian Treasury, indicating that there was new national political interest. Catalysed by commissioned working papers—that came together in a Sydney conference—the World Bank published *At Home and Away. Expanding Job Opportunities for Pacific Islanders Through Labour Mobility* in 2006. Nic Maclellan and Peter Mares contributed an overview of labour mobility in the Pacific, followed by an economic analysis of remittances by Richard Brown and others, that became the core of its policy directions and recommendations (Maclellan & Mares, 2006; Brown et al., 2006). The report pointed to the conjuncture of elusive economic growth and a growing youth population in the PICs, emphasised that 'finding productive employment is becoming more challenging' and argued that 'greater labour mobility would expand the employment options available to Pacific islanders, but it is currently limited and skewed in favour of skilled workers' and 'motivated by the need for jobs for the Pacific islanders who cannot source them domestically' (World Bank, 2006b, pp. vii–viii). It suggested that migration might even outweigh the gains from further trade liberalisation. If not the first report to have similar objectives and recommendations, it was unusually comprehensive, came from a respected international organisation and was supported by academic research that focused on three themes: the extent of demographic pressures facing the Pacific (in a chapter entitled 'The Young and the Restless'), the impact of remittances on sending households and communities (especially in Fiji and Tonga) and the possibility of 'designing migration programs that balance the benefits of migration with the concerns of sending and receiving countries'. It proved timely.

The report recommended 'the exploration of pilot schemes between Pacific populations and other nations with an economic need for unskilled labour and a development interest in the Pacific region' (p. ix). The 'other nations' were unspecified, but the implication was obvious since the report used Australia and New Zealand as examples of where demographic trends pointed to future labour shortages being likely. The ten World Bank member PICs—all the independent PICs (except Tuvalu and Nauru, that had not yet joined the Bank)—were the likely source countries, but the report mainly utilised available data from Fiji and Tonga.

In addition, the report pointed to the value of migration as a means of fostering 'greater regional stability' and achieving poverty alleviation and 'pro-poor outcomes' which meant that 'migration options need to be extended beyond the skilled and elite to the poor and unskilled who are unlikely to find such opportunities domestically' (p. ix). It went on to emphasise that in the PICs strong evidence existed that remittances reduced both poverty and relative deprivation, contributed to investments in education, and provided social protection, but this would be enhanced were migration opportunities to be extended to the poorest. Directed by Manjula Luthria, the senior regional economist of the World Bank, and the report's editor, it developed Devesh Kapur's general statements for a particular region.

The Bank suggested cornerstone '4-Cs': choice of workers with appropriate skills (who were not overqualified), circular movement to allow good employees to return in a second year, cost-sharing of travel expenses between employers and migrants, and commercial viability (so that the scheme reflected labour market conditions rather than having fixed quotas). While it examined horticulture in Australia as a potential arena where temporary work programmes might be established, it stepped back from suggesting where regional labour might be deployed, arguing that this was best left to the private sector to identify, but it did indicate that there were 'likely to be several sectors in other neighbouring countries where such programs could be beneficial' (p. x). Agriculture was not the only possibility. Nor were Australia and New Zealand the only possible destinations; South Korea and Malaysia were also regarded as possible subsequent labour markets.

The report revived and generated debate. In parts of the Pacific pessimism remained. A leading article in Papua New Guinea's main newspaper advised PNG and other PICs to abandon trying to obtain seasonal employment since Australia would never give way, but was instead promising a technical college for skills that would be locally useful: 'We appreciate the World Bank's report but we must say that report will not be enough to change the Australian government mindset' (*Post-Courier*, 16 August 2006). Other PICs with fewer development opportunities were more accepting. Helen Hughes remained a fierce opponent, arguing that the World Bank's 'pressure on developed countries to accept more immigrants … follows its abandonment of the key role of growth in development in favour of welfare' (Hughes & Sodhi, 2006, p. 8). However, as the ABC reported, the notion of a Pacific guestworker scheme was stirring some unusual alliances, underpinned by those between the Labor government and the National Farmers Federation (NFF), whereas the Construction Forestry Mining and Energy Union (CFMEU) expressed concerns over labour standards and the exploitation of migrants. Particularly hostile to the proposals was Bill Shorten, then

National Secretary of the Australian Workers Union (AWU), who claimed that fruit growers in Mildura (Victoria) were 'plotting' to bring in 10,000 Chinese workers to pick fruit and vegetables for four months each year:

> This is a race to the bottom. A significant portion of Victoria's fresh fruit crop is picked by undocumented workers who are highly vulnerable to exploitation and in some cases are offered wages as low as $3 per hour. Let's call it the return of the Kanak culture. (Quoted in *The Fiji Times*, 30 May 2006)

But PICs continued to press, OXFAM called for a pilot scheme, key Australian organisations such as the NFF now supported such a scheme, thousands of temporary skilled and semi-skilled workers were entering Australia on temporary 457 visas and Australia welcomed skilled—but less needy—Pacific island doctors, nurses and rugby players. Ironically, some South Sea Islanders saw a modern organised scheme as partial 'restitution' for blackbirding.

While Australian debate focused primarily on the economic need for workers in Australia, as debate shifted towards the role of seasonal migration as both aid and trade, and as a factor in potentially reducing instability in the region, opinion moved further towards support. Australia carefully watched New Zealand's RSE scheme that had begun in 2007 and perceived it as generally positive. In 2005 the NFF had developed their own *Labour Shortage Action Plan* and were arguing in 2008 that there was a shortage of as many as 100,000 workers, and so supported a Workforce from Abroad Employment Scheme similar to that in Canada.

The Commissioner of Labour from Vanuatu, Lionel Kaluat, who had already sent 800 workers to New Zealand through the RSE scheme (Chap. 5), was particularly enthusiastic and sought to send 5000 workers to Australia (ABC Radio, 15 April 2008). Alexander Downer had been converted and was arguing that Nauru especially desperately needed such opportunities. In August 2008, Kevin Rudd's new Labor government formally announced at the Pacific Islands Forum a pilot scheme for four PICs (Vanuatu, Tonga, Kiribati and PNG) in what was initially somewhat tentatively called the Pacific Seasonal Worker Pilot Scheme (PSWPS). Significantly the scheme was to be organised through the

Department of Education, Employment and Workplace Relations rather than through Foreign Affairs (and therefore through its then aid arm, AusAID).

The CFMEU still opposed any scheme; as its national secretary, John Sutton argued: 'It would be the thin end of the wedge … it would spread out to other industries. What is most difficult with guestworker arrangements is that the visas are temporary and contingent and there's a substantial onus on workers not to complain' (quoted in Morris, 2008, p. 69). By contrast, the AWU, which covered Australian agricultural workers, had finally lent the scheme support, while most labour unions instinctively argued against migration, that the scheme would displace Australians, and seasonal workers would not get the range of welfare provisions and entitlements of Australian workers, that ignored demand. Caught up in complex arguments about skilled workers, inadequate data existed on demand; there were no effective local assessments and no feasible national perspective. It had long been pointed out that employment conditions on many farms—often distant from towns—were unattractive. A 1992 study of fruit pickers in the Goulburn Valley (Victoria) concluded that 'Picking is undisputedly an unattractive job … work is hard and low paid [with] inadequate accommodation, lack of efficient transport, poor working conditions and deficient work organisation' (Ruben, quoted in Hanson & Bell, 2007, p. 102). Fragmented data suggested significant regional shortages, the unwillingness of Australians to meet these shortages, discontent with backpacker labour and the need for new mobility to remedy the shortages.

Almost immediately after the Bank's *At Home and Away*, the Australian Senate's Standing Committee on Employment, Workplace Relations and Education completed its own inquiry and produced a much less detailed report entitled 'Perspectives on the future of the harvest labour force' (Australia, 2006). Although the Committee reviewed some of the same evidence and referred to the Bank's report, it did not perceive that the labour shortage was then significant enough to justify contracting workers from PICs, pointed to a lack of accommodation and recognised existing exploitation of illegal workers (a rather prescient observation). It made no recommendations, not even for a trial programme, but reviewed the Canadian experience and concluded that were there to be an Australian scheme, the Seasonal Agricultural Worker Program (SAWP) should be the model. It further observed that were such a scheme to go ahead in the future that would be a response to Australian foreign policy concerns, pointing to an emerging arc of instability and a problem of failed states, suggesting that 'in the near future an Australian government may need to put aside purely domestic considerations in order to address this problem' (2006b, p. ix). It went further to observe that 'migration was the only route to viability for Pacific micro-states' (2006b, p. 36), but argued that this would be better resolved by skilled labour migration supported by a new training facility. While it reflected opinion in the Liberal-National Party coalition government, it was the last formal opposition to Pacific guestworkers at the national level, as foreign policy and employment policy finally converged.

The Canadian Model

As the World Bank observed: 'Australia and New Zealand may be the only industrialised nations that do not import temporary labour for agriculture' (2006b, p. 116), but 'compensated' for this by using backpackers and WHMs from several European and Asian countries. Perhaps it was a foregone conclusion that the report's findings would be accepted, and policy implementation would follow, as it concluded by modelling various assumptions about wage rates, taxes, paid accommodation and transport costs, derived from the Canadian SAWP experience, and described as 'world's best practice in the importation of temporary foreign labour for agriculture' (2006b, p. 127). Although it was not regarded as without flaws, the Bank argued that 'well-managed programs can be designed to overcome most concerns quite successfully' and that the Canadian scheme was superior to that in the UK by 'rewarding' good workers with re-entry the next year (2000, pp. ix, 21–22). The SAWP had begun in 1966 when Jamaica signed the initial agreement with Canada and immediately sent 264 workers. In 1976, the Canadian government extended the programme to include Mexico and the smaller Eastern Caribbean states—Antigua and Barbuda, Dominica, Grenada, St. Kitts-Nevis, St. Lucia, St. Vincent and the Grenadines, and also the small British colony of Montserrat, all more similar in size and economic structure to the PICs. The scheme offered four decades of experience. In its early years, it faced political resistance due to high unemployment in Canada and perceptions that Caribbean migrants were taking jobs from locals. As elsewhere that proved unfounded, and the scheme gradually grew. The SAWP was perceived as a valuable model since Canada was an 'immigration country' much like Australia and New Zealand, it lacked local labour and half the workers came from small island developing states (SIDS).

Under the SAWP, workers could stay in Canada for a maximum of eight months, were tied to a single employer with whom they shared the transport costs, were given free accommodation and 85 per cent returned to the same employer in subsequent years while only 1.5 per cent overstayed. This success was attributed to transport cost-sharing with employers, workers having the right of return and their being low-skilled workers from rural areas with few alternative employment prospects. Workers were guaranteed a minimum of 240 hours of work over six weeks at or above the prevailing minimum wage rate. By the time of the World Bank report, the SAWP scheme was hosting about 20,000 workers annually, many of whom were returnees and almost all were men. Nearly half came from Caribbean island states.

The overall benefits for Canada were seen as a reliable labour force, with increased productivity where workers returned, enabling farms to stay operational (with downstream benefits for transport and food processing industries) while the spending of seasonal workers benefited stores and services in small towns. For the workers the benefits were secure incomes substantially greater than those at home, the use of these incomes for welfare gains (including housing, nutrition and health care) and the access of the very poor to these schemes, that suggested a degree of equity and a reduction in inequality (although no data were presented to demonstrate that this was true other than in Mexico).

The Bank did however recognise distinct problems with the Canadian scheme. Academic studies had suggested it was exploitative—indeed a 'revolving door of exploitation' (Sharma, 2001)—and migrant workers had been involved in protests and strikes, alongside cases of abuse, substandard accommodation (since it was provided free), industrial accidents (some the result of inadequate training) and overlong working hours.

Seasonal workers were discouraged from joining unions and were excluded from workplace health and safety legislation. Many were too isolated and remote to have much social contact with either compatriots or local communities. Although the scheme was subject to annual review, through consular liaison officers, these were 'seen to be too remote from the workers and to suffer from a conflict of interest—maintaining good relations with Canada and the smooth operation of the scheme versus taking up the fight on behalf of individual workers' (World Bank, 2006, p. 122) that might have later disadvantaged their own country's participation. Government officials in both Canada and Mexico were dismayed at the amount of bureaucracy required for a programme involving relatively few people.

Most academic studies of the SAWP focused on Mexican workers and whether they experienced more disadvantage than Caribbean islanders because of cultural or linguistic differences is impossible to determine. Tanya Basok (2002, p. 144) argued that Mexican workers under the SAWP (and, here, by extension Caribbean islanders) had become the preferred and predominant form of agricultural labour in Ontario at least partly because of their malleability and limited freedoms under the programme: 'it is not the vulnerability of farming that makes growers dependent on offshore labour but the need to secure labour that is not only unfree to change jobs but that is available for work on demand. Since such unfree labour does not exist in Canada, it is vital for Canadian agriculture to have access to unfree foreign workers'. The kind of worker was thus as significant as any actual labour shortage, and implicit within this was a regime of 'discipline, power and control', or, minimally, one that was paternalistic and inequitable. Such cautionary notes were never then made apparent, perhaps unknown or presumed irrelevant, or outweighed by enthusiasm for imminent implementation. Other concerns have subsequently been raised over the Canadian 'model' (Chap. 11). While the Bank referred to similar schemes in Germany, the United States and the UK, these were quickly dismissed as being of less relevance to the Pacific than the SAWP, mainly because the SAWP focused on low-skilled rural workers (whereas the UK scheme involved more skilled workers), while the United States H-2A scheme was more exploitative of workers (e.g. in drawing up-front 'deposits' from them).

The Bank certainly recognised that all labour migration involves risks and social costs, pointed to concerns over children being reared with one or both parents absent, marriages possibly breaking down and the potential spread of sexually transmitted diseases (including HIV/AIDS). These were evident in the context of Tuvaluan and i-Kiribati seafarers (Dennis, 2003) and balanced the substantial economic gains for a considerable proportion of both nations' populations, against the problems of maintaining contact with families, increases in sexually transmitted disease and domestic violence (on return), and complex gender and childhood development issues. A new scheme might have extended such concerns in the Pacific, alongside the extra burdens that were likely to be placed on women particularly, the possible health risks attached to working away from home and the vulnerability of women to sexual harassment.

Regulation and pre-departure training (modelled on the two-day home country programme of SAWP—although it was regarded as inadequate) was regarded as crucial. The Bank expressed concern over the challenges of ensuring equitable opportunities for islanders from remote parts of countries through their limited access to bureaucratic requirements (such as health checks, passports and visas), available only in the capital, and of recruiting unskilled workers rather than skilled trades or professional staff, and the careful regulation of private sector recruiters who had already exploited workers in World Bank (2006b, p. 131). Home countries were encouraged to develop policies that would encourage remittances alongside setting up bank accounts, but the ultimately complex issues of working out who would bear various costs, such as transport and accommodation, levels of taxation, insurance and health benefits, were not discussed. Who would undertake training, regulation and monitoring, and how existing Pacific communities and various social organisations might be involved, was also left for another time and to the particular countries. Nonetheless, the Bank's report was not merely a review of migration in the Pacific and of guestworker schemes elsewhere but was very close to a blueprint, and a set of ground rules, that would quite quickly be adopted.

Labour Shortages and Development Directions

As one more basis for the World Bank's report, a study was undertaken of the Australian horticultural industry to ascertain whether there were labour shortages, how and why these might occur, and the attitude of horticulturalists to migrant labour. There had been considerable, more anecdotal reporting of labour shortages, such as Kathryn Shine's 'Fruit rots as workers pick and choose' (2005), and raids by immigration officials to catch illegal workers had become a feature of harvest periods, that had also pointed to the underpayment of the illegal workers. A high proportion of the workers were backpackers (WHMs), and many growers had concerns that they were less productive, more injury-prone and unlikely to return subsequently. Overall growers claimed before parliamentary committees and elsewhere that labour shortages and inadequacies had resulted in direct and indirect crop losses.

The study undertaken in the Swan Hill/Mildura area revealed that half the farmers found it difficult to recruit an adequate labour supply—which had also constrained their expansion—and a quarter (probably an underestimate) admitted to having employed illegal workers. However, a third of the farmers were uninterested in hiring overseas workers, especially if there were unemployed local people available. Those who were interested expressed no preference for region of origin or gender, preferred some ability to speak English, were willing to pay above award rates and had a clear preference for employing those who would return. Farmers were rather less enthusiastic about sharing any costs, most feeling that fixed costs, such as air fares, should be recovered from wages. Most were unwilling to provide on-site board and lodging to workers—a partial reflection of ongoing shortages of adequate reasonably priced accommodation in the area. Farmers with greater labour requirements were more enthusiastic. None found it easy to estimate what their future labour requirements would be, either over the following season or over the long term as their enterprises evolved. The survey concluded that a pilot project was required to educate sceptical farmers.

Curiously what amounted to a little-known pilot programme in Victoria occurred in the same year, when a small group of ten Fijian men were brought to work for four months on a tobacco farm in Victoria, organised through the British-American Tobacco company. The farmer was enthusiastic about the programme, disappointed only that the workers had to return home after merely four months: 'They're very reliable workers. They're polite people, they're willing to work, they turn up every day and never cause any problem in the community'. The workers were equally enthusiastic, pointing out that their earnings were roughly ten times those in Fiji and would be used for their children's education (*The Fiji Times*, 30 May 2006). No reference seems to have subsequently been made to this modest exploratory programme, but it set the pace for what was to follow.

In effect the real pilot project was in New Zealand and the first year of the RSE programme in 2007–2008 was extensively reviewed, with particular reference to Tonga and Vanuatu. The review concluded that a comparable Australian scheme was certainly feasible and would be widely welcomed, but cautioned that there were 'significant dangers in regarding seasonal workers as commodities to be traded between countries, and the heritage of the blackbirding era should be warning enough for policy-makers' (Maclellan, 2008b, p. 31). At much the same time, Manjula Luthria, noting that large pools of unemployed Pacific youth were young, fit and anxious to work, and that islanders could return with savings of more than A$5000 a year, several times what they could earn at home, stressed:

> In New Zealand the seasonal worker program has built hope into island communities. It has brought a real pride of labour. It's something we as outsiders in the aid community can never build. The scheme is the shortest way to achieve the largest amount of development …The best and the brightest were leaving there and the rest were trapped there. (Quoted in Stapleton, 2008, p. 9)

Her enthusiasm was soon vindicated, and it was a theme the World Bank returned to more than a decade later, almost unchanged and no less enthusiastic, in its endorsement and publication of *Pacific Possible. The Million Dollar Prize* (Curtain et al., 2016).

Australia cautiously watched the New Zealand experience and was eventually convinced, uneasily embarking on a new structure of migration and shifting away from a migration policy intended to be oriented to nation-building and permanence rather than alleviating possible temporary labour shortages (Millbank, 2006). The new direction appeared more as a response to development disadvantages in Australia rather than to those in PICs. Much as the SAWP, both the RSE and the SWP (like other precursors) were designed with an explicit focus on gaining a cheap and reliable source of labour for national agricultural industries, as much as linking migration to development in the PICs. It is a matter of ideology, debate and argument which objectives had priority, but the timing indicates the former. The scheme was necessarily and unsurprisingly advertised by the Australian government as having as its key objective that of 'demonstrably boosting the development of Pacific island communities through new employment opportunities, increased remittances incomes, and options of upskilling of Pacific island seasonal workers' (quoted in Opeskin & MacDermott, 2009, p. 368). Almost a century after the end of blackbirding another phase in the migration of short-term agricultural workers was about to begin, modelled on the SAWP and with multiple echoes from it. That new phase is the subject of the following chapters.

Bibliography

Australia. (2006). *Perspectives on the future of the harvest labour force*. Senate Standing Committee on Employment, Workplace Relations and Education.

Australian Government. (2003). A Pacific Engaged. Australia's relations with Papua New Guinea and the Island states of the Southwest Pacific. .

Bakewell, O. (2008). 'Keeping them in their place': The ambivalent relationship between development and migration in Africa. *Third World Quarterly, 29*(7), 1341–1358.

Basok, T. (2002). *Tortillas and tomatoes: Transmigrant Mexican Harvesters in Canada*. McGill-Queen's University Press.

Bertram, G. (1986). 'Sustainable development' in Pacific micro-economies. *World Development, 14*(7), 809–822.

Brown, R., Connell, J., Jimenez, E., & Leeves, G. (2006). Cents and sensibility: The economic benefits of remittances, in World Bank (2006b). In *At home and away* (pp. 47–99). World Bank.

Castles, S. (1998). Globalization and migration: Some pressing contradictions. *International Social Science Journal, 50*(2), 179–186.

Connell, J. (1983). *Migration, employment and development in the South Pacific. Country Report No. 18. Tonga*. SPC and ILO.

Connell, J., & Brown, R. (2005). *Remittances in the Pacific: An overview*. ADB.

Curtain, R., Dornan, M., Doyle, J., & Howes, S. (2016). *Pacific possible. Labour mobility: The ten million dollar prize*. World Bank and ANU.

De Bres, J., Campbell, R., & Harris, P. (1974). *Migrant labour in the Pacific*. CORSO.

de Haas, H. (2010). Migration and development: A theoretical perspective. *International Migration Review, 44*(1), 227–264.

de Haas, H. (2012). The migration and development pendulum: A critical view on research and policy. *International Migration, 50*(3), 8–25.

de Haas, H., Castles, S., & Miller, M. (2020). *The age of Migration: International population movements in the modern world*. London: Red Globe Press.

Dennis, J. (2003). *Pacific island seafarers: A study of the economic and social implications of seafaring on dependents and communities*. South Pacific Commission.

Ellerman, D. (2005). Labour migration: A developmental path or a low-level trap? *Development in Practice, 15*(5), 616–630.

Galbraith, J. (1979). *The nature of mass poverty*. Harvard University Press.

Hanson, J., & Bell, M. (2007). Harvest trails in Australia: Patterns of seasonal migration in the fruit and vegetable industry. *Journal of Rural Studies, 23*, 101–117.

Hayes, G. (1982). Migration, metascience and development policy in island Polynesia. *The Contemporary Pacific, 3*(1), 1–58.

Hughes, H. (2003). Helping the islands to help themselves. *Quadrant, 47*(7), 1–3.

Hughes, H., & Sodhi, G. (2006). *Should Australia and New Zealand open their doors to guest workers from the Pacific? Costs and benefits*. Centre for Independent Studies Policy Monograph 72.

Hugo, G. (1982). Circular migration in Indonesia. *Population and Development Review, 8*(1), 59–83.

Kapur, D. (2004). *Remittances: The new development Mantra?* UNCTAD G24 Discussion Paper No. 29.

Kite, S. (1975). Overseas employment in New Zealand for Tongan workers. Noumea: Unpublished paper, Regional Conference on Manpower Planning, Labour Recruitment and Migration Policies.

Levick, W., & Bedford, R. (1988). Fiji labour migration to New Zealand in the 1980s. *New Zealand Geographer, 44*(2), 14–21.

Maclellan, N., & Mares, P. (2006). *Remittances and labour mobility in the Pacific.* Swinburne Institute for Social Research.

Maclellan, N. (2008b). *Workers for all Seasons? Issues from New Zealand's Recognised Seasonal Employer (RSE) program.* Swinburne Institute for Social Research.

Macpherson, C. (1981). Guest-worker movements and their consequences for donor and recipient countries: A case study. In G. Jones & H. Richter (Eds.), *Population mobility and development: Southeast Asia and the Pacific* (pp. 257–277). ANU Development Studies Centre Monograph No. 27.

Mares, P. (2007). Objections to Pacific Seasonal Work Programs in rural Australia. *Public Policy, 2*(1), 68–87.

Massey, D., Arango, J., Hugo, G., Kouaouci, A., Pellegrino, A., & Taylor, E. (1993). Theories of international migration: A review and appraisal. *Population and Development Review, 19*(3), 431–466.

Millbank, A. (2006). *A seasonal guest-worker program for Australia?* Parliament of Australia Research Brief No. 16.

Morris, S. (2008, April 8). Fertile soil for guest workers, *Australian Financial Review*, 69.

Opeskin, B., & MacDermott, T. (2009). Resources, population and migration in the Pacific: Connecting islands and rims. *Asia Pacific Viewpoint, 50*(3), 353–373.

Poirine, B. (1998). Should we hate or love MIRAB? *The Contemporary Pacific, 10*(1), 65–105.

Sharma, N. (2001). On being *not* Canadian: The social organisation of 'migrant workers' in Canada. *Canadian Review of Sociology and Anthropology, 38*(4), 415–439.

Shine, K. (2005, March 5). Fruit rots as workers pick and choose. *The Australian*, 5.

Stapleton, J. (2008, June 21). Fruit rots on the vine for want of seasonal workers. *The Australian*, 9.

Wickramasekera, P. (2011). *Circular migration: A triple win or a dead end?* ILO.

World Bank. (2006b). *At home and away. Expanding job opportunities for Pacific Islanders through labour mobility*. World Bank.

5

Early Days

After years of hesitation and opposition Australia eventually introduced its pilot guestworker scheme in 2009. It was the start of a new era, but an era that carefully and closely followed the Canadian Seasonal Agricultural Worker Program (SAWP) and the New Zealand Recognised Seasonal Employer (RSE) scheme that had begun two years earlier. New Zealand had announced at the Pacific Islands Forum in Fiji in October 2006 that it would introduce a new seasonal migration scheme prioritising workers from Pacific Island Countries (PICs). The RSE scheme duly began six months later in April 2007, quickly proving a considerable boost to the horticulture and viticulture industries (almost necessitating that Australia would have to follow suit). Australia's own announcement came in 2008; a year later its Pacific Seasonal Worker Pilot Scheme (PSWPS) began.

Labour migration evolved quickly in the first decade of the twenty-first century. After a few small-scale schemes in New Zealand ended, PIC pressure, World Bank advocacy, agricultural labour shortages and a convergence of theory and practice that had suggested 'triple wins' elsewhere led to the establishment of the new guestworker schemes. Potential benefits to Australian farmers underpinned the PSWPS, despite its being advertised and promoted in the Pacific as an aid project. With some

K. Petrou, J. Connell, *Pacific Islands Guestworkers in Australia*,
https://doi.org/10.1007/978-981-19-5387-3_5

rather later differences (as the Australian scheme extended northwards into regions of tropical crops—sweet potatoes near Bundaberg, and bananas in Tully—but never apparently sugar cane), the SWP and RSE schemes were so similar that most of the conclusions and experiences from one were equally applicable to the other. Positive early outcomes in New Zealand, and general enthusiasm in the PICs, brought Australia's participation. This chapter therefore briefly examines the renaissance of guestwork in the southern hemisphere, and the circumstances of the early days in New Zealand and Australia, as the experiment crossed the Tasman Sea.

New Zealand and the Recognised Seasonal Employer Scheme

The 'new age' of Antipodean guestwork that began in New Zealand was quickly promoted as an anticipated 'triple win' scenario. The RSE scheme was first implemented for five PICs (Kiribati, Samoa, Tonga, Tuvalu and Vanuatu—the so-called kickstart states). Fiji was excluded because of the 2006 military coup. Other PICs were recognised as potentially eligible for the scheme, just as soon as the 'kickstart' had been implemented, notably the Federated States of Micronesia, Marshall Islands, Nauru, Palau, Papua New Guinea (PNG) and Solomon Islands. Solomon Islands and, rather tentatively, Nauru formally entered the scheme in 2008, PNG joined after 2010 (with small numbers), but the others never did. It was capped at 5000 people in the first year, with Vanuatu, Tonga and Samoa dominating numbers. Like the Canadian SAWP scheme, it centred on workers being overseas for up to nine months and then returning home (Ramasamy et al., 2008). Overseas recruitment was only permitted if employers were unable to recruit New Zealanders. Since recruitment could scarcely respond to sudden unanticipated shortages, as the arrival of labour would have been too late, quotas had to be developed based on past experience.

For most PICs, New Zealand employers paid half the return air fare between New Zealand and the home country. However, for i-Kiribati and Tuvaluan workers, employers only contributed half the return airfare between New Zealand and Nadi (Fiji): these workers had to pay much more to enter the scheme, and there was no particular incentive for employers to hire them. Unsurprisingly their numbers were relatively small (Table 5.1). Nauruans were disadvantaged by having no agricultural skills, relevant work experience or great interest in participating, but began going in small numbers from 2014. While the scheme was something of a trial, numbers were certainly parsimonious relative to the needs of the PICs. In the first year barely 3500 RSE visas were issued to PICs, and nearly a thousand went to five Asian countries, led by Malaysia. Visa numbers were far below demand with 4500 Tongans and 4000 Samoans applying in that single year (Maclellan, 2008b). Despite New Zealand's close ties with the PICs, the inclusion of larger Asian states, and 'Employer' rather than 'Employee' in the title, emphasised that the scheme was primarily oriented to the needs of New Zealand farmers.

The New Zealand Ministry of Social Development monitored the labour market to identify regions with seasonal labour shortages where there were not enough New Zealanders to fill the vacancies. Employers believed that it was better to participate in the RSE scheme and have a secure source of labour rather than use irregular, if often cheaper, undocumented workers (Ball, 2010). Potential employers applied to the New Zealand Department of Labour to recruit Pacific workers, and had to register and be approved as a Recognised Seasonal Employer, which required demonstrating they were financially viable, could pay workers at least the minimum wage for 30 hours a week, provide accommodation and 'pastoral care' to workers, and have a dispute resolution process in place. Effectively that meant that most were large employers rather than family farms. The employers were then assessed by the New Zealand Department of Labour to ensure that the facilities provided for the migrant workers, notably reasonably priced accommodation, were of appropriate standard. By early 2008, within the first year, some 75 employers had been given official RSE status. Employers could recruit directly in the PICs or could accept nominated workers from those states.

RSE workers enjoyed the same labour rights and protections as New Zealanders, received wages no less than the statutory minimum wage (and usually exactly that minimum, initially NZ$12 [A$11] per hour) and paid non-refundable taxes The minimum hourly salary was roughly

Table 5.1 Number of RSE visa grants by financial year

Country	2007/ 2008	2008/ 2009	2009/ 2010	2010/ 2011	2011/ 2012	2012/ 2013	2013/ 2014	2014/ 2015	2015/ 2016	2016/ 2017	2017/ 2018	2018/ 2019	2019/ 2020
Brazil		3											
Czech Republic	1												
Fiji								30	92	355	359	444	487
India	41	77	25	40	15	25	28	27	35	35	35	35	35
Indonesia	249	271	271	304	299	303	305	313	321	323	325	326	263
Kiribati	69	38	48	149	142	138	127	136	162	189	231	263	289
Malaysia	364	374	406	375	317	273	308	333	335	321	329	335	120
Nauru								20	20	17	14	19	5
Papua New Guinea				6	6	31	58	96	68	121	124	172	134
Philippines	80	76	75	74	74	68	68	67	67	66	67	66	67
Samoa	647	1228	1021	1219	1162	1137	1169	1238	1454	1690	1878	2315	2409
Solomon Islands	238	311	256	252	407	423	491	511	590	593	643	696	756
Taiwan		12	54	31	31	34	34	20	20	20	20	23	10
Thailand	195	684	727	827	658	565	588	624	637	634	629	628	418
Tonga	805	1355	1142	1411	1398	1573	1538	1563	1687	1822	1899	2037	1807
Tuvalu	99	49	54	51	88	56	71	70	64	80	80	92	115
Vanuatu	1698	2342	2137	2352	2412	2829	3070	3435	3726	4171	4445	5130	4237
Vietnam		1				1							
Grand Total	4486	6821	6216	7091	7009	7456	7855	8483	9278	10,437	11,078	12,581	11,152

Source: Ministry of Business Innovation and Employment, New Zealand

equal to the unskilled daily wage in Vanuatu, above that in Tuvalu and Kiribati and slightly below that in Tonga and Samoa. Sick leave accrued to workers after six months. Workers could remain in New Zealand for up to seven months at a time (or nine months in the case of the more distant Kiribati and Tuvalu). However, employers could potentially employ the same workers year after year, so providing both an incentive for them to work well and be law-abiding, if they sought to return, and for New Zealand employers to establish long-term relationships and invest in training and skill development. These contexts and circumstances were almost identical to those that applied in Australia when its own scheme began.

At the end of the first year the pilot scheme had employed people from six PICs, and five Asian states, and more than 3000 Pacific islanders were employed in orchards and vineyards in several parts of the country from the Bay of Plenty to Otago. Some of the earliest workers however went into jobs that were far from fruit picking. A group of 24 Tongan women, recruited for the RSE scheme in 2007, were intended to work as forklift drivers, mechanics, computer operators and supervisors (PIR, 7 November 2007). The RSE scheme grew by 50 per cent in its second year and then more steadily, taking a decade before the numbers of RSE workers had doubled to pass 10,000 in a year (Table 5.1). By then Fiji had joined in, after political barriers were removed, and PNG numbers were beginning to grow, but otherwise Tonga, Samoa and Vanuatu remained the principal PICs involved.

Australia and the Pacific Seasonal Worker Pilot Scheme

Australia initially resisted the introduction of guestworkers, firstly, because there were concerns that temporary workers would overstay and become illegal migrants; secondly, because of arguments that Australians were available for such jobs and that too little had been done to attract them into the industry; thirdly, that migrant workers would be exploited by unscrupulous employers; and, fourthly, that low wages paid to migrant workers, and full employment, would discourage improved wages for local workers in the industry (Maclellan & Mares, 2007). Most objections had little substance where recurrent labour shortages existed and where there was a history of employment of illegal workers at very low wages. The successful adoption of the RSE in New Zealand, and incessant pressure from the Pacific, proved catalysts for change in Australia, and signalled a new era in its economic relationships with PICs.

Australia introduced its PSWPS two years after the RSE, modelled on the same principles (and therefore on the SAWP) and with similarly selected PICs (initially Kiribati, PNG, Tonga and Vanuatu) but without Asian participation. Priority had to be given to available Australian workers, and the more conservative scheme was expected to initially allocate 2500 temporary visas per year, little more than half the number of RSE visas granted annually. As in New Zealand, workers could work for up to seven months in any 12-month period, on horticultural enterprises or farms. They had to be aged between 21 and 45 at the time of visa application, be healthy, fit and of good character (having undergone a police check). Australia also specified that workers should not have participated in the New Zealand RSE scheme, to broaden the base of opportunity. Owing partly to worsening economic conditions (with the establishment of the scheme effectively coinciding with the Global Financial Crisis [GFC]), implementation of the programme proceeded slowly. The GFC meant that although demand for agricultural workers was unchanged, local supply increased somewhat as unemployed students, backpackers and others became available. Floods and drought in 2010, especially in Queensland, further hampered PSWPS uptake, and it barely got off the ground.

The PSWPS, like the RSE, began cautiously with small numbers. Between July 2009 and June 2012 some 2400 visas were made available, but they were issued quite slowly. By February 2011, after 18 months, only 214 workers had come to Australia under the scheme, mainly from Tonga (181) but with small groups from Kiribati (19) and Vanuatu (14). Arrangements for recruitment from PNG were finalised in 2010 and recruitment began in 2011, with just seven Papua New Guineans coming in that year. Host employers sought to reject some of the Tongan and i-Kiribati workers because of a limited demand for labour and what was said to be their lack of experience and motivation (Callick, 2010). Most workers were employed picking citrus fruit, pruning vines or harvesting almonds, all activities without home country parallels, hence basic new skills had to be learned. Unlike in New Zealand, the early days of Australia's new labour mobility pilot were widely seen as a disappointment, employers found the costs and bureaucracy of recruitment considerable (being particularly critical of the process of recruitment mediated by three labour hire firms) and PICs were concerned at Australia's apparent lack of commitment. Such small participant numbers could make no real contribution to development needs in the region.

The Pilot Schemes in Operation

Despite the difference participant numbers between Australia and New Zealand, the experience of workers from the PICs that were involved early in the RSE (and subsequently the PSWPS), demonstrate some of the successes and failures of the earliest years, and indicate that similar problems arose in both countries. Remarkably, most of the policies and practices established in these formative years barely changed in later years, and even the outcomes remained strikingly similar. We consider this here in the context of two of the countries most involved in the two schemes both then, and more recently, Tonga and Vanuatu.

Vanuatu

Vanuatu, which became independent in 1980, was classified by the United Nations as a least developed country (until late in 2020 when it graduated from least developed status) and has a population of just over 300,000 scattered through 65 populated islands in a complex archipelago. Its economy is based on agriculture (with kava exports being significant) and on tourism (until its borders closed in 2020 against COVID-19). It has exhibited only slight economic growth in this century, per capita income levels have not significantly increased, more than 40 per cent of the population live below the poverty level, and both unemployment and under-employment are rising. On the smaller islands especially many people are semi-subsistence agriculturalists, planting root crops and with a limited engagement in fishing. It is regarded as the most hazard-prone country in the world and was badly affected by both Cyclones Pam and Harold in the 2010s. Vanuatu particularly enthusiastically sought access to overseas employment programmes to reduce unemployment and boost national income. It was involved in the pilot RSE scheme and in numerical terms was one of the principal beneficiaries of both the early RSE and PSWPS schemes. It was thus a 'first mover', which proved a major subsequent benefit for negotiating opportunities and marketing experience.

In New Zealand, the preference for ni-Vanuatu workers was a deliberate decision by employers who, at the start of the RSE, preferred to employ workers who did not have connections to a significant diaspora population in New Zealand. Employers believed that this lack of diaspora connections meant ni-Vanuatu workers were less likely to abscond and hence the scheme would be more likely to succeed. In the RSE's first year, the World Bank facilitated recruitment by connecting New Zealand employers with local chiefs from rural villages in Vanuatu (Curtain, 2018). Good reports from the New Zealand experience proved valuable for Vanuatu's early entry into the PSWPS (Fig. 5.1).

The Vanuatu Department of Labour licensed local employment agents who were the contact points and recruiting agents for New Zealand employers. The agents negotiated fees from the employers for finding workers, but charged no fees to the workers. Agents were responsible for pre-departure briefings (covering such topics as health insurance, trade unions and means of remitting income) and were also expected to compile lists of 'work-ready' people. Such people were usually aged between 21 and 55, with police and medical certificates and the support of a community leader, such as a chief or pastor, confirming that they were of 'good character'. A few agents proved bad choices and cheated some ni-Vanuatu applicants or gave them unrealistic expectations, and others were biased towards their own family members, but generally appropriate selections were made.

It was initially intended that workers be contracted from outside the two main islands, Santo and Efate. The Vanuatu Commissioner of Labour, Lionel Kaluat, stated in 2007:

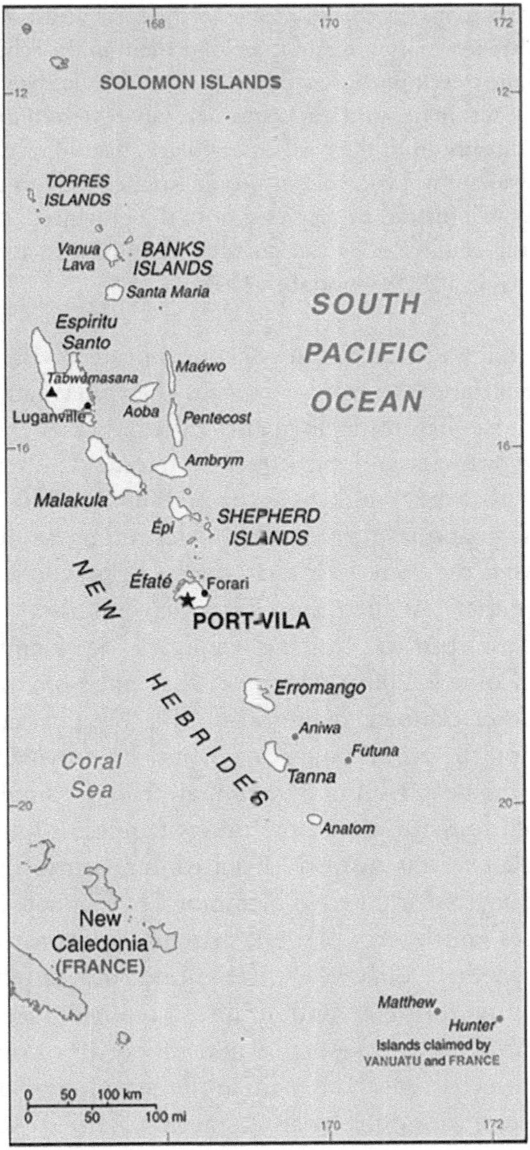

Fig. 5.1 Vanuatu. (Source: CIA World Factbook)

> Our objective is to give access to people in the rural areas. We believe the best workers are in the islands and to involve them in the scheme will help boost economic development in the islands, and to address the issue of urban drift. In future we want a system where laborers from the islands will be recruited directly from their island or village, instead of coming to the towns. That way we will control our people from coming to town looking for jobs. They will instead be recruited from the provinces, and once they are back in the country, they can go back and invest in their respective places. (Quoted in PIR, 1 November 2007)

Workers paid for a medical and police certificate, a visa fee, a passport and half their airfare (the employers paid the other half), resulting in large upfront costs before deployment. In some cases these costs were advanced by employers and subsequently refunded. Intending participants were told that they would bring back about A$4000—roughly four times the average annual rural income—depending on how hard they worked and what they did with their wages. In practice, workers saved and remitted rather less than this. Work was not always available and local expenditure proved enticing (Connell & Hammond, 2009; Hammond & Connell, 2009). Many recruits came from Tanna where, as one NGO worker claimed, it proved both 'a big physical mental and emotional test on the people going down' but also provided 'real changes for the police, the hospital, immigration, all the government offices that have never been tested to do all the things required to send such a big group of people overseas to work. It jolted Tanna into the twenty-first century in one big step' (quoted in Hammond & Connell, 2009, p. 205).

Communities and leaders played a part in selection. In Lamen Bay (Epi), at least in these early years, the village council played a role in screening and monitoring migration, forced or encouraged migrants to remit to a community fund—that would support the community hall, school and health centre—and remain in the community 'once they have had five consecutive seasons away' (Craven, 2013, p. 49). In Tanna, younger married men with no children were chosen, as it was thought RSE participation would provide them with 'a good start' in life. However, older men, who were thought to be more reliable and more likely to return home with money, were also chosen. Professional and skilled workers were not selected. At least some workers were sponsored by their communities so that their income might support community needs such as shared water tanks or village halls. As in the earlier Fijian scheme with New Zealand, this tended to result in the selection of 'worthy' recruits who were either needy or regarded as good workers who would not be troublemakers. Those first involved in the scheme were particularly motivated (as expected in a first cohort) and in some respects were 'target migrants' making calculations of just how much income was needed for particular objectives such as building a house, setting up a small business or getting children through high school.

Recruiting from Tanna, Vanuatu

Many early recruits came from the large, densely populated southern island of Tanna, where internal rural-urban migration had long been common and domestic remittances were a significant source of income. Unlike any other previous migration experience that Tannese (and most other ni-Vanuatu or Melanesians generally) might have been involved in, people were not allowed to sign up or migrate purely as individuals, but had to be involved in community activities and take part in preliminary briefings. Participants had to set up bank accounts in Tanna (so that the income would return there), donate NZ$250 (A$232) to a community fund, use group flights which would return people directly to Tanna rather than through Port Vila (where the pilot programme suggested that a proportion of funds were absorbed or 'lost') and abide by a local prescribed Code of Conduct with various precepts that focused on abstinence from alcohol and sexual activity and respect for leaders (Connell & Hammond, 2009) and that partly replicated a Tannese moral order.

Workers had to be married, as it was assumed that married couples had clear needs and a measure of stability. One partner would stay behind. Since New Zealand employers sought equal female participation (with women's skills preferred for activities such as grading and packing), the scheme offered an opportunity for women to be more independent. However, women had less familiarity with the outside world, tended to be more conservative and less likely to put their names forward or have them put forward. About 20 per cent of the first group of RSE workers from Tanna were women. By contrast on Ambrym island women were not selected in the first year; as the Chair of the local Council said: 'We want the women in the villages to look after the children, the [food] garden and the livestock' (quoted in Bailey, 2019, p. 29). Men proved more expendable.

Recruits came from throughout Tanna and were relatively evenly distributed in particular areas, so there was no obvious spatial bias in selection; at the national scale, a broad spatial equity involved most parts of Vanuatu. Workers came from the many different religious denominations of Vanuatu, including 'kastom' religions. Most were not particularly young—the average age being 35 for women and 36 for men—and were therefore from established households.

Almost all workers were semi-subsistence agriculturalists, with market sales their only source of income. A small minority had wage or salary employment. Overall three quarters (73 per cent) of the households had incomes less than 50,000 vatu (about A$580). Two-thirds (65 per cent) had never gone beyond primary school. The recruits were not well educated, had little facility in English and came from low-income households: a reasonable cross-section of the population of Tanna, demonstrating the considerable degree of income poverty stemming from a widespread poverty of opportunity. Despite Tannese having a local reputation for being difficult, they were seen by the Vanuatu Department of Labour as 'very dynamic, willing to try new things, and take the risk', but also as experiencing land shortages, hence having some need for migration.

The underlying rationale for migration was invariably income generation. Any general interest, excitement and novelty were at best secondary. Income generation had three key objectives: notably paying school fees (which were roughly 40,000 vatu [about A$490] per year, plus transport costs, for high school, and 9000 vatu [about A$110] for primary school). Education offered the possibility of jobs in the public service, or in tourism, and an opportunity to break away from reliance on subsistence agriculture. More than half (61 per cent) of the recruits (from a sample of more than 300 workers) listed school fees as their first priority. Small business development (centred on a small store) was the second most important category (25 per cent), while improved housing (13 per cent) came third. In practice, the principal direct material goal was house construction, with permanent materials that needed less frequent replacement, would be dry in wet weather and allowed gutters and a tank for collecting drinking water.

Elsewhere in Vanuatu too, income was used for improved housing, water tanks, lighting (usually through solar panels) and education fees. Much was spent on mobile phones, since telecommunications were extending at the same time as the introduction of the RSE scheme. On Epi, iPads, MP3 players, CD players phones and other electronic gadgets were soon important, alongside solar panels (Kumar, 2012). While, unsurprisingly, not all expenditure was directed to valuable welfare gains, there is substantial evidence of the schemes making a contribution to basic needs and to moving rural areas of small island states a little closer to achieving some of the Sustainable Development Goals (SDGs). More obviously modern houses and solar panels had changed the landscape; some stores, meeting houses and kava bars (*nakamals*) were more likely to have lights; no longer was Lamen Bay (Epi) just dark at night and without permanent buildings (Kumar, 2012, p. 74). Lamen Bay was however atypical.

Although Tanna was a prime source of workers, the rationale behind recruitment was much the same elsewhere in Vanuatu. On Epi too, the reasons for participation were dominated by the desire for a better house, followed by the need to pay children's school fees; as one worker stated 'We need money to build good houses for ourselves … because during cyclone season we have to move our kids to stronger houses' (quoted in Kumar, 2012, p. 56). Some desire to travel overseas was never too far away. Most were primarily subsistence farmers with limited cash incomes from local food sales. Alternative income-generating opportunities in rural Vanuatu were few; as a Lamen Bay resident stated bluntly, 'Who would want to work in the copra fields anyway? It's hard work for little money. When you can earn so much by going to New Zealand, it makes no sense to waste our time on other things' (quoted in Craven, 2013, p. 52). That was a poor omen for future local agricultural development.

Certain gains were more likely to be apparent on different islands. Thus, RSE workers from Mele, the closest village to Port Vila and already relatively well-off, were more likely to invest in a significant local store than workers from Epi for whom building a modern house was more important (Kumar, 2012). Strategies were often carefully considered. After a first RSE experience, one ni-Vanuatu from Epi stated:

> I bought a solar panel, two bikes and a chainsaw [from New Zealand] which are more expensive here in Vanuatu. I'm going to use the remainder of the money to buy cement and bricks here and start building my house. I will go back to New Zealand again to earn some money and then come back and complete the house. (Quoted in Kumar, 2012, p. 70)

Very early on workers recognised the value of return, to realise greater earnings and savings the second time around.

It was not long however before people pointed to inequities in selection. On Epi in 2010 workers tended to be chosen from Lamen Bay, the main island centre and the residence of the Member of Parliament, whereas men travelled from the eastern side of the island—a three-hour walk away—to be interviewed for the RSE but were not selected. As one worker complained, 'Our island is divided into four parts but most people going to New Zealand are only from Lamen Bay. It is because the agent is from there and he is doing "family business"' (quoted in Kumar, 2012, p. 68). Individuals distant from Lamen Bay were further disadvantaged by the lack of formal employment and markets that limited their ability to generate income to cover the basic costs of participation.

From Vanuatu to New Zealand and Australia

Both in New Zealand, and later Australia, fewer work hours were available than promised and anticipated, despite guarantees of minimum hours; unexpected pay deductions occurred (mainly for accommodation) and piece work rates did not equate with the minimum wage. Labour requirements proved hard to estimate. Employers were sometimes exploitative or failed to meet adequate standards (especially for accommodation), and costs were more than workers anticipated, for food, warm clothes and recreation. Some workers found access to towns for food and other basic goods irregular, difficult and expensive. Many complained about accommodation costs, but they had little choice and were generally paying standard Australian or New Zealand rates, albeit seemingly enormous to ni-Vanuatu workers (Kumar, 2012). Accommodation could leak, be overcrowded and cold and local churches sometimes had to intervene to ensure more adequate food and accommodation. Workers missed local food and generally made do on a basic diet of rice, bread and meat. Purchasing such food collectively minimised costs. The moral code was not enough to prevent some problematic emotional entanglements, and several workers were sent back early for drunkenness or not turning up for work.

In most cases ni-Vanuatu workers remitted money and brought back significant sums, but almost all returned with less than they anticipated. Downtime was a problem; as one ni-Vanuatu man stated,

> I haven't been able to work enough. Sometimes even when it is sunny they don't have work. I want one day off a week, but no more, as it says in our contract. We don't want to be on holiday … the amount of work hasn't been enough and we came here to save but we struggle in this situation. (Quoted in Williams, 2010, p. 74)

Numerous workers were frustrated being well aware that neighbouring farms had work that needed doing, but they were legally unable to pick up work from other employers. Some managed to supplement their agricultural incomes in other ways. One group of workers from an Ambrym village reportedly raised NZ$10,000 (A$9300) from busking with ukelele, tea-chest bass and massed voices outside the farmers market in the small town of Cromwell (McDonald, 2008). Others busked elsewhere—an enjoyable way of supplementing incomes.

Income was generally used for their specified initial objectives, including communal projects, and mobile phones were acquired by many, reflecting a familiar pattern of remittance use in the Pacific, initially on education, consumption and then on welfare gains (such as improved housing), eventually followed by investment in small-scale business (Connell & Brown, 2005). Within a couple of years, in villages such as Whitesands (Tanna), several new small stores had been constructed within a hundred-metre radius, raising questions about limitations to the only kind of business development with which most people were familiar and were able to establish (Connell & Hammond, 2009).

Some workers who had seized the opportunity to take part in the first scheme were far from unambitious; as one 27-year-old Tannese, who believed he would save about NZ$4200 (A$3900), and who was looking forward to coming for another five seasons, stated:

> I want to build a good house for my family and put in a solar panel so my kids can have lights to study and read books. After that I want to start a small business, maybe buy a rental house in Port Vila. (Quoted in McDonald, 2008, p. 2)

Since workers were chosen (or chose themselves) according to perceived needs, the scheme made a further contribution to local and regional equity. Despite frustrations, about half of all those who had migrated sought to return for a second phase. On Epi too many were enthusiastic to return having learned the art of money management and were able to work faster because they now knew what was required (Kumar, 2012). Women and older men however found the work hard and were less likely to wish to return. In Tanna those who remained at home absorbed extra work and responsibilities without obvious problems or rancour.

The experience of early labour migration from Vanuatu was overall largely positive for New Zealand, Australia and Vanuatu. The ni-Vanuatu workers, available for seven months, were a more reliable source of labour than backpackers, so reducing training times and costs. Migrant workers filled vacancies, did not displace local workers, made some social and economic contribution to local society, and universally returned to Vanuatu at the end of the contract period, avoiding concerns about overstaying. A minority returned early with relatively little to show, mainly because of dislike of work disciplines, the cold or frustration over lack of work, while a few were deported for breaches of conduct. These early days suggested that, despite diverse concerns, the schemes could benefit participating countries, agricultural businesses and workers, and even that it might serve the needs of the poor more effectively than many forms of aid. Nonetheless, the two schemes certainly served the needs of agricultural businesses more than the needs of workers and their families. Fewer than 3000 workers were guestworkers at any one time—just 1 per cent of the national population—so that few households benefited directly. The overall contribution to the Vanuatu economy was limited, but the first three years had offered a promising start.

Tonga

Rather like Vanuatu, Tonga is an archipelago of about 40 populated islands, but dominated by the main island of Tongatapu, with the capital Nuku'alofa, where 70 per cent of the Polynesian population of 100,000 live. The national economy is centred on agriculture, and in many of the smaller outer islands, a semi-subsistence economy prevails. Tonga has however long been dependent on international migration and remittances and more than half of all ethnic Tongans live overseas, notably in New Zealand, and also in the United States and Australia. Many Tongans have travelled widely and are relatively familiar with New Zealand and Australia unlike residents of Vanuatu.

Like ni-Vanuatu, Tongans participated in the earliest phases of both the RSE and the PSWPS and absolutely dominated the latter. The first migrants participating in the Australian scheme were 50 Tongan workers who arrived in February 2009. Just six ni-Vanuatu workers followed in the same year, as lack of demand for workers meant that the remainder of an anticipated 100 visas were not issued (Fig. 5.2).

As it had done for the RSE, the Tongan government had put together a work-ready pool of potential seasonal workers, although the initial Australian hiring company both took workers from this list and hired directly (through a network of Australian-Tongan contacts). Direct hiring led to some tensions, since it produced a particular kind of selectivity by religion. The Tongan group put together by the Ministry of Labour was recruited throughout Tonga, according to the population of particular districts, although employers could choose workers from a single district or from across the country. Selection was biased towards those whom the government regarded as 'good reliable people' who were unlikely to overstay (being married with children at home) and also targeted the 'grassroots of Tongan society: the poor and the unemployed' (quoted in Blanco, 2009, p. 45).

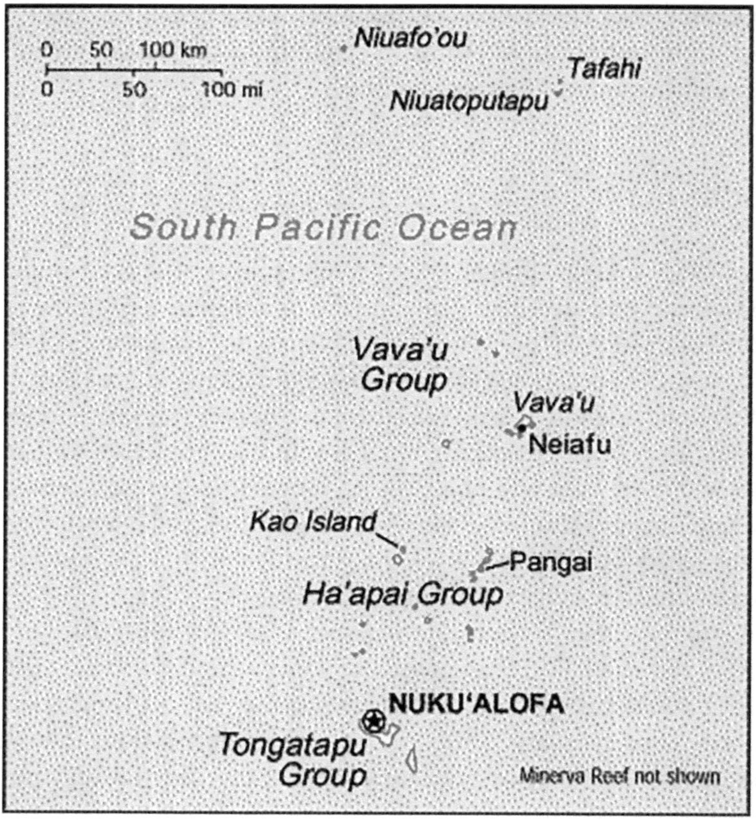

Fig. 5.2 Tonga. (Source: CIA World Factbook)

The Tongan participants were recruited for agricultural work in the Riverina (Victoria and NSW) and in southern Queensland. They were mainly married men aged between 26 and 40, from households of six people, and came from each of the four main island groups according to their population size. Almost all were either semi-subsistence agricultural workers or earned incomes from some kind of unskilled wage employment, and came from poorer households than non-participant households. They were not well educated, since the recruitment firm, Tree Minders, was more concerned with workplace skills. English literacy later proved a problem for skill acquisition. Like workers from Vanuatu, they had gone through briefing sessions in Tonga.

Tongan workers were first employed in Robinvale (Victoria), but went through a period of unemployment when work dried up and were later deployed elsewhere in Victoria and in Queensland. Being employed as casual workers they received no benefits during the period of unemployment. Productivity increased quickly after unfamiliar skills (almond picking and machinery operation) were acquired. The Tongans believed that acquired skills with chainsaws, for which they gained a certificate, would be useful in Tonga. Younger workers were more productive, especially where workers were paid a standard minimum hourly wage. Hours could be long; as one observed, 'In Tonga I would sleep for 12 hours a day; here I am working for 12 hours' (quoted in Blanco, 2009, p. 57). Incomes were boosted and assimilation made easier where workers were able to stay with permanent resident Tongans in the neighbourhood, although that was unusual. The labour hire company organised participation in church choirs and rugby games, which provided welcome social opportunities. During periods of downtime with relatively little work, boredom and alcohol consumption became more common.

Despite periods without work, the Tongan workers grossed more than A$10,000 per head, substantially more than they would earn in a year in Tonga. Costs were kept down by sharing accommodation and basic, bulk food purchasing. Many commented that they were earning more than they had expected. Most sent money to just their own household, but some remitted to as many as five households. Remittances averaged about A$358 per week, in the first fully employed phase, with workers with larger families tending to remit more. All the workers intended their remittances to be used for paying bills (electricity and telephone), home renovation and their children's education, with lesser amounts sent for churches, medical bills and vehicle or land purchase. A quarter of recipient households used remittances to pay for local agricultural workers to tend their own crops during their absence. Those who had never worked overseas before tended to be more oriented to using remittances for immediate consumption objectives while those with a previous migration history, who had achieved basic material goals, were more interested in seeking to establish or develop a small business, including painting and fishing businesses (Blanco, 2009). Almost all the workers sought to return for a more lucrative second phase.

A sample of the households of seasonal workers in the two Tongan islands of 'Eua and Tongatapu indicated that they were poorer than non-PSWPS households in terms of incomes and in access to services (such as piped water and electricity). Poorer, more rural workers were more likely to participate in both the early RSE and PSWPS schemes (Blanco, 2009; Gibson et al., 2008). The remittances to households of migrant PSWPS workers were said to constitute as much as 95 per cent of their incomes, while workers saved assiduously and denied themselves foods, that is unlikely. One worker commented that he was 'always hungry, because we don't spend enough on food' (quoted in Blanco, 2009, p. 71). Certainly, Tongans worked and saved hard to meet their objectives. For those working in the Riverina, cigarettes and phone credit were the only non-essential items that they purchased (and almost all the phone credit was used in talking to families in Tonga).

In Tonga, remittances were used mainly for consumption (bills and education fees) though all households first donated something to churches. Diets changed slightly as food was more likely to be purchased. Some households hired labour in the absence of male workers, while others without labour experienced declining productivity, but were unconcerned, perceiving the benefits of remittances to be considerably in excess of any short-term lost production. Strong social and family networks were usually adequate to reduce any negative costs of the absence of workers (Blanco, 2009; Rohorua et al., 2009). Remittances were also used to pay off debts on housing materials and car purchase. In this early phase, only one household had been able to invest in business development. Households welcomed the financial security and freedom that the scheme had given them as the best outcomes of the PSWPS, while the most negative social outcome was the absence of the husband and father. Benefits were particularly strongly perceived on the much smaller island of 'Eua, south of Tongatapu, where economic opportunities were unusually limited.

Despite the various problems associated with limited employment in Australia, the workers in Australia and their households in Tonga regarded the first phase of the scheme as successful. Those who had gone to New Zealand recognised this as a learning experience for Australia; as one Tongan ministry official observed,

> Overall the RSE is going very well. You can see tangible things, like buildings going up for young families. The first year has been a very good learning experience—we've learned a lot from the boys from Lapaha and Kolovai [Tongatapu villages] who first went on RSE. New Zealand is really doing all the hard work for Australia! (Quoted in Maclellan, 2008b, p. 2)

Workers from Tonga constituted mainly unskilled workers with limited incomes and education, and the labour mobility schemes thus opened up opportunities that would not otherwise have existed for poor rural households (Gibson et al., 2008) with fewer diasporic links in New Zealand. Broadly, the same proportions of Tongans as ni-Vanuatu had worked on these early schemes, and, as in Vanuatu, the initial years held out considerable promise for the future.

First Movers

The seasonal employment trajectories of other smaller national groups in these early years have largely escaped analysis, but there is little reason to think that they differed much from those of the Tongan and ni-Vanuatu workers, in terms of their employment, remittances, attitudes and aspirations. Samoans were similar to Tongans in background, experience and achievements, but initially favoured New Zealand's RSE. PNG, unfamiliar with such schemes, responded belatedly and overoptimistically; when its first small group of workers left in 2011, the Chairman of its Seasonal Worker Taskforce believed that workers would return with about A$20,000 each and that this would inspire rural business development (ABC Radio, 23 May 2011). Tuvaluans, on the RSE scheme, were however somewhat different coming from remote, dry coral atolls where agriculture had no resemblance to that in Australia or New Zealand. With a population of about 10,000 people scattered across nine coral atolls, Tuvalu had regularly sought new migration opportunities overseas and was the smallest country participating in the scheme. Workers were recruited from each of its islands, following the long-established national principle of enabling equitable access to overseas job opportunities. However, unlike early days elsewhere, geographical disadvantage and the very high initial costs of participation meant that only relatively wealthy Tuvaluans were able to be involved. Although workers were generally positive about the work experience itself, most found the long working hours difficult and found it hard to adjust to the cold. While some returned with savings in excess of NZ$5000 (A$4600) after 28 weeks' work, many returned with very little money or were in debt, unusual for workers from other PICs. Nonetheless, such disappointments did not prevent new and repeat flows in subsequent years (Bedford et al., 2010). Workers from Kiribati had broadly similar experiences but few if any Tuvaluans or i-Kiribati ever participated in the PSWPS.

Overall, within the PICs the schemes proved to be both generally pro-poor, because of domestic selection procedures, and were not localised in particular regions, but rather with opportunities spread widely in the sending countries (Gibson et al., 2008; McKenzie et al., 2008; Blanco, 2009). In Vanuatu the average household sending workers to the RSE scheme, while poor by regional standards was richer than the average ni-Vanuatu household not sending workers (Gibson & McKenzie, 2011). However, every country sought to achieve some version of geographical equity. Migrants moved out of relative poverty. Governments, recruiters and local agents preferred agricultural workers—and sought to exclude educated, or even skilled, workers (and the urban unemployed) reducing the costs and increasing the benefits to sending countries. The guest-worker schemes ruled out the rich and skilled, and provided a rare and important outlet for the relatively and absolutely poor.

At the local scale village councils and leaders were involved in recruitment and selection—seemingly in favour of the most 'needy and deserving' and workers were expected to contribute to village projects as well as their own. Such procedures had a long history in Kiribati and Tuvalu and were not unusual elsewhere. At much the same time, village-based selection schemes for Fijian workers recruited for the Middle East appear to have had a similar balanced and ethical basis (Connell, 2006), as they were in earlier labour migration to Nauru from Kiribati and Tuvalu, and from Fiji and Samoa to New Zealand (Levick & Bedford, 1988; Macpherson, 1981). Local and regional equity and both household and community benefits characterised early temporary migration schemes.

After the first year in Vanuatu, women were not obviously discouraged from participating.

While Tuvaluans and i-Kiribati were particularly disadvantaged and earned less than other national groups, the schemes were not without other problems. Many workers struggled with the cold, unfamiliar technology (and even appliances such as toasters), and more critically with tax returns, ATMs, scarce work and too much downtime. Workers had little engagement with unions that might have resolved unemployment and downtime problems. Both knew little of the other. Limited working hours—but also long hours—posed problems, while wage calculations and deductions sometimes lacked transparency (Maclellan, 2008a). Tongans in the Riverina were able to benefit from the presence of established islander communities that assisted with welfare services and broad pastoral care, linking people into local activities and minimising boredom, frustration and the resort to alcohol. Any possible social costs of workers being absent, while children and old people remained, or acquiring new attitudes to work and life overseas, were yet to occur.

Some evidence suggested that the withdrawal of labour meant lost productivity in some labour-sending communities, including Vanuatu, which may have had negative impacts on diet and health (Rohorua et al., 2009) but that was almost certainly counteracted by income gains. Remittances reduced both the incidence as well as the depth of poverty (Gibson et al., 2008; Blanco, 2009; Brown et al., 2014). Savings supported welfare gains and material wellbeing through improved housing and education. Most of the early seasonal workers regarded the experience as positive and felt they had earned a satisfactory income. Farmers too recognised the benefits, especially as the schemes moved through second and third years and workers were more experienced. The PICs were generally enthusiastic despite the small number of participants, a function of demand in New Zealand and Australia. Despite frustration and privations, islanders were usually willing to persevere because of the significant earnings that could be made. Even in these early years, as workers could be taken advantage of in an unfamiliar social and economic world (just as employers too were unfamiliar with the aptitudes and quirks of their new labour force and a new employment system), a large number—probably the majority—sought to return.

An Early Triple Win?

The new schemes had got off to a reasonable start despite the small numbers. They probably initially attracted the best workers and more well-organised employers, and there were real incentives for both groups, and for the participating governments to ensure that the schemes worked successfully. Early assessments largely focused on measures of income and material wellbeing rather than non-income aspects of wellbeing and vulnerability. Social costs later became more significant, including the emotional stresses of separation, but these were not nearly so certain or obvious as the material gains. In that regard, both schemes achieved early 'triple wins', if on a small scale.

Numerous teething problems occurred—including inadequate (and expensive) housing and facilities, lack of work at certain times (but continued costs), lack of awareness of certain requirements and local conditions (including the need for warm clothing), unfamiliarity with the kinds of agricultural work and labour organisation that were required, and frivolous expenditure of 'new money'. Bank accounts, the English language and cold weather were novelties to many. Minor car accidents occurred (partly a function of ni-Vanuatu experiencing driving on the 'wrong' side of the road) with one woman so badly injured that her husband had to fly from Vanuatu to help her return home. Some workers returned home after barely two months (*Fiji Times*, 30 August 2008). Uncertainty existed over hourly rates and piecework, with most Pacific islanders preferring standard hourly rates for all workers, and over deductions from wages for various reasons. There was scope for mistreatment. One contractor provided a three-bedroom house to be shared by 22 workers. Most workers found that the costs of accommodation were greater than anticipated—especially since they were sharing quite basic facilities and often faced additional costs for electricity and water (Bedford et al., 2010). This necessarily reduced their earnings, and some felt they had not been able to accumulate the kinds of sums they had anticipated and expected. The Australian scheme officially guaranteed six months' work. Employers, used to a rapid turnover of casual workers, found it difficult to commit to six months' employment or guarantee 30 hours work per week for the contract period (TNS, 2011). It also proved difficult to forecast and predict agricultural labour requirements far in advance, especially in uncertain climatic conditions. This proved the greatest problem in both destinations as workers did not gain the work they expected or were repeatedly and uncertainly shuffled around between employers.

A lack of leadership within groups of workers in a new social context made both any regulation and social cohesion difficult and resulted in periods without an obvious sense of direction and purpose. The provisions for pastoral care were vague and uneven since it was devolved to employers rather than regulated by Ministries of Labour. Source countries had inadequate resources to monitor the wellbeing of the migrant workers. Where there was considerable downtime workers were more likely to spend some of their savings, especially on alcohol, since there was little else to do in unfamiliar rural areas and within a dominantly male environment.

Over the first few years of both guestworker schemes, participant expectations around earnings reduced and workers, especially returnees, developed a greater familiarity with what was expected, what was available and how they might conserve savings. The PICs learned relatively quickly who would turn out to be the more effective workers, and what kinds of pre-departure training were most useful. Employers learned how to accommodate new workers and valued returnees who could explain particular tasks and objectives to newcomers in their own language, take on more sophisticated roles such as driving and supervising, had already acquired basic skills and knowledge, and were familiar with the neighbourhood and the social environment. Most workers favoured Australia since it was assumed that wages would be higher and it would be warmer there, hence it was a particular disappointment that it offered few opportunities. As workers gained experience of working in different conditions (machinery, techniques, labour routines, grading and packing, etc.), amidst quite unfamiliar plants, trees and temperatures, there was a strong incentive for employers to re-hire those who wished to return. Return migration began to challenge aspirations to equity. Correspondingly, few useful skills were acquired. Skills that were acquired were rarely transferable to quite different PIC agricultural systems, as was true in the earlier

Samoan episode (Macpherson, 1981). An early predilection—that never waned—for purchasing chainsaws suggested potential downstream environmental issues.

That the main concern in both New Zealand and Australia was over the difficulty of securing a reliable workforce, emphasised that both schemes were structured around the needs of Australasian farmers. Pacific economic and social development was a secondary aim, especially in Australia, and the schemes were linked to Ministries of Labour/Employment and not to NZAid or AusAID. In Australia the scheme developed slowly having been inadequately publicised. As one farmer told Christine Brickenstein:

> Another key failure [of the pilot scheme] is the lead agency's inability to market the scheme to growers. That comes from a lack of understanding what marketing means but also an inability arising from the fact that DEEWR [Department of Education, Employment and Workplace Relations] mainly deals with domestic labour market issues rather than international migration (quoted in Brickenstein, 2014, p. 246)

Marketing and migration had yet to be reconciled. PIC governments and households were interested in the schemes for different reasons. Nonetheless, in these early years, both schemes benefited the PICs (through human capital development and the flow of remittances), the destination countries (through access to valuable labour to fill gaps), and the migrants themselves and their households (through the acquisition of capital). Beyond these gains the schemes had also served local and regional equity objectives. In a small way a triple win had been achieved and expansion beckoned. In a few islands, notably Epi and Tanna, the schemes had made a very real impact.

Already a new pattern of migration was becoming engrained in local societies—even as a normative rite of passage and an emerging culture of migration—that seemingly brought widespread benefits. Yet, despite significant welfare gains for the households of those employed, the limitations of short-term labour migration were evident in the minimal structural changes, the difficulty of converting household gains into sustainable development (hence the desire to return) and the inability of small island states to translate household gains into more effective structures of national development. Nonetheless, both schemes largely confirmed much earlier conclusions from another hemisphere that gains were invariably most obvious for recipient countries, who controlled the flow and recruitment (and most obviously for the farmers since the numbers were small) and benefits were less obvious for source countries and for the migrants themselves (e.g. Wickramasekera, 2011). By mid-2011, only 560 workers had been recruited by Australia over the two-year period since the PSWPS began. It was an inauspicious, small and largely positive start, but enough for the pilot schemes to be over and for Australia to welcome larger numbers. From 2012 onwards, numbers grew, the geographical extent of the scheme broadened to incorporate workers from other PICs, while it spread further within Australia. In the next chapter, therefore, we turn to the contemporary Australian scheme and what has been a more expansionist phase, that at the end of the 2010s extended the context of seasonal work so that it might be feasible for longer time periods and into industries beyond agriculture and horticulture.

Bibliography

Bailey, R. (2019). *New Zealand's Recognised Employer Scheme (RSE): 10 year longitudinal case study*. ANU Department of Pacific Affairs.

Ball, R. (2010). Australia's Pacific Seasonal Worker Pilot Scheme and its interface with the Australian horticultural labour market: Is it time to refine the policy? *Pacific Economic Bulletin, 25*(1), 114–130.

Bedford, R., Bedford, C., & Ho, E. (2010). Engaging with New Zealand's recognised seasonal employer work policy: The case of Tuvalu. *Asian and Pacific Migration Journal, 19*, 421–445.

Blanco, C. (2009). *Temporary migration, remittances and development: Early impressions of the Australian Pacific seasonal worker pilot scheme*. Honours thesis, University of Sydney, School of Geosciences.

Brickenstein, C. (2014). *Foreign Seasonal Workers in Agriculture; a comparison of regulations and practices in the Pacific and in Europe*. PhD thesis, Macquarie University, Sydney.

Brown, R., Connell, J., & Jimenez-Soto, E. (2014). Migrants' remittances, poverty and social protection in the South Pacific: Fiji and Tonga. *Population, Space and Place, 20*(5), 434–454.

Callick, R. (2010, March 2). Pacific scheme withers on the vine. *The Australian*.
Connell, J. (2006). Migration, dependency and inequality in the Pacific: Old wine in bigger bottles? In S. Firth (Ed.), *Globalisation and governance in the Pacific Islands* (pp. 59–106). ANU E-Press.
Connell, J., & Brown, R. (2005). *Remittances in the Pacific: An overview*. ADB.
Connell, J., & Hammond, J. (2009). *Iso Iso*: The first ni-Vanuatu guest workers in New Zealand. *Pacific Economic Bulletin, 24*(1), 83–95.
Craven, L. (2013). *The relationship between temporary migration and vulnerability to climate change in rural Vanuatu*. Honours thesis, University of Sydney Department of Government.
Curtain, R. (2018). What can Papua New Guinea do to lift its numbers in the seasonal worker programs of Australia and New Zealand? ANU Development Policy Centre Discussion Paper 70.
Gibson, J., & McKenzie, D. (2011). Australia's PSWPS: Development impacts in the first two years. *Asia Pacific Viewpoint, 52*(3), 361–370.
Gibson, J., McKenzie, D., & Rohorua, H. (2008). How pro-poor is the selection of seasonal migrant workers from Tonga under New Zealand's Recognised Seasonal Employer Program? *Pacific Economic Bulletin, 23*(3), 187–204.
Hammond, J., & Connell, J. (2009). The new blackbirds? Vanuatu guestworkers in New Zealand. *New Zealand Geographer, 65*(3), 201–210.
Kumar, R. (2012). *Migration and development: The development impacts of short-term labour migration of Ni-Vanuatu workers to New Zealand*. MA thesis, University of the South Pacific.
Levick, W., & Bedford, R. (1988). Fiji labour migration to New Zealand in the 1980s. *New Zealand Geographer, 44*(2), 14–21.
Maclellan, N. (2008a). Seasonal workers for Australia: Lessons for New Zealand. *Farm Policy Journal, 5*(3), 43–53.
Maclellan, N. (2008b). *Workers for all Seasons? Issues from New Zealand's Recognised Seasonal Employer (RSE) program*. Swinburne Institute for Social Research.
Maclellan, N., & Mares, P. (2007). Labour mobility in the Pacific: Creating seasonal work programs in Australia. In S. Firth (Ed.), *Globalisation and governance in the Pacific Islands* (pp. 137–172). ANU E-Press.
Macpherson, C. (1981). Guest-worker movements and their consequences for donor and recipient countries: A case study. In G. Jones & H. Richter (Eds.), *Population mobility and development: Southeast Asia and the Pacific* (pp. 257–277). ANU Development Studies Centre Monograph No. 27.

McDonald, H. (2008, April 28). Rich and poor share a harvest of plenty. *The Australian*, 1–2.

McKenzie, D., Garcia Martinez, P., & Winters, A. (2008). *Who is coming to New Zealand under the new Recognised Seasonal Employer program?* World Bank.

Ramasamy, S., Krishnan, V., Bedford, R., & Bedford, C. (2008). The recognised seasonal employer policy: Seeking the elusive triple wins for development through international migration. *Pacific Economic Bulletin, 23*(3), 171–186.

Rohorua, H. T., Gibson, J., McKenzie, D., & Martinez, P. G. (2009). How do Pacific island households and communities cope with seasonally absent members? *Pacific Economic Bulletin, 24*(3), 19–38.

TNS. (2011). *Final evaluation of the Pacific Seasonal Worker Pilot Scheme*. TNS.

Wickramasekera, P. (2011). *Circular migration: A triple win or a dead end?* ILO.

Williams, A. (2010). *The experiences of the recognised seasonal employer scheme and Ni-Vanuatu temporary workers in Te Puke, New Zealand.* Master of Planning thesis, University of Otago.

6

Taking Part

Australia has fewer socio-economic ties with most Pacific Island Countries (PICs) than New Zealand, despite having been the colonial power in Papua New Guinea (PNG) (and then hurriedly departing) and exercising trusteeship of Nauru. A consequence has been a much smaller Pacific diaspora population—both proportionately and absolutely—with Pacific islanders often only migrating to Australia indirectly via New Zealand. As a result, Australia has proportionately fewer 'human' ties to the islands than does New Zealand. Despite being a major aid donor to the Pacific, Australia has been slower to respond to PIC needs and wanted little to do with Pacific Islander migration. Nonetheless, and urged on by agricultural business interests, Australia belatedly adopted the Seasonal Worker Programme (SWP) scheme in 2012, after a tentative trial programme, opening it up to nine PICs and Timor-Leste.

After a slow start, hindered by lack of publicity, high compliance costs ('red tape'), the Global Financial Crisis (GFC) and the persistence of illegal labour, SWP participation gradually began to grow, albeit with two PICs—Vanuatu and Tonga—dominating numbers. As numbers increased, an Australian push sought to recruit more workers from a wider variety of PICs and ensure greater participation of women, who

K. Petrou, J. Connell, *Pacific Islands Guestworkers in Australia*,
https://doi.org/10.1007/978-981-19-5387-3_6

made up only a small proportion of early recruits. The scheme received a boost when, in 2016, Australia announced its 'Pacific Step-up' plan to prioritise engagement with the Pacific Islands, which was later grandiosely described by the Prime Minister, Scott Morrison, as 'a new chapter in relations with our Pacific family'. Expanding Australia's Pacific Labour Mobility (PLM) programmes, with eventual movement into industries beyond agriculture, was identified as a key means of fostering economic prosperity within the region.

From here on, we focus on labour mobility between the PICs and Australia, but draw on examples from the Recognised Seasonal Employer (RSE) where it is useful to compare trends between the two schemes, or where research on the SWP is lacking. This chapter describes the growth of the SWP scheme from 2012, considers PIC participation rates, recruitment practices and how effectively workers are prepared for the realities of life in Australia through pre-departure and arrival briefings. In so doing, it examines issues of equity around recruitment within and between PICs and considers who misses out and who benefits from PLM participation. In subsequent chapters we examine the impact of the scheme in Australia and in the PICs.

Where Do Workers Come From?

The SWP and RSE schemes were designed with an explicit focus on gaining a cheap and reliable labour source for national agricultural industries while simultaneously facilitating economic development in participating PICs. Participation rates were primarily a function of demand in New Zealand and Australia, but were also influenced by home country governance and recruitment practices (see below). Numbers of participants did not reflect domestic populations and varied markedly between the PICs (Table 6.1). Tonga and Vanuatu have numerically been the primary beneficiaries of both the RSE and SWP schemes since their establishment. Both were early entrants to the schemes, and so had the advantage of 'first mover' status, whereby employers were keen to re-employ workers who understood the work and had performed well in previous years. At the end of the 2010s, some 14.7 and 8.1 per cent of the Tongan and Vanuatu

Table 6.1 Number of SWP visa grants by country, 2012–2020

Pacific Island Country	Financial Year of Visa Grant							
	2012/2013	2013/2014	2014/2015	2015/2016	2016/2017	2017/2018	2018/2019	2019/2020
Fiji	0	0	<5	160	190	247	436	227
Kiribati	34	14	11	20	124	364	377	146
Nauru	10	0	0	17	0	0	0	0
Papua New Guinea	26	26	35	42	139	92	128	116
Samoa	22	162	185	140	309	527	677	543
Solomon Islands	42	9	21	61	87	175	314	228
Timor-Leste	21	74	168	224	477	915	1567	1526
Tonga	1200	1497	2179	2624	2691	2790	3737	2217
Tuvalu	0	20	7	<5	0	0	0	0
Vanuatu	119	212	567	1198	2149	3349	4964	4821
TOTAL	1474	2014	3177	4490	6166	8459	12,200	9824

Source: Department of Home Affairs, Australia

workforce, respectively, were employed as guestworkers, either in Australia or New Zealand, compared with a mere 0.01 per cent of the PNG workforce (World Bank, 2021).

Crucially involvement was a competitive process, becoming part of a marketing exercise; when PNG sent only its second group of workers its government office proclaimed that

> the participation of 12 energetic young able bodied men and women from various economic and cultural backgrounds in the Pacific seasonal workers scheme was crucial to opening the doors for more Papua New Guinean participation in a competitive labour mobility program that continues to be dominated by other Pacific islanders. This group of 12 PNG seasonal workers will work towards further promoting and consolidating the hopes of millions of young Papua New Guineans; their very presence in Australia will market, promote and further showcase the potential of the PNG labour market, thus reinforcing the confidence of Australian employers/farmers in sourcing more Papua New Guineans. (Quoted in PIR, 31 October 2011)

Early success in the RSE assisted Vanuatu's entry into the Pacific Seasonal Worker Pilot Scheme (PSWPS), and then the SWP. Nonetheless, Tongan workers initially dominated the Australian scheme, thanks to several Tongan residents in Australia who owned two large labour hire firms which employed SWP workers; a third large labour hire firm also had a close relationship with Tongan-Australian growers. In addition to employing many Tongans themselves, these firms promoted Tongan workers to other Australian growers (Curtain & Howes, 2019). Word of mouth about workers' reliability could be invaluable in influencing recruitment practices, but it could also deter employers from hiring workers from certain PICs. Indeed, an initial preference for Tongan SWP workers waned after worker productivity declined and incidents of bad behaviour increased. Assisted by the Vanuatu Government's Employment Services Unit (ESU), which worked closely with Australian employers, ni-Vanuatu workers eventually outnumbered Tongans in the SWP. Tongans and ni-Vanuatu continued to numerically dominate both the RSE and SWP until at least 2020.

Despite the large proportion of ni-Vanuatu and Tongans participating in the schemes, other PICs gradually became more active participants. Fiji, with the most vibrant agricultural economy in the region, was a late arrival to both schemes, while Timor-Leste sent a substantial number of workers to Australia, largely due to its effective recruitment systems. By contrast, and despite being much the largest PIC in terms of population, very few guestworkers from PNG travelled to Australia. This lack of participation resulted from the absence of intermediaries to broker work arrangements, an over-reliance on recruitment via work-ready pools (Curtain, 2018), and a recent past where international migration from PNG was almost non-existent. Much the same was true of Solomon Islands. The remote nations of Kiribati and Tuvalu, with much smaller national populations, also had limited participation in the SWP. Indeed, Tuvaluans travelled almost exclusively to New Zealand where a growing expatriate community existed, having experienced difficulties in securing places with unfamiliar Australian employers (Falefou, 2017, p. 259). In Kiribati, interest in the schemes grew over time, and in 2019, the government called for greater equity in recruitment between PICs. One i-Kiribati MP, Pinto Katia, pointed out that while Tonga and Kiribati had similar sized populations, Tonga sent roughly 20 times more workers to New Zealand each year than Kiribati (RNZ, 12 April 2019). Nauru, with no past experience of migration, no recent history of manual employment and almost full employment through hosting Australia's refugee holding centre, expressed no great interest in the SWP. However, participation rates fluctuated, and recruitment patterns changed in response to employer preferences and demands, and PLM governance arrangements. The participant numbers from different PICs were not therefore the outcome of particular Australian preferences and choices, but reflected the practical operations of the PICs themselves and of labour hire firms and companies.

Recruitment

While the Australian government oversaw and administered the SWP, the recruitment of workers was largely in the hands of participating PICs and Australian employers. Many PICs, including Vanuatu and Tonga, were eager to be involved in the SWP and developed (relatively) user-friendly recruitment systems, but others had more cumbersome procedures. Recruitment models could be broadly classified as either 'government-light' or 'government-central' (Table 6.2). In numerical terms, 'government-light' proved the more successful model and was the strategy employed by Tonga and Vanuatu. Under this approach, recruitment was handled by employers and their agents (and thus entirely in the private sector). In Vanuatu, agents were formally licensed, whereas in Tonga the process was more informal and relied upon intermediaries such as SWP returnees and Tongan agents resident in Australia. Indeed over 63 per cent of Tongan workers had some existing connections in Australia: more than double that of any other country (World Bank, 2017).

Agents were involved in visa processing and pre-departure briefings and played an important role in dealing with administrative hurdles such as medical and police checks for workers. Agents and employers were key in promoting their workers and providing pastoral care once workers were in Australia. This government-light model was less burdensome for participating governments and was generally preferred by employers who valued the ability to select new workers through returnees. Nonetheless, reliance on agents could be risky, and in Vanuatu various reports existed

Table 6.2 PIC labour mobility governance styles

Country	Classification
Fiji	In between
Kiribati	In between
PNG	Government-central
Samoa	In between
Solomon Islands	In between
Timor-Leste	Government-central
Tonga	Government-light
Vanuatu	Government-light

Source: Curtain and Howes (2020)

of agents charging illegal fees to workers recruited into the programme. The use of agents could also make it more difficult for governments to manage workers once they arrived in Australia. Consequently, the government-light approach could carry reputational risks when agents 'misbehaved', or when workers were not fully briefed or absconded (Curtain & Howes, 2020).

PIC governments implicitly played a more prominent role in the 'government-central' recruitment model and did most of the work when it came to recruiting workers and overseeing the scheme. This worked best when governments were responsive to issues that arose. When they were not, as was the case in PNG, participation suffered. Timor-Leste provided a good example of an effective government-central approach, stationing two Timorese labour attachés in Australia to provide pastoral care, promote the scheme and assist with worker selection and visa processing from the Australian side. This made it easier to solve any problems workers might face as they arose. The workers themselves were selected from a work-ready pool (Curtain & Howes, 2020; Wu, 2019). This 'pool' comprised workers who were selected by governments, had met all the participation requirements and were ready to be deployed. Governments approached work-ready selection in various ways. In Fiji, for example, in 2016, the Fiji Minister for Employment personally travelled to 12 rural areas with a team of government officials in tow, to recruit work-ready workers, and consulted with local communities to discuss expectations around the scheme and provide in-person support and guidance. The Fijian government specifically targeted more remote rural communities where people had existing relevant agricultural experience and were more likely to be living in poverty. By contrast, in PNG where the government also ran a work-ready pool, there was limited outreach to rural communities and little support provided for potential workers (Curtain & Sherrell, 2017). Effective government involvement at different scales was the key to success under the government-central recruitment model.

While work-ready pools could be effective, most Australian employers disliked this system, expressed concern over nepotism in selecting workers and preferred to do their own recruitment. Many employers found it difficult to obtain information about workers in work-ready pools, an issue that they felt could lead to subsequent problems around health and performance in the workplace, preferring more hands-on recruitment methods. As employers established relationships with their workers and agents, engagement with work-ready pools generally declined over time. Employers' choice of PIC and workers was influenced by various factors including the reputation and experience of workers from different PICs, the ease of navigating and working with PIC governments and agencies, and employers' understanding of protocols and culture from their chosen PIC. Most employers generally preferred to recruit workers from a single PIC and tended not to recruit from countries where travel connections were expensive or infrequent, or where workers had a poor reputation or might abscond (often through diaspora contacts in Australia). For PICs, worker reputation, personal connections to employers and user-friendly recruitment systems were integral to success.

Pacific Labour Migration Policies and Legislation

Despite the growing importance of PLM in the region, by 2021 Vanuatu was the only country to introduce legislation around seasonal work, passing the Seasonal Employment Act in 2007 around the time the RSE began. This Act remained in place until late 2019 when a new labour mobility policy was adopted to cater to the growth of labour migration in the intervening years, and the new opportunities, such as in the broader Pacific Labour Scheme (PLS) (Chap. 10), that had become available. The new policy recognised and sought to rectify the lack of government support for workers while overseas, highlighted the need for a strategy to help workers invest their savings and better manage the impacts of labour mobility on sending households and communities. It summarised these needs through six key aims:

1. Increase labour mobility opportunities and make them available to a wider range of people.
2. Enhance Vanuatu's skill base and better match skills development to Vanuatu's labour needs.

3. Improve how labour mobility is governed and operated.
4. Strengthen the social and economic benefits of labour mobility.
5. Maintain Vanuatu's attractiveness and trust of the receiving country as a major source of workers for the international labour mobility market.
6. Reduce costs for workers and ensure their wellbeing and protection. (Government of Vanuatu, 2019)

The new policy thus aimed to improve and maintain Vanuatu's participation and primacy in the scheme while ensuring better, more equitable access and outcomes for the workers. In response to these key aims, by late 2020, the Government of Vanuatu had announced plans to move away from the agent model of recruitment (see below) to a decentralised, government-controlled recruitment system that was intended to increase geographical equity of access to the schemes. In early 2021, after several years in preparation, a worker re-integration strategy was launched by the Government of Vanuatu with support from the Department of Foreign Affairs and Trade (DFAT) and the New Zealand Ministry of Foreign Affairs and Trade (MFAT). The strategy aimed to support workers and their families throughout the life cycle of labour mobility participation through three specific objectives:

1. Support skills development and transfer to increase human capital
2. Increase financial investment into agriculture
3. Identify strategies to support seasonal workers and their families to increase agricultural production, food security and income security, and climate change and disaster preparedness training to sustain an agricultural business. (DARD and DoL, 2020, p. 10)

It was proposed that these objectives be met via mentoring, training and support centred on financial management, soft skills acquisition and agricultural skills development, but it was couched in such general terms that it was difficult to envisage how it might be translated into useful action.

Other countries did not follow Vanuatu's practical concern, but by 2019 several PICs were preparing their own labour mobility policies. In Samoa, legislation was drafted in 2008, but was yet to be passed. Kiribati had prepared a broader National Labour Migration Policy in late 2016 that covered temporary labour migration and aimed to ensure i-Kiribati could migrate 'with dignity'. Recognising existing limitations, PNG devised a new Labour Mobility Policy in 2021 to decentralise the management and governance on the national scheme. The lack of specific policies in most other PICs indicates some lack of national interest in worker welfare, while lower rates of participation in some countries meant that temporary labour migration schemes were of more limited national significance and hence targeted policies seemed unnecessary.

Agents: Advantages and Pitfalls

Agents were a key ingredient in both Vanuatu and Tonga's high levels of participation in the SWP (Curtain & Howes, 2020), but they were also a source of controversy. In Vanuatu, applications to become an agent were processed annually, and screened by the Department of Labour, with licensed agents bound by various conditions including the need to 'conduct a fair and transparent selection process' (Republic of Vanuatu, 2007, p. 10) and to provide pre-departure training. After remaining relatively constant for many years, the number of agents expanded significantly around 2018 when the government relaxed registration requirements, and in 2019 there were 107 and 23 licensed agents recruiting for the SWP and RSE schemes, respectively (Bedford et al., 2020). Overseeing the practices of so many agents proved difficult, and problems increased as some unscrupulous agents chose not to follow official recruitment guidelines. In response, the Vanuatu government increased scrutiny of applications, and by 2021 the number of licensed agents had fallen to a more manageable 69, 13 being RSE agents with the rest recruiting for the SWP, where worker numbers were not capped.

Charging illegal fees became a key ongoing problem within the agent system. In Vanuatu, agents were not allowed to charge fees to workers, and instead recouped their costs from employers. However, no guidelines existed over how much agents should charge. For the RSE, agent fees averaged NZ$45–50 (A$42–47) per worker, an amount that was not enough to cover recruitment costs, let alone generate a profit (Bedford et al., 2020). For SWP workers, the fee hovered around A$100–150, so that only a handful of agents operated at a scale that made business economically 'worthwhile': agents did much work for little remuneration. As well as taking care of workers' passports and other application documents, agents were required to provide pre-departure training, transport workers to the airport and remain in constant contact with them (generally via the internet) while they were in Australia to keep up morale and deal with any issues that might arise. Agents generally did not invoice employers for their services until after workers arrived in Australia, and while by 2020 supplier agreements were required between agents and employers, in practice they were rarely used (Wade, 2020). Codes of conduct were non-existent, but if agents failed to meet the standards set out in the Seasonal Employment Act, their licences could be cancelled.

Australian employers repeatedly expressed concern over agents illegally charging recruitment fees to workers. However, while some agents operated illegally, not every monetary transaction between workers and their agent was coercive. Rather, following traditional practices of exchange, ni-Vanuatu workers sometimes presented their agent with a small (approximately A$200–300) monetary gift of thanks when they returned home—a large amount by local standards but only a small percentage of their total savings. Such cultural nuances made it difficult to 'solve' the problem of recruitment fees (Bedford et al., 2020), and by late 2020 the issue had become political. In September 2020, the Deputy Prime Minister of Vanuatu, Ishmael Kalsakau, used Facebook to accuse opposition MP, John Salong, a well-known recruitment agent, of charging workers fees. At about the same time, Australian media reports suggested that licensed agents who ran for office in Vanuatu might be promising workers a place in PLM schemes in exchange for votes. With only two former agents running for office, such a practice could not have been widespread.

Alongside accredited agents who did not follow the rules, ongoing issues also surrounded 'fake agents' who extorted money from would-be labour migrants (RNZ, 2019). Warnings and news about fake agents, as in a 2019 Facebook post, were not uncommon on social media:

> The police today arrested a young boy from Ambae who was using the name [XXX] to purposely steal money from people. He was arrested because he was lying to people all over town, telling them he was an agent for the SWP. If you've given money to him, you can go to the police station to claim it or take information to the police. Thanks to the Vanuatu police force.

To add legitimacy to their claims, some scammers used the names of registered agents and pretended to be acting on their behalf. Another Facebook post from 2019 warned,

> If someone comes and tries to trick you on the island or in Vila by collecting money in my name, don't give it to them. So far I've already received reports of two groups of people who have been lying to everyone and collecting money from them using my name. I never sent these people or asked them to collect money. Some of these people are using old contracts to trick people. This is a really serious issue and we're going to follow up on it. Thanks.

According to Vanuatu's 2007 Seasonal Employment Act, masquerading as a fake agent could result in a fine of up to 2,000,000 vatu (A$20,000) or a jail sentence of up to two years. Vanuatu's Commissioner of Labour, Murielle Meltenoven, announced in 2019 that organisations that falsely claimed to handle SWP applications could be fined up to A$55,000. However, at least until then, agents were regulated but not monitored, making scams relatively easy to pull off.

Elsewhere too, the widespread enthusiasm to take part in seasonal labour created opportunities for fraud, but not necessarily from official agents. In Samoa, spurious agents defrauded applicants of 'consultation' fees and fees for passports, and medical or police clearance certificates that never eventuated. One man had applied for a scheme offering seasonal work in American Samoa, unaware that no such scheme existed (Fatupaito et al., 2021). The degree of shame and embarrassment attached to being tricked suggested that fraud was probably under-reported.

Citing ongoing issues with the agent model, in late 2020, the Government of Vanuatu announced plans to move to a government-controlled work-ready pool, which would have potentially increased government revenue from PLM. The proposal was approved by the Council of Ministers and was planned to be taken to parliament in 2021. Were the legislation to pass, a 12-month transition period would ensue, during which time all agents would be phased out, so removing both corrupt and ethical agents. In early 2021, the Government of Vanuatu announced that agents would have the year to 'prove themselves' and show that they were following the rules and benefitting (rather than ripping off) participants. Curtain and Howes (2019) have directly linked Vanuatu's success in the scheme to their use of agents, and government intervention could reduce ni-Vanuatu participation rates.

Employer Involvement in Recruitment

In addition to recruiting via agents, some employers travelled to PICs to interview and recruit workers directly. For example, the manager and pastoral care workers from 'Appleseed', the largest RSE employer recruiting from villages in Lamen Bay, Epi, Vanuatu, travelled to the island annually to interview workers. Rachel Smith (2016) describes the visit that took place in 2012.

> Appleseed's manager and pastoral care workers had been visiting Epi to meet and interview workers since 2008. In 2012, the interviews took place in a small building owned by a local recruiter. Potential workers were required to meet the usual employment criteria such as physical fitness, but were also spoken to about Appleseed's expectations around motivations and behaviour. During the interviews, the Appleseed team scored applicants on criteria including reasons for migration, family situation, and goals and objectives for savings. While standards of work in previous seasons were the most important criteria for return workers, these 'other' factors were also taken into consideration. The Appleseed manager kept a record of employee aspirations and achievements, while the team from Appleseed often visited workers' homes and took notes about and photos of workers' projects such as home improvements and small businesses. The Appleseed team were proud of the 'development' their involvement in the RSE scheme had enabled.

In this case, somewhat exceptionally, recruitment was about more than just performing work in New Zealand, and employers took on a degree of moral responsibility for the development impacts of the RSE in 'their' chosen villages. Some other employers travelled to workers' home villages specifically to see how PLM had impacted local lives and livelihoods. Joe, a ni-Vanuatu man employed in rural Victoria in 2012, described how his employers, their children and a supervisor from his workplace travelled to his village on Santo:

> They [the Australians] came and Ned and Robert showed them around. We made a big feast according to our *kastom* [customs]. If a stranger comes, you have to make a feast for them.

Travelling to villages such as this provided one means for helping employers to understand workers' customs and contexts. Here too, employment was much more than a mere commercial arrangement, even as visits could also encourage new recruits.

During their visit, some employers used the opportunity to 'test' workers' suitability for horticultural employment. Victoria Stead (2021, p. 2) describes how representatives from a packing shed in rural Victoria selected ni-Vanuatu workers on a trip to Port Vila:

> In the absence in Port Vila of both apples and mechanized packing lines, the hopefuls had been put through a series of exercises with golf balls, meant to simulate the processes of sorting and classing fruit into Class 1, Class 2, and juice quality grades. Describing the process to me subsequently, in her office at the shed in rural Victoria, the manager of the packing shed had explained that though it wasn't ideal, it was nevertheless the best available way to get a sense for who would work well. 'Obviously, though', she had added, 'we don't always get it right'.

Getting it 'wrong' in these auditions meant that workers who were 'not quick enough' or 'not discerning enough' in their fruit sorting could be sent home early, a loss for both employers and workers who would not be able to meet the savings goals they had planned. Ensuring the most appropriate workers were chosen was a complex and critical process in both sources and destinations.

Direct employer recruitment and visits to home villages were relatively straightforward in countries such as Vanuatu, geographically close and with regular direct travel connections to Australia, but remote PICs, such as Tuvalu and Kiribati, were much more costly and difficult to visit, had fewer links with Australia and lacked direct contacts with employers. It was more difficult for them to benefit from these kinds of relationships.

While growers accounted for 89 of the 125 approved employers (AEs) operating in Australia between July 2012 and February 2020, labour hire companies brought in far more workers; whereas the average grower employed 161 workers over this period, the average labour hire firm employed 1013 workers, nearly ten times as many. The top four SWP employers were all labour hire companies and accounted for roughly half (48 per cent) of all positions. The dominance of large labour hire companies was mainly due to the high transaction costs of participating in the scheme; workers had to be brought in from overseas; and pastoral care and accommodation were required. Labour hire companies were better placed than smaller employers to understand and comply with the complex SWP government regulations (Curtain & Howes, 2020), had a wider knowledge of regional labour markets and were thus more able to ensure that workers were deployed to appropriate locations. These firms were also expected to play a broad role in providing pastoral care—broadly social protection—and ensure that growers provided adequate facilities for workers. However, the largest labour hire companies were distant from both farmers and workers, sometimes a state and several hundred kilometres apart, creating a gulf and a lack of pastoral care that could contribute to accommodation and employment problems for both groups (Chap. 7).

In addition to these 'formal' recruitment strategies, some Pacific islanders actively sought out opportunities to participate in guestwork by drawing upon personal networks. Rachel Smith (2016) describes how a group of ni-Vanuatu from Lamen Bay, Epi, became friends with an Australian couple who visited their village as tourists, sometimes for several weeks at a time. The villagers began asking the Australians if they knew of any opportunities to participate in the SWP. The couple found a farmer in Victoria who was willing to take on some ni-Vanuatu workers as a trial, and they took on the responsibility of recruitment and pastoral care for the workers while they were in Australia. As we discuss in the next section, personal contacts such as this could be instrumental in gaining access to the scheme and indicate the interest and agency of potential workers. The context of participation took many forms.

Worker Selection: Knowing the Right People

In Vanuatu, the increased number of licensed agents led not only to unscrupulous behaviour but also to growing competition to place workers in Australia (and New Zealand). For SWP agents to be successful, they needed to forge a good relationship with Australian employers, who ultimately decided who was hired and for how long. Some agents, from both Vanuatu and elsewhere, were able to draw on pre-existing contacts in Australia. This worked against PICs, such as PNG, where such contacts were scarce (Curtain, 2018). Agents who could provide reliable workers were likely to be given further work, so they tended to recruit 'trusted' workers who were often friends or relatives. While this may seem like nepotism, it mirrored traditional practices of finding domestic employment through kin networks (Petrou, 2020), and it was expected that those with access to opportunities would share them with family. Many of the ni-Vanuatu workers we met over the years had personal connections to their agent.

> Keith lived in Port Vila where he had a small business owning and driving buses. Keith took part in the PSWPS in 2011, and in 2012 had returned to Australia for the second time. He was able to participate in the scheme because his uncle worked as an agent and made arrangements for Keith to travel to Australia.
>
> Vanessa lived in Port Vila where she worked as a radio announcer. In 2012 she took leave from her job so that she could participate in the SWP. Her husband had already participated in the RSE scheme three times. Vanessa's brother, who worked as a team leader in the SWP heard that women were to be included in the next round of recruitment, and suggested she should participate. After her brother encouraged her, Vanessa spoke with her uncle, who was an agent, and he arranged for her to travel to Australia.

> Edna lived in Luganville, Santo where she was employed in housekeeping at a local hotel. She met her agent when he was staying at the hotel as a guest. Edna told her agent that she was interested in seasonal work, so he took her name and when he needed workers, he contacted her and asked if she was ready to participate. She told him that she was, and not long after, he arranged for her to travel to Australia.

Personal connections could be vital both in learning about the SWP and in finding a place in it. Across the PICs, most participating workers first heard about the scheme through personal networks including friends (45 per cent), family (24 per cent) and community leaders (4 per cent), and were primarily selected into the scheme through these friends (32 per cent) (World Bank, 2017). Consequently, communities with few connections to agents or workers were less likely to participate, so contributing to inequities around access to the scheme as opportunities were concentrated within some villages and extended families.

Finding employment for family members might have been expected, but it was not always straightforward. Terry, a Tongan-Australian working in agriculture, explained to Kirstie in 2020:

> It's a hard decision when you bring workers, it's very hard. Because you don't know what every person can do. Some who come, they only want to come and see Australia. They don't worry about money. Some people they know [they are] coming to work. Some people they think [they're] only coming for holiday […] some people only work to pay their rent and they're happy. The young one[s]. But the people coming, they got family in Tonga, they know why they come, to work hard.

Should a worker prove unreliable, it not only reflected badly on that worker but could also end further employment opportunities for friends and relatives. Participating in the SWP was widely considered a valuable and much sought-after opportunity, and the pool of potential workers far exceeded the number of available places, not only in Tonga and Vanuatu, but across the PICs. Personal connections were consequently invaluable.

Selecting Workers: Community Involvement?

Recruitment was not only about individual workers, but often involved community input and participation. In Vanuatu, hopeful workers needed to provide a letter of permission to participate from their spouse or family members, which was then endorsed by a village leader. Letters of support from communities were also required in Tonga, while in Fiji, where recruitment was focused on rural areas, a letter from the District Officer or Advisory Council was required (Bedford et al., 2020). In the early days, in rural areas of Vanuatu and Tonga, communities often played a role in selecting those who were considered trustworthy and reliable workers (Bailey, 2014; Blanco, 2009; Hammond & Connell, 2009). Moral precepts were involved. In the Torres Islands older men were regarded as the most reliable and it was believed they would bring the most money back. In Ambrym, people were chosen who worked hard, could follow instructions and were not 'overly dependent' on cigarettes, alcohol and kava (McKenzie et al., 2008). Such workers would be more likely to remit money to family members (rather than 'wasting' it), would return to the village (rather than linger in town) and would build an ongoing relationship with Australian employers, with the obvious implication that communities would benefit more. It was also anticipated that these workers would not misbehave or abscond, behaviour that could sully the village's reputation.

As the schemes expanded, communities were not always as involved in selection as they would have liked; in Vanuatu in 2015–2016, only 23 per cent of communities surveyed by the World Bank (2017) played a role in worker selection, while in Tonga 34 per cent of communities were involved. Some of these communities thought recruitment practices discriminated against younger people and those with disabilities, and many regarded women as disadvantaged in selection processes. By contrast, in urban areas, workers were more likely to self-select for participation and in Vanuatu many were 'walk-ins' who had applied through the Department of Labour (Gibson & McKenzie, 2014). That was also so in Samoa, where applicants had to meet quite standard criteria, including being between 21 and 40, literate in English and able to meet high fitness levels. When the SWP scheme eventually extended beyond agriculture, Samoa's Ministry of Commerce, Industry and Labour stated that 'previous experience with animals or knives is an advantage'. Applicants had to have a photo ID (a passport or drivers' licence) and were expected to have a phone number, an email account or a Facebook Messenger account (Fatupaito et al., 2021), characteristics probably biased against some rural residents. It was not always evident, at least to those rejected, why they had missed out. Some blamed village tensions. A Samoan bus driver who had been rejected twice by the Ministry stated:

> I look like a failure and a loser and that really saddens me … I have waited for this moment for so long … I look at the future of my children and to buy a family vehicle and a small shop, but all these dreams disappeared because I wasn't picked … the villagers mocked me for not going … Unfortunately the village decision is very biased … The village mayor said that I can't go because I might get into trouble and tarnish the village name. (Quoted in Fatupaito et al., 2021, p. 60)

While in small states individual characteristics might well be known to those involved in the selection process, there are clear indications that local authorities retained some input into the process. However, as the schemes became more established, PICs tended to move further away from community involvement in, and towards more institutional control of the process and more individual choice.

Geographies of Participation

Many Pacific islanders wanted to participate in the SWP, but, beyond personal constraints, opportunities were influenced by home country recruitment practices which often targeted certain areas or villages. While never rigorously enforced (and indeed unenforceable), the SWP was established with the broad aim of recruiting from rural populations where income earning opportunities were limited. Consequently, at the outset of the schemes, seasonal workers from Tonga (Blanco, 2009) and Vanuatu tended to come from relatively poor, rural villages (Connell & Hammond,

2009; McKenzie et al., 2008; see Chap. 5) despite the first group being selected from peri-urban Mele, perhaps the richest village in the country. Governments in Fiji, Timor-Leste and Tonga continued to officially prioritise recruitment from rural agricultural areas (C. Bedford, 2013; Curtain, 2018), although these were not necessarily the poorest households. In Timor-Leste, for example, workers tended to be selected from the south-east, where populations had more experience of lowland agriculture, rather than from the more impoverished western mountains, although most were actually recruited in the capital, Dili (Wu, 2019). By contrast, Kiribati and Tuvalu employed a quota system based on island or area of origin to spread opportunities equitably and ensure workers came from a variety of islands and villages (Bedford et al., 2010, 2020). Yet, despite this policy of seeming geographical equity, 60 per cent of Tuvaluan workers stated that migrating internally—invariably from the outer islands to the capital, Funafuti—greatly improved their chances of participating in the SWP. The same was true of 43 and 26 per cent of i-Kiribati and ni-Vanuatu workers, respectively (World Bank, 2017). Predictably, workers from less geographically dispersed countries did not consider location to be as important in influencing participation.

Government policies sometimes unintentionally skewed recruitment towards urban areas, as occurred when PNG enforced educational and English language requirements. Indeed, urban residents have probably accounted for a large and growing proportion of seasonal workers; 40 per cent of the one hundred ni-Vanuatu workers interviewed by the Labour Mobility Assistance Program (LMAP) (2017a) in 2017 resided in Shefa province, where the national capital, Port Vila, is located. Census data similarly indicate that a slightly higher proportion of urban (4.4 per cent) than rural (3.7 per cent) residents had ever participated in seasonal employment (VNSO, 2017) (Table 6.3). This need not necessarily represent a problem, where urban areas are increasingly characterised by poverty and hardship (Abbott & Pollard, 2004), so that recruiting urban workers could potentially resolve some poverty issues. But there is little evidence that this occurred. As we discuss later, many PLM participants held employment prior to participating in the schemes.

That workers have been recruited from every province in Vanuatu suggests at least a degree of equity. A 2018 sample study (of just under half

Table 6.3 Vanuatu population that has ever participated in the SWP or RSE by sex and location, 2016

Region		Total Population			SWP Participants			RSE Participants			Total PLM Participation (per cent of population)		
		Total	Male	Female	Total	Male	Female	Total	Male	Female	Total	Male	Female
VANUATU		144,175	71,870	72,305	1009	763	246	4537	3782	755	3.8	6.3	1.4
	Urban	35,027	18,134	16,893	367	252	115	1186	912	274	4.4	6.4	2.3
	Rural	109,148	53,736	55,412	642	511	131	3351	2870	481	3.7	6.3	1.1
TORBA		5342	2643	2699	16	11	5	129	120	9	2.7	5.0	0.5
SANMA		28,777	14,484	14,293	123	97	26	609	529	80	2.5	4.3	0.7
PENAMA		17,081	8436	8645	40	36	4	199	180	19	1.4	2.6	0.3
MALAMPA		22,370	11,009	11,361	83	76	7	454	411	43	2.4	4.4	0.4
SHEFA		52,584	26,864	25,720	685	497	188	2521	2010	511	6.1	9.3	2.7
TAFEA		18,021	8434	9587	62	46	16	625	532	93	3.8	6.9	1.1

Source: VNSO (2017)

of all workers) showed that there was a relatively even distribution throughout the country, in terms of the ratio of workers to total population (Table 6.4). Larger islands were however the main source of workers; Efate (442 workers), the site of Port Vila, and on average the wealthiest island in the country, and Tanna (388 workers) led the way. However, several islands with populations of less than a thousand were proportionately better represented; Mere Lava, a remote and isolated volcanic island in the north, and densely populated Tongoa, had 10 per cent or more of their population away as workers (Bailey & Rereman, 2019). However, since data were collected by 'island of origin', several workers may have been living in urban centres: a disguised bias towards recruitment from Port Vila. Whereas RSE employers tended to recruit from areas beyond Efate (possibly a function of their having long recruited ni-Vanuatu workers), SWP labour hire firms and their recruitment agents were more likely to recruit from Efate. Occasional suggestions of an unusual degree of equity and altruism existed; one Epi village chief stated that he had told the recruiting agent:

> You send the people from the small island [Lamen] that's all well. If you go and take people from Paama, that's good. Then you send people from Tongoa and the Shepherds, because they are many but they don't have land…that is fine. Then us who have a lot of land, we won't go. Us that have land, we plant kava, and we have the chance to find money. (Quoted in Smith, 2016, p. 226)

Table 6.4 Participation rates in SWP and RSE work by province, Vanuatu 2017–2018

Province	RSE	SWP	Total	Per cent of Population
Malampa	347	135	482	1.2
Penama	180	54	234	0.7
Sanma	298	55	353	0.6
Shefa	530	261	791	0.8
Tafea	256	166	422	1.1
Torba	208	21	229	2.2
Total	1819	692	2511	0.9

Source: Bailey and Rereman (2019)

Recognising the needs of even nearby others was not usual, even in early days.

Whether high rates of 'absenteeism' brought problems to the similarly small islands of Mere Lava and Tongoa or boosted island development is impossible to ascertain. Otherwise, the impact at this small scale probably reflected the success of recruitment strategies in different islands, and uneven access to information about the schemes. Even in 2018 many remote communities in Vanuatu had no idea such schemes existed (Dornan, 2018); rather earlier, many Paamese were interested in participating in seasonal work, but because they had no connections to an agent, they knew little about how to access these opportunities and were far from services that might have assisted them (Petrou, 2020). As agents focused on the larger, accessible islands, it became apparent that 'many of the rural and poorest areas of Vanuatu [we]re experiencing extreme barriers to participation' (Bailey, 2019a, p. 28). Recruitment skewed towards urban areas may exacerbate inequality of opportunities between rural and urban areas and raise questions about equity at a different scale.

For some ni-Vanuatu urban residents, participation in the PLM schemes was their first experience of agricultural labour. Alice, a ni-Vanuatu seasonal worker, provided an example from her own circumstances. Although her family originated from the island of Ambrym, Alice was born on Santo and moved to Vila when she was two years old where she had lived ever since. Before participating in the SWP, Alice was employed as an accountant and knew nothing of horticulture; as she told Kirstie,

> I've worked my whole life [in town], I grew up, I came, I went to school … I went straight to working in an ofice […] Because I grew up in town, you know what life in town is like, we're not like everyone who grows up on the island where they make gardens and do all of those things […] I came here [to Australia] and I just realised that I could do different things, like I can do other things which I thought I never could do.

For Alice, the opportunity to earn a significantly higher wage than was possible in Vanuatu—even in a much sought-after, skilled office job—was a key motivator to her participation in the scheme.

Where choice was available, recruiters and employers generally preferred to recruit directly from rural areas since, in the words of one employer for Vanuatu:

> rural recruits are more job focused and productive than those that live in a city area. Our workers live every day growing, gardening, building and walking—this gives them a wonderful mindset for the work on our farm. They are already used to labouring to feed their families, if they don't grow it or hunt it they don't eat. These are the best workers; they are already fit and healthy. (Quoted by Richard Curtain, in comments on Bailey & Rereman, 2019)

While this is a crude essentialist typecasting of rural life in the Pacific, it is indicative of the perceptions and orientations of many employers, and it is certainly true that many long-term urban residents, like Alice, had limited experience of agricultural work compared to their rural counterparts. Some ni-Vanuatu had somewhat similar views of rural life. As a 25-year-old urban tourism worker told John in 2019:

> In villages life is easy. They don't need to go. In town we have to pay for everything so it is more useful for us. We need to establish a business and ensure our children get a proper education so they can get a good job

If scarcely an accurate summary of urban (or rural) life, it reflected perceptions of a modernising nation and of alternative priorities.

Employers often preferred to recruit groups from particular villages or islands, so that they would form a small community overseas. In such instances, existing social relationships could play some part in regulating behaviour, maintaining wellbeing, and supporting and reminding each other of the employment and income rationale for participation. As one employer, speaking of ni-Vanuatu workers said: 'I definitely think recruiting from a community is better as they all know each other … [and they] feel that by coming here … they are representing their village so they will work together to ensure there are fewer problems' (quoted in Williams, 2010, p. 92). Speaking about a group of SWP workers in 2020, Andrew, a Pacific islander resident in Australia, explained to Kirstie the potential difficulties of regulating behavioural norms in a new environment and therefore the value of localised recruitment:

> There's been a growing issue of just you know the general drinking problems, clashes within the [community], because you've got the Tongans, you've got Samoan boys, there's about 27 Samoan boys and 20 Fijian boys and they're bringing the Fijian girls, 15 of them, and now they've got the Vanuatu girls or Solomon Islands—one of the two. And so that creates quite a bit of dynamic [...] one of the issues is that, which most islanders would know this, when you live in the islands, the fence is very clear, boundaries are very clear and you have somebody to enforce it within your community. So when you leave that and you come over here it, there's no boundaries, there's no fence, you know, apart from the rules, the policies that you've got to follow [...] So when they are being exposed to this kind of freedom for the first time, they, some of them, don't know how to handle it. So that's been an ongoing issue.

Recruiting a coherent group of workers from a single village or cultural group could minimise these issues.

Record-keeping systems are poor or non-existent both in the PICs and Australia, making it extremely difficult to know exactly where workers were recruited from, or why particular islands, villages or communities were chosen either by governments, Australian employers or other recruiters. Accessibility played a role, particularly in archipelagic states like Vanuatu, where populations were highly dispersed, or in Tonga where most people lived on the main island; it was simply faster, easier and cheaper to recruit from regions that were more accessible (Rockell, 2015; Calnitsky, 2016). Most Solomon Islands participants came from the main island of Guadalcanal. Early on, unable to find a way into the programmes from distant villages and islands, many young men migrated to their nation's capital, in the hope of finding recruiters there (see above) (Cummings, 2016; World Bank, 2017). At an international scale, it was far more difficult (and expensive) to recruit from less easily accessible Kiribati and Tuvalu. Personal networks also influenced who was chosen and, as we described above, those who had some connection to a recruiting agent were far more likely to participate (Petrou & Connell, 2018; Smith, 2016; World Bank, 2017). Based on the limited data available, we can tentatively conclude that there has usually been a bias towards larger more accessible islands, where the capitals are located, most of which are, on average, economically better off than smaller islands, and that this bias may have increased over time. Such a bias seems particularly true of more recently participating PNG and Solomon Islands, where the schemes are unknown in many regions.

Return Workers

Trustworthy and reliable workers were often invited back to Australia for subsequent seasons and were an invaluable resource for employers. Return workers often took on roles with increased responsibility as team leaders, assisted with pastoral care for newer recruits and provided on the job training for first-time workers in their own language. Some workers had changed their attitude after their first trip. Maggie Cummings (2016, p. 29) describes the case of Willy, who had made little money in the first year by picking apples too slowly and then spending the income on clothes, alcohol and convenience food, only to return as a church-going worker who no longer drank and who had become a shift manager, to the extent that some members of his work team accused him of being 'in the pocket of the white man [the farmer] and in danger of becoming too much like one, stingy, lonely and bossy'. Experience was invaluable, promotion was feasible and earnings increased.

On average, SWP workers across all PICs participated in the scheme for two seasons; however, some had participated as many as nine times (up to and including 2019). Return migration became more common as the scheme expanded. Once a worker had participated in the scheme twice, they were 70–80 per cent likely to continue participating for up to six years. Nonetheless, the likelihood of return varied somewhat between countries. Timor-Leste and Vanuatu provided many return workers, other countries—perhaps less impoverished—were less likely to have returnees. Workers in both the RSE and SWP schemes averaged 2.4 visits (Howes, 2018). Curtain and Howes (2020) argue that return was linked to the success of Vanuatu and Timor-Leste as sending countries; return workers created opportunities for new workers to join the scheme as their employers were keen to continue engaging similar SWP workers. The workers themselves were able to recruit additional workers into the scheme either directly or by recommending them to others. But it posed problems where a limited number of new jobs were created; in Timor-Leste, with the greatest proportion of returnees, a steadily growing number of potential workers the government's work ready pool grew in size annually (Rose, 2019). By 2018 the extent of return migration was being recognised as a problem in some PICs.

While not all workers returned or saw participation as a long-term livelihood strategy, by 2018 in the more established RSE scheme some ni-Vanuatu workers had participated in seasonal work for as many as ten years. Indeed, by 2019 at least one ni-Vanuatu RSE worker was on his 13th season, and by then had built more than one house (Corlett, 2019). Rochelle Bailey followed a group of 22 ni-Vanuatu seasonal workers from 2007 onwards. Seven had participated in the scheme every year; two participated most years, but took short breaks. When asked if they would continue participating, these long-term workers explained, 'It is my job now', 'I hope my job continues so I can keep helping my community' and 'I need it to continue paying for school fees' (quoted in Bailey, 2019b, p. 42). John was one such long-term worker.

> John from Sanesup village is married with three children. He came to New Zealand due to limited opportunities to earn wages at home. Since he began he has been able to support two of his children through their education, purchase additional cattle and fencing for his land, build a house and a church, purchase two buses and land in Port Vila, and build another house in Efate so his children had a home to reside in while undertaking tertiary studies. He is also currently building a rental property to generate income for his family. John has a son who also joined RSE in 2009, but works in a different Central Otago township than his father. His son exited in 2013 when he acquired enough funds to get married and purchase a bus. Recently his son has reapplied to RSE to earn money for his children's school fees. (Bailey, 2019b, p. 43)

Returnees argued that they needed to keep participating; otherwise, their goals would not be met. Yet 14 of the 20 workers had made at least one attempt to establish a business (as their goals were usually revised upwards), in what Bailey (2018) describes as 'entrepreneurial audacity'. Similar trends and patterns of continuity emerged from Australia, pointing to both the benefits and shortfalls of seasonal migration.

Such long-term engagement in seasonal work departed from the initial intention of the schemes. At its inception the RSE was envisaged as a three-stage plan that would engage workers for seven years. It was thought that workers would initially participate in the scheme to meet basic needs such as food and superior housing for their families. After three to five years they would then progress to pursuing goals such as education and meeting other large expenses. The final stage of participation was meant to result in small businesses or similar investments and secure a place for workers in the local economy, at which point workers would withdraw from the scheme and surrender their place to sons and nephews. Even this gendered pattern, where kin might take over, denied a broader equity where only a few new jobs were at stake each year. Thus, in PNG, with a population of 9 million, but where participation in the SWP was limited, some participants had already been involved at least four times (Dun & Klocker, 2017). By 2020, one of the perceived emerging 'problems' with the RSE was that withdrawal after seven years had not happened, and workers were participating for more than seven years to meet evolving aspirations. In Vanuatu, suggestions were made that limits be set on the number of returns to reduce dependency on both guestworker schemes and ensure that long-term participants did not abandon agriculture (Dornan, 2018). However that did not transpire—nor did attempts to deter participation of urban residents—since it was feared that term limits would mean fewer skilled return workers would become available and demand might shift elsewhere. At the same time, the Australian government introduced a three-year multiple entry visa that cut costs and encouraged return.

Remarkably, the vast majority of men (90 per cent) and women (75 per cent) surveyed by the World Bank (2017) wanted to continue participating in the SWP for as long as possible, while none of the ten-year veterans of the RSE had any intention of stopping in the next five years, rather seeing the work as their career (Bailey, 2018). While the implications of such long-term engagement, and the associated ageing of the seasonal workforce, remain poorly understood, it emphasises that goals cannot easily be achieved and that they are constantly being revised upwards.

Demographic Trends and Structures

Employers naturally preferred to recruit workers who were young and healthy (but also experienced). Officially only minor variations in the minimum age existed: ni-Vanuatu had to be 21, and Tongans and Samoans could participate at 18. Men overwhelmingly dominated both schemes, even as the proportion of men and women varied between PICs (Table 6.5). The 'average' Pacific guestworker was a married (or partnered) man aged in his early 30s. These attributes—age and marriage—were regarded as proxies for maturity (younger workers would not take work seriously) and responsibility. Workers with families were considered more stable by home communities, less likely to abscond and more likely to return and use their money for 'productive' purposes like establishing small businesses, rather than fritter away the profits on consumables and kava (Hammond & Connell, 2009). Such preferences were far from unique to Australia, and recruiters for Canada's Seasonal Agricultural Worker Program (SAWP) chose workers with the same attributes (Bailey, 2014). Discussions on ni-Vanuatu Facebook pages also reflected these normative assumptions. One 2019 post from a ni-Vanuatu worker observed:

> 21, 22, 23, 24 year olds don't know how to take work seriously yet. Unless they live in rural areas where work is in their blood. If you send workers who are too young, they're just going because they want to enjoy themselves. Their attitude and their behaviour is a headache.

As we noted earlier, Terry in the Riverina had a similar view of Tongans. Workers were thus likely to be young, male, able-bodied and in the 'prime' of their working lives. That led to a loss of manpower at home in these population groups. It also led to a highly idealised, and patronising, perception by one Queensland employer of Solomon Islands workers:

Table 6.5 Female participation numbers in the SWP, 2012–2020

Financial Year of Visa Grant	Gender	Fiji	Kiribati	Nauru	Papua New Guinea	Samoa	Solomon Islands	Timor-Leste	Tonga	Tuvalu	Vanuatu	Total
2012–2013	Female	0	10	<5	7	<5	13	10	138	0	30	**214**
	Male	0	24	8	19	18	29	11	1061	0	89	**1259**
	2012–2013 Total	**0**	**34**	**10**	**26**	**22**	**42**	**21**	**1199**	**0**	**119**	**1473**
2013–2014	Female	0	0	0	6	6	0	19	177	13	32	**253**
	Male	0	14	0	20	156	9	55	1320	7	180	**1761**
	2013–2014 Total	**0**	**14**	**0**	**26**	**162**	**9**	**74**	**1497**	**20**	**212**	**2014**
2014–2015	Female	0	0	0	8	<5	0	39	261	0	94	**405**
	Male	<5	11	0	27	182	21	129	1918	7	473	**2772**
	2014–2015 Total	**<5**	**11**	**0**	**35**	**185**	**21**	**168**	**2179**	**7**	**567**	**3177**
2015–2016	Female	39	0	<5	<5	5	11	46	327	0	190	**626**
	Male	121	20	13	38	135	50	178	2297	<5	1007	**3863**
	Not Specified	0	0	0	0	0	0	0	0	0	<5	**<5**
	2015–2016 Total	**160**	**20**	**17**	**42**	**140**	**61**	**224**	**2624**	**<5**	**1198**	**4490**
2016–2017	Female	15	14	0	40	8	0	125	301	0	393	**896**
	Male	175	110	0	99	301	87	352	2390	0	1755	**5269**
	Not Specified	0	0	0	0	0	0	0	0	0	<5	**<5**
	2016–2017 Total	**190**	**124**	**0**	**139**	**309**	**87**	**477**	**2691**	**0**	**2149**	**6166**
2017–2018	Female	30	58	0	10	33	9	292	293	0	521	**1246**

	Male	217	306	0	82	494	166	623	2497	0	2828	**7213**
	2017–2018 Total	**247**	**364**	**0**	**92**	**527**	**175**	**915**	**2790**	**0**	**3349**	**8459**
2018–2019	Female	25	47	0	31	58	42	447	650	0	890	**2190**
	Male	411	330	0	97	619	272	1120	3088	0	4075	**10,012**
	2018–2019 Total	**436**	**377**	**0**	**128**	**677**	**314**	**1567**	**3738**	**0**	**4965**	**12,202**
2019–2020	Female	23	18	0	27	36	28	412	578	0	881	**2003**
	Male	204	128	0	89	507	200	1114	1639	0	3940	**7821**
	2019–2020 Total	**227**	**146**	**0**	**116**	**543**	**228**	**1526**	**2217**	**0**	**4821**	**9824**

Source: Department of Home Affairs, Australia

> Our workers come from a remote rural village in the Solomon Islands … They have to grow their own food or they don't eat … Job opportunities are non-existent. This shapes their positive work ethic. They arrive in Australia physically and mentally fit. They are here to work. They want to work. They are enthusiastic … They want to learn and they want to be appreciated for the job that they perform. They are energetic, easily trained, honest, happy and reliable, and they are grateful. They really are every employer's dream. (McCarthy, 2017)

Few employers can have been so fortuitous.

Selective recruitment had obvious effects on domestic labour forces. In 2018–2019 as many as a third (33 per cent) and a quarter (24 per cent) of rural Tongan and ni-Vanuatu men aged 20–49, respectively, were participating in either the SWP or RSE scheme and were away for at least six months of the year (Bedford et al., 2020). Other PICs had lower participation rates; in Fiji and Kiribati, the proportion was much smaller at roughly 1 per cent and 5 per cent, respectively, and in Samoa it was just under 10 per cent. Throughout the PICs, however, this was the same population that was generally relied upon to perform the hardest labour such as clearing and planting food gardens, so their absence was potentially significant for village livelihoods and social lives (see Chap. 9).

Overall, few women participated in the SWP (Table 6.5), but their participation rates varied. Women accounted for a notably higher proportion of workers from Kiribati, PNG, Solomon Islands, Timor-Leste and Tonga. However, other than in Tonga, overall participation rates from many of these PICs were low so that these statistics probably reflected choices made by a small number of employers for particular roles. In one example from PNG, a group of five women was recruited from a single village because a trusted intermediary was able to spend time in their village explaining the scheme and gaining the confidence of their families and the community (Curtain, 2018). Women's participation in the SWP sat at around 13 per cent until 2018–2019, when it rose due to an increase in the number of Tongan women participating. Women's participation in the RSE was slightly lower at around 11 per cent in 2016–2017. This was partly due to strict requirements by the New Zealand government that RSE workers not be recruited to undertake tasks New Zealanders were willing to do. That made it harder to recruit RSE women for packhouses, the 'preferred' work for Pacific women (Bedford et al., 2020). Employer hiring preferences further entrenched male dominance; employers preferred not to hire women for physically demanding roles, but valued their input for such tasks as mushroom or berry picking and for overnight shift-work in packhouses. Such perceptions were widely shared. In 2019 John was told by the Australian High Commissioner in Kiribati, with reference to mushroom pickers, that 'women are better at fragile tasks' and were successful in the tourism industry on Hayman Island since i-Kiribati were 'very smiley people'. Such archaic views represented Pacific versions of the Asian 'nimble fingers' syndrome where women were assigned to particular tasks because of their assumed 'natural' attributes.

In Melanesia, the high proportion of male workers reflected gender norms and historical precedents. In Vanuatu and elsewhere, men dominated labour mobility flows during the blackbirding era and for contemporary employment in New Caledonian mines in the 1960s and 1970s (Chap. 3). Internal migration streams to towns were also initially male dominated, reflecting traditional norms around who 'should' travel (Bedford, 1973b; Bonnemaison, 1985, Haberkorn, 1989). Twentieth-century plantation and mining work in PNG and Solomon Islands was similarly male dominated. Polynesian migration was more evenly balanced, but men from Samoa and Tonga continued to outnumber women. Focusing on Tonga and Vanuatu, the World Bank (2018a) suggested that women's relative lack of participation in the SWP was influenced by three factors: selection processes at the local level, gender biases of Australian employers and local agents, and gender norms around women's role in the home. Much of this related back to women's traditional role as caregivers within families and communities, despite women being the main domestic agricultural workers in most parts of Melanesia, a double burden that freed men to travel (Chap. 5). In Vanuatu, gender played a role in where workers were recruited from. Women participants were much more likely to have come from urban areas than men; almost 40 per cent of ni-Vanuatu women compared with a quarter of men were 'urban' based (VNSO, 2017). Women in Vanuatu were expected to be more culturally conservative than men, and so those with existing experience of wage labour in towns were both the most likely to apply and the most likely to be able to escape cultural constraints.

Despite an official Australian government preference for a greater gender balance, women's participation is unlikely to increase without attitudes around women's role in home communities shifting, and Australian employers seeing more jobs as appropriate to women. Men in most PICs were reluctant to take over women's domestic responsibilities when their wives and partners were away in the SWP; hence, single women and those who could find family members to replace their lost domestic labour were most likely to be involved. In 2016–2017 less than half of female SWP workers were married compared with just over two-thirds of men (Chattier, 2019; World Bank, 2018). Many of the female SWP workers we met over the years fitted this description.

> Mary was thirty-three years old when she participated in seasonal work for the first time. Originally from Malakula, she had been living in Vila for a long time. Mary had four children, but had separated from their father some time ago. Normally, the children lived with Mary, and visited their father occasionally. However, she had organised for them to stay with their father while she was working in Australia. Mary was participating in the SWP so that she could earn enough to rent a stall in Port Vila's Centrepoint market house where she could sell baskets and other handicrafts. She also wanted to use her stall to sew and sell traditional island dresses, something she had already been doing on a smaller scale from her house.

Husbands, wives and partners of SWP workers were often required to take on additional domestic or caring tasks, just as Mary's ex-husband had done. Whereas men usually disdained such new tasks when women were absent, when men departed women acquired more roles and tasks. As one ni-Vanuatu woman said, 'when my husband is away I have additional farm activities as well as looking after the kids and the garden'. Another ni-Vanuatu woman commented, 'My husband went last year and it left me very busy but now for the first time I'm going on RSE. I'm excited about it; more women are doing this now' (quoted in Addinsall et al., 2020, p. 112). For women, as much as for men, migration could be a novel and exciting prospect, as well as a means of generating income.

While many women wanted the opportunity to participate in the SWP, not all women chose to do so when the chance arose. In the early days of the RSE, women from Ambrym were unsure of travelling to 'unknown' places, so many were unwilling to participate in the scheme (Bailey, 2014). Some women, more than men, were simply uninterested in participating (World Bank, 2018). While the Australian government sought gender equality, disability and social inclusion (GEDSI) in recruitment practices, without a better understanding of the impacts and needs of female workers, it remained difficult to provide more opportunities for women's participation in the SWP, as long as employers determined what workers were required.

The High Cost of Labour Migration

Applying for the SWP was an expensive process, and the costs alone could provide a barrier to participation. Potential workers had to cover costs of passports, internal and international travel, visas, and medical and police checks. Not all expenses were paid upfront, and employers could cover initial airfares. Workers had then to repay all but the employer's contribution (A$500). Yet, as most services (passport offices, etc.) were in PIC capital cities, prospective workers had to meet costs relating to domestic travel, food and accommodation themselves. Costs varied, but could amount to several hundred (and sometimes even thousands) Australian dollars (C. Bedford, 2013). Thus, prospective SWP workers from the PNG highlands needed to cover three weeks of accommodation costs and living expenses in Port Moresby (K300, A$120) while they dealt with bureaucracy and awaited deployment (Curtain, 2018), although many probably lived with *wantoks* (kin) during this period. The World Bank (2017) suggested that in 2015/2016 the average total cost of SWP participation across the PICs was A$1474, a considerable sum for rural villagers without regular income-earning opportunities (Table 6.6). Some workers applied for micro-finance loans to meet application costs, but applicants without regular or reliable income—a category many rural residents fell into—were not always granted loans (Bailey, 2014). It was

Table 6.6 Pre-departure costs for SWP participation (A$)

	Samoa	Timor-Leste	Tonga	Vanuatu	Other PICs	Pacific Average
Passport	42	59	38	94	56	67
Internal travel	54	51	119	320	23	152
Police clearance	26	1	86	31	38	27
Medical	33	50	3	84	22	54
Visa	169	240	119	139	188	167
International airfare	754	1284	951	769	1228	892
Work clothing	108	87	122	60	47	80
Other costs	13	50	–	45	–	35
Total pre-departure	1199	1822	1438	1542	1602	1474

Source: World Bank (2017, p. 35)

therefore rather more difficult for relatively poor people in remote areas to participate.

Some prospective workers from rural areas lodged with family in town while they awaited departure, but this merely shifted the economic burden to their urban relatives. In 2011 Kirstie met a small group of Banks Islanders who were staying with a relative's family at an informal settlement in Port Vila while they waited for their flight to New Zealand.

> Julie and Ben both held steady jobs in hospitality at one of Vila's many tourist restaurants. Together, they lived in an informal settlement with their two children, Julie's parents, her sister, Lucy, and Lucy's teenage son. Lucy was employed in housekeeping at a local hotel, but even with three steady incomes, money was tight. When Ben's three male relatives travelled to Vila to await their flight to New Zealand, they were welcomed into the household. However, after several weeks, the financial pressure became too much and Julie asked her Banks Islander guests to contribute to household expenses. Financial pressure or not, this was considered an inappropriate request, and Julie was quickly criticised by her neighbours for being 'stingy'.

Here and elsewhere it was culturally expected that urban households would cover expenses for workers as they transited through town, yet such extended household costs were often an unrecognised and unexpected cost.

The costs, borne both by workers and their families, partly explain why urban residents, who did not need to travel to apply for passports, could attend (and knew about) recruitment drives and easily access other services, were at a distinct advantage. Speaking about the RSE scheme, Billy, a well-connected local businessman resident on the tiny island of Paama, explained to Kirstie in 2011:

> There aren't many people from this island [who have gone to New Zealand]. For those of us on the island, sometimes it is hard to find the money … to pay for passports and other things. Those who live in town, they find it easier because they work and earn money, so they can pay their way to go. (Billy, quoted in Petrou, 2020, p. 95)

This alone explained a bias towards urban participation in Vanuatu, and probably in other PICs.

In recognition of the costs associated with long stays in Port Vila to obtain medical checks and other documents, the Government of Vanuatu trialled a mobile medical service in mid-2017. It was intended that a doctor would travel to various locations throughout the archipelago, spending a week at a time performing medical checks on prospective workers. Problems soon arose however, as workers, agents and employers failed to make appointments and, where appointments were made, workers often did not attend. The programme was quickly redesigned so that a permanent facility offering medical checks was established on Santo by April 2018. Medical checks were more costly on Santo (12,000 vatu or A$155) than in Port Vila (8500 vatu or A$110), but cheaper travel and accommodation costs in Santo meant that this decentralised facility became a popular option; over a 12-month period more than 800 workers completed medical checks there (Bailey & Sorensen, 2019). Nonetheless, even travelling to Santo from the northern islands of Vanuatu remained extremely expensive. For very remote islands in every PIC, everywhere but Nauru and Samoa, costs were more prohibitive.

The costs associated with application meant that those who held formal employment were best placed to finance participation. Indeed, perhaps surprisingly, seasonal workers were much more likely to hold formal employment than the national average in their home PIC. Some 59 per cent of SWP workers surveyed by the World Bank (2017) were in formal employment prior to participating in the SWP, so departing from the notion that the jobs should be taken by relatively poor people in rural areas. Employment status prior to participation varied between countries. The majority of ni-Vanuatu and Tongan workers were unemployed before participating in the SWP. Ni-Vanuatu workers who were employed were more likely to be informally employed or performing unpaid work for their families than those from other PICs.

Across the PICs, the most common type of formal employment previously held by SWP workers was in the service sector, and many had left skilled or semi-skilled employment in accounting, hospitality and engineering. Other workers 'shared' jobs with kin; one family member participated in guestwork while another worked at home, and when the first returned from Australia or New Zealand, they swapped places. Such arrangements ensured that opportunities for domestic and international employment were shared between kin (World Bank, 2017). In Vanuatu at least, that represented something of a revival of past 'job rotation' practices amongst ni-Vanuatu men during the early years of rural-urban labour migration. Taking it in turns within villages or families to travel to town and work meant that no one man benefitted more than others from urban employment, so maintaining a rough level of village socio-economic equality (Haberkorn, 1989).

Among the many ni-Vanuatu SWP workers we met, it was more common to resign from formal employment to take up seasonal employment, although many had the option to be re-hired upon their return home, while a few were able to take leave.

Freddy was participating in the SWP for the first time in 2020. His family was originally from Paama, but Freddy lived on Santo where he was employed as a petrol station attendant. When asked about his job in Vanuatu, Freddy explained: 'For me, before I came, I made a resignation letter for the boss of my company to tell him that I was coming to Australia. So he understood why I came. He understood my reasons, so he told me that it was OK for me to go, and if I came back and wanted to work there again I could. I could reapply again'.

Lydia lived in Port Vila where she was employed as a supervisor at Bon Marché, one of the largest supermarkets. She explained: 'When I left, I

> wrote a resignation letter, so Bon Marché, when you write a resignation letter you can apply to work there again. But if you don't write a letter and you just walk out, it's hard to work there again. I don't know, when I get back to Vila I'll just decide if I want to go back to work or go to my [home] island. Bon Marché is a big supermarket, but the salary isn't good'.
>
> Marinet was participating in the SWP for the first time. She lived in Port Vila where she had been working at a takeaway food shop to earn money so that she could apply for her passport and travel to Australia. Marinet's boss was happy to have her back once she returned: 'For me, my boss will welcome me back any time. He told me that if I wanted to work there again when I get back then I can. It's up to me, I get to choose. If I'm tired, I'll have a break.'
>
> Vanessa worked as a radio announcer in Port Vila, and was participating in the SWP for the first time. She had taken leave from her job so that she could work in Australia: 'I took leave without pay for seven months [...] I wrote a letter to my boss and he accepted my request for leave without pay for seven months. When I go back [to Vila], I can work again'.

No clear indication existed that migrant workers were leaving 'key' jobs—such as in teaching or health care—that would result in a temporary skill drain (World Bank, 2017; see Chap. 9). It is even more difficult to assess whether youth were being dissuaded from pursuing further study or skilled domestic occupations in favour of the 'fast money' possible overseas. However, intergenerational transfer of guestwork opportunities (Bedford et al., 2020) indicates that for some at least, there was an expectation that they pursue a career in temporary labour mobility.

Prospective workers who travelled to Australia (or New Zealand) were able to recoup domestic travel and other costs once they started work. Indeed, this was often their first objective (Blanco, 2009). Others, rejected from the scheme at the last hurdle, could potentially face long-lasting financial debts, although how often this occurred was unknown. Within the PICs rejection was relatively straightforward and resulted from such things as missing documents, blacklisting, poor behaviour in past years or criminal convictions (Bailey, 2014). However, how many workers were rejected at what stage and at what cost is impossible to discern.

Pre-departure and Arrival Training: On to the Job

Before leaving their home countries, SWP workers received pre-departure briefings to prepare them for their time in Australia. These were provided by PIC governments and recruitment agents, but the accessibility and quality of training varied markedly. On average, most workers (97 per cent) from Samoa, Tonga, Vanuatu and Timor-Leste had access to pre-departure briefings, but workers from some other PICs—particularly Solomon Islands—had more limited access. The sessions covered information about wages and pay deductions; other topics such as gender roles and potential health issues while in Australia were less common (World Bank, 2017). The pre-departure trainer's manual designed for Nauru provided a largely generic programme (Table 6.7) and recommended that training not exceed six hours per day so that workers were not overwhelmed. Trainers were advised that each session should have three main components:

> 1. Welcome, introduction and objectives—Always briefly explain the objectives for each session to participants and outline how long the session will go for.
> 2. Actual training—this should be based on exercises and short 'lectures' which summarize, add to, and reinforce what was discussed in the exercises. There will be some content that will be taught more in a lecture style (e.g. payslip deductions), but there will be other content that can be led by the participants, with the trainer providing more of a guiding role (e.g. what makes a good worker).
> 3. Conclusion—Always finish a session with the key messages: What participants must know. (ILO, 2015, p. 14)

Trainers were encouraged to use various techniques such as role plays, short quizzes and other activities to engage workers and test their knowledge. Examples of activities included the following:

> Prepare a 'fake payslip' which has unnecessary deductions (Training Aide 7). Get participants to work in groups to identify what is wrong with each

Table 6.7 Sample pre-departure training plan for Nauru

Day 1	Topic	Session time
Session 1	Introduction: What is it all about?	9:30 am–10:55 am (5 min break)
Session 2	What to expect as a seasonal worker	11 am–12:30 pm (30 min break)
Session 3	Skills for being a good worker	1 pm–2:25 am (5 min break)
Session 4	Goal setting and participating in training	2:30 pm–3:30 pm
Day 2		
Session 5	Letter of offer	9:30 am–10:55 am (5 min break)
Session 6	Payslip	11 am–12:30 pm (30 min break)
Session 7	Budgeting	1 pm–2:25 am (5 min break)
Session 8	Remittances	2:30 pm–3:30 pm
Day 3		
Session 9	Australian life/culture	9:30 am–10:55 am (5 min break)
Session 10	Australian laws: Rights	11 am–12:30 pm (30 min break)
Session 11	Australian laws: Responsibilities	1 pm–2:25 am (5 min break)
Session 12	Keeping in touch; health hygiene and recreation	2:30 pm–3:30 pm
Day 4		
Session 13	Coming to Australia (logistics)	9:30 am–10:55 am (5 min break)
	Summary	11 am–12:30 pm (30 min break)
	Opportunity for individual questions/ appointment	1 pm–3:30 pm

Source: ILO (2015, p. 11)

payslip. You can make this easy or challenging depending on the understanding of the workers. Provide a calculator to each group. Make sure you go through the answers carefully. (op.cit., p. 32)

Needs and Wants. Distribute a list or pictures of different goods and services such as: a bottle of wine, picture of a classroom, fancy car, etc. Get participants to work in groups to decide on which goods and services are

> needs and which are wants. Ask them to calculate how much money on 'little wants' might cost over time—for example a few beers a night at the pub will cost around $4–6 each. How much will that amount to over 6 months? (op. cit., p. 33)

Other activities covered topics including how to appropriately address situations where employers had breached their obligations, how to stay healthy and how to navigate cultural differences, particularly around gender norms.

While this 'model' programme potentially took four days, the Australian government recommended that pre-departure training take two full days. However, in some cases the content—or a distilled version of it—has been delivered in as little as half a day (World Bank, 2017). In 2017, about half (55 per cent) of the ni-Vanuatu workers surveyed by LMAP (2017a) had received their pre-departure training over two days, about a quarter received training over a shorter period, while the remainder attended training that lasted for three days (7 per cent) or even longer (14 per cent). While most workers (80 per cent) attended training run by Vanuatu's Employment Services Unit (ESU), other training was delivered by agents. The varied quality and accessibility of pre-departure briefings across the PICs largely depended on PIC budgetary issues (World Bank, 2017). Alongside briefings, workers had access to a pre-departure handbook and short videos produced by the Australian government. The handbook was available in workers' home languages or the lingua franca of Melanesian countries, providing a reference guide for issues around payslips, contracts, unions, tax and Australian culture.

The varied content, cursory delivery and lack of access to briefings meant some workers were ill-prepared for their time in Australia, and faced additional cultural, climatic and linguistic challenges, which briefings could not easily prepare them for. Writing about the experiences of Samoan SWP workers, one community leader observed:

> Everyone talks about how working and living in Australia is nothing like they were told. It is not really a training programme but a lot of talking and it does not help with the practical issues that the workers encounter. They say that the key message is from the Government that they must be disci-

> plined. They understand that the job is to work hard and follow the instructions but there are times when they deal with practical problems and lack the language and support to talk about them. Sometimes they need help to speak to their employer and other times they might need non-work support. (Quoted in Union Aid Abroad APHEDA, 2020, p. 11)

Not all workers were unprepared. Ni-Vanuatu workers surveyed by LMAP (2017a) in 2017 felt that pre-departure training did an adequate job of covering topics such as how to stay physically fit in Australia, how to remain healthy and how to understand their rate of pay. However, other aspects such as Australian laws, pay deductions and the rights and responsibilities of employers were not so well understood. Women generally felt less well prepared than men, particularly around topics such as culture and gender norms in Australia, Australian laws, and worker and employer rights and responsibilities. Just under half (44 per cent) of all workers felt at least quite well prepared for their time in Australia. Similarly, the ni-Vanuatu SWP workers we met in rural South Australia in 2020 believed their pre-departure and arrival training was adequate. The presence of many returned workers was also beneficial as they were able to help 'new' workers navigate the practicalities of life and work in a new environment.

Simon: We did some training, but it was just something we did at [our workplace]. But for understanding life here, like to tell us what life is like in Australia, there wasn't anything. Our agent just did a briefing and then we came. But there wasn't anyone else who spoke to us, it was just our agent who went through everything with us.

Kirstie: Were you well prepared?

Simon: I think everything was OK.

Jenny: It was OK because there were some of us who had travelled already, they already knew some things, so we helped each other with things like living in a house. Because back home, some of us live on the island and it's a different style of living. And the places where to go, we were all new here, but we managed to travel around and learn the places [...] At work, every-

> thing is good. I think in terms of things here, getting around, everything is good. Anything we didn't know, we learned when we got here.

Returning workers proved invaluable when it came to filling in inevitable gaps in formal training and that lesson was eventually learned. In late 2020, the Vanuatu government, along with World Vision, began recruiting experienced workers to help deliver pre-departure training. This provided a valuable perspective for first-time workers in helping them to navigate some of the common problems—such as strange accommodation and difficulties around pay rates and deductions—that we discuss in the next chapter.

In Australia, workers were provided with arrival training; 79 per cent of ni-Vanuatu workers received this training from their employer in Australia and most (92 per cent of women and 80 per cent of men) felt this had prepared them quite well or very well (LMAP, 2017b) But even so, arrival training varied in quality and, by design or accident, briefings often omitted presentations by unions or the Fair Work Ombudsman (Howe et al., 2019). These omissions posed problems when workers needed to complain or understand their rights. In 2012, Kirstie met a group of ni-Vanuatu workers in rural Victoria who recounted how they had experienced information overload during their arrival briefing. The group subsequently befriended a local family who took them through various activities ranging from visiting and using the local bank to tutorials on how to clean the cabins they were staying in. For some workers, the learning curve about 'life in Australia' was steep, and it was easy to feel overwhelmed. Informal connections to a local community, where workers were fortunate enough to have them, were important in bridging gaps in formal briefings.

Winners and Losers? Uneven Beginnings

While most eligible PICs sent workers to Australia, some were far more 'successful' than others, even though both schemes responded to a widespread poverty of opportunity. Vanuatu and Tonga benefitted from first mover status, but employer preferences, geographical accessibility and PIC recruitment strategies were important in determining which countries, and which parts of countries, sent the most workers. A bias towards large islands and urban centres appeared almost ubiquitous, whereas remote and more impoverished areas tended to be disadvantaged, despite initial intentions. Numerical success was influenced by both home and host country factors, but participation rates evolved and changed over time, as happened when Tongan recruitment decreased in response to concerns over 'misbehaving' workers. Demand invariably increased. Other than New Zealand's inclusion of Asian workers, in almost every way the patterns of recruiting and selection were virtually the same in the two countries.

Enthusiasm at both national and local levels was important, ranging from stolid and less visible government organisation to Australian employers travelling to PICs to select workers and then monitor their development. Having locally resident kin, as many Tongans did in the Riverina, could encourage and enable a successful transition to seasonal work. As with all employment opportunities, some places and people were simply not interested in participating or were quite unfamiliar with the schemes (as in most parts of PNG). Some women fell into this category, while others were discouraged from participating, even as Australia officially sought greater gender equity. By contrast, many workers returned over and over again, ever seeking to embellish their household's development.

Within the PICs themselves several factors influenced opportunities for participation. Government policies (targeting rural areas or not), connections to agents, gender and geography all played a role in who was chosen to participate. The most likely participants were young (often urban) men raising some issues around the uneven impact of the scheme. Compared to the early days of the schemes (Chap. 5), there was a noticeable shift from the focus on equitable recruitment as participation rates increased and opportunities became (theoretically) more accessible. To achieve greater equity—even assuming that national commitment existed—would have required a more hands-on government role, but scarce human resources made that difficult. Nonetheless, guestworker (and domestic) jobs were often shared between kin, and many families and communities benefitted from the SWP. The following chapter turns to the workers' varied experiences of work and life in Australia.

Bibliography

Abbott, D., & Pollard, S. (2004). *Hardship and poverty in the Pacific: Strengthening poverty analysis and strategies in the Pacific*. Asian Development Bank.

Addinsall, C., Rihai, N., & Nasse, A. (2020). Agroecology and sustainable livelihoods as a framework to empower rural Ni-Vanuatu women. In L. Singh-Peterson & M. Carnegie (Eds.), *Integrating gender in agricultural development. Learnings from South Pacific contexts* (pp. 105–120). Emerald.

Bailey, R. (2014). *Working the Vines: Seasonal migration, money, and development in New Zealand and Ambrym, Vanuatu*. PhD thesis, University of Otago, Dunedin.

Bailey, R. (2018, April 9). RSE workers ten years on. ANU DevPolicyBlog.

Bailey, R. (2019a, April 26). *Pacific workers in rural Victoria*. ANU DevPolicyBlog.

Bailey, R. (2019b). *New Zealand's Recognised Employer Scheme (RSE): 10 year longitudinal case study*. ANU Department of Pacific Affairs.

Bailey, R., & Rereman, J. (2019, January 8). Vanuatu's seasonal workers: Where are they from? ANU DevPolicyBlog.

Bailey, R., & Sorensen, M. (2019, January 31). Labour mobility and medical mobility: Lessons from Vanuatu. ANU DevPolicyBlog.

Bedford, C. (2013). *Picking winners?: New Zealand's recognised seasonal employer (RSE) policy and its impacts on employers, Pacific workers and their island-based communities*. PhD thesis, The University of Adelaide.

Bedford, R. (1973b). *New Hebridean mobility: A study of circular migration*. ANU Department of Human Geography Publication HG/9.

Bedford, R., Bedford, C., & Ho, E. (2010). Engaging with New Zealand's recognised seasonal employer work policy: The case of Tuvalu. *Asian and Pacific Migration Journal, 19*, 421–445.

Bedford, C., Bedford, R. & Nunns, H. (2020). *RSE Impact Study: Pacific stream report*. Bedford Consulting.

Blanco, C. (2009). *Temporary migration, remittances and development: Early impressions of the Australian Pacific seasonal worker pilot scheme*. Honours thesis, University of Sydney, School of Geosciences.

Bonnemaison, J. (1985). Territorial control and mobility within ni-Vanuatu societies. In M. Chapman & R. Prothero (Eds.), *Circulation in population movement: Substance and concepts from the Melanesian case* (pp. 57–79). Routledge and Kegan Paul.

Calnitsky, N. (2016). Grape vines and orchard lines. *Left History: An Interdisciplinary Journal of Historical Inquiry and Debate, 20*(1), 11–35.

Chattier, P. (2019). Beyond development impact: Gender and care in the Pacific Seasonal Worker Programme. *Gender and Development, 27*(1), 49–65.

Connell, J., & Hammond, J. (2009). *Iso Iso*: The first ni-Vanuatu guest workers in New Zealand. *Pacific Economic Bulletin, 24*(1), 83–95.

Corlett, E. (2019, December 12). RSE Scheme transformed the New Zealand fruit growing industry. *Radio New Zealand*. https://www.rnz.co.nz/national/programmes/eyewitness/audio/2018726025/.

Cummings, M. (2016). Uncertain belongings: Relationships, money and returned migrant workers in Port Vila, Vanuatu. *Journal of Pacific Studies, 35*(3), 21–33.

Curtain, R. (2018). What can Papua New Guinea do to lift its numbers in the seasonal worker programs of Australia and New Zealand? ANU Development Policy Centre Discussion Paper 70.

Curtain, R., & Howes, S. (2019, June 28). *Tiny Tonga's SWP success*. ANU DevPolicyBlog.

Curtain, R., & Howes, S. (2020). *Governance of the seasonal worker programme in Australia and sending countries*. ANU Development Policy Centre.

Curtain, R., & Sherrell, H. (2017, February 2). *How recruitment and selection can shape seasonal work programs: Comparing Fiji and Papua*. ANU: DevPolicy Blog.

Department of Agriculture & Rural Development and Department of Labour & Employment Services. (2020). *Framework for the reintegration of agricultural workers in Labour Mobility Programs 2021–2023*. Vanuatu Government.

Dornan, M. (2018, March 16). *Vanuatu grapples with seasonal worker success*. ANU DevPolicyBlog.

Dun, O., & Klocker, N. (2017). The migration of horticultural knowledge: Pacific Island seasonal workers in rural Australia—A missed opportunity? *Australian Geographer, 48*(1), 27–36.

Falefou, T. (2017). *TOKU TIA: Tuvalu and the impacts of climate change*. PhD thesis, University of Waikato, Hamilton.

Fatupaito, A., Utuva, L., Tauave, S., Alofipo, A., Meleisia, M., Schoeffel, P., Arthur, T., & Alexeyeff, K. (2021). Samoa's New Labour Trade. *Journal of Samoan Studies, 11*(1), 51–63.

Gibson, J., & McKenzie, D. (2014). The development impact of a best practice seasonal worker policy. *Review of Economics and Statistics, 96*(2), 229–243.

Government of Vanuatu. (2019). *National Labour Mobility Policy*. Government of Vanuatu.

Haberkorn, G. (1989). *Port Vila: Transit station or final stop?* Australian National University.

Hammond, J., & Connell, J. (2009). The new blackbirds? Vanuatu guestworkers in New Zealand. *New Zealand Geographer, 65*(3), 201–210.

Howe, J., Clibborn, S., Reilly, A., van den Broek, D., & Wright, C. (2019). *Towards a durable future: Tackling labour challenges in the Australian Horticulture Industry*. University of Adelaide.

Howes, S. (2018, August 6 and 8). ANZ Seasonal labour programs are not too reliant on return workers. ANU DevPolicyBlog.

International Labour Organisation. (2015). *Pre-departure training for Nauru seasonal workers: Trainer's manual*. ILO.

LMAP. (2017a). *LMAP Vanuatu reintegration and tracer study report for management*. LMAP.

LMAP. (2017b). *Approved employer survey and focus group discussion*. CARDNO.

McCarthy, K. (2017, December 1). *The seasonal worker program: A personal story*. ANU DevPolicyBlog.

McKenzie, D., Garcia Martinez, P., & Winters, A. (2008). *Who is coming to New Zealand under the new Recognised Seasonal Employer program?* World Bank.

Petrou, K. (2020). *If everyone returned, the Island Would Sink: Urbanisation and migration in Vanuatu*. Berghahn.

Petrou, K., & Connell, J. (2018). 'We don't feel free at all': Temporary ni-Vanuatu workers in the Riverina, Australia. *Rural Society, 27*(1), 66–79.

Radio New Zealand. (2019, April 23). Vanuatu rogue agents threatened with jail. *RadioNZ*. https://www.radionz.co.nz/international/pacific-news/387565/vanuatu-rogue-agents-threatened-with-jail

Republic of Vanuatu. (2007). Seasonal Employment Act. Government of Vanuatu.

Rockell, D. G. (2015). *Pacific Island labour programmes in New Zealand: An aid to Pacific Island development?: A critical lens on the recognised seasonal employer policy*. PhD thesis, Massey University, Palmerston North.

Rose, M. (2019, October 11). Seasonal workers and rising oceans: McCormack needs to understand reality. ANU DevPolicyBlog.

Smith, R. (2016). *The goal of the good house: Seasonal work and seeking a good life in Lamen and Lamen Bay, Epi, Vanuatu*. Unpublished PhD Thesis, University of Manchester, Manchester.

Stead, V. (2021). Precarity's reach: Intersections of history, life, and labour in the Australian horticultural industry. *Journal of the Royal Anthropological Institute, 27*(2), 303–320.

Union Aid Abroad APHEDA. (2020). Inquiry into strengthening Australia's relationships with countries in the Pacific region: Submission 20. Submission to Joint Standing Committee on Foreign Affairs, Defence and Trade—Foreign Affairs and Aid Sub-Committee: Canberra.

Vanuatu National Statistics Office. (2017). *2016 Post-TC Pam Mini-Census Report: Basic Table Volume 1*. Vanuatu Government.
Wade, J. (2020, September 1). *Licensed agents: The key to Vanuatu's SWP success.* DevPolicy Blog.
Williams, A. (2010). *The experiences of the recognised seasonal employer scheme and Ni-Vanuatu temporary workers in Te Puke, New Zealand.* Master of Planning thesis, University of Otago.
World Bank. (2018a). *The social impacts of seasonal migration: Lessons from Australia's Seasonal Worker Programme for Pacific Islanders*. World Bank.
World Bank. (2017). *Maximizing the development impacts from temporary migration*. World Bank.
World Bank. (2021). *Pacific labor mobility, migration and remittances in Times of COVID-19*. World Bank.
Wu, A. (2019). *Timorese participants in the Australian seasonal workers programme: Migration, remittances and development*. PhD thesis, University of Sydney.

7

Destination Australia

The Seasonal Worker Program (SWP) was designed to meet Australian agricultural labour needs, and Pacific guestworkers eventually proved invaluable to the industry. In this chapter we consider the context of SWP workers' employment in Australia, the manner in which labour needs have been conceptualised, and how the COVID-19 pandemic has influenced understandings of Australia's agricultural labour requirements, which, as we stress, are very far from uniform. Subsequently, we explore why farmers chose to participate in the SWP, and the extent to which the scheme met and satisfied labour shortages. Such shortages were usually, but not always, located in more remote areas where local workers were less likely to be available and Australian citizens unwilling to travel for employment. Consequently, SWP workers were often employed in out-of-the-way places, and we therefore examine some of the problems that could arise around accommodation, wages, access to transport and workers' health. Some of these issues were longstanding, and symptomatic of wider problems within the Australian horticultural industry. This chapter thus recognises that SWP workers could sometimes work and live in poor

K. Petrou, J. Connell, *Pacific Islands Guestworkers in Australia*,
https://doi.org/10.1007/978-981-19-5337-3_7

conditions, facing problems even as they were valued by their employers and performed an essential role in the Australian food supply chain. At the same time, we emphasise the impossibility of quantifying the extent of the issues discussed here; many guestworkers experienced no serious problems during their stay in Australia and most wanted to return again.

Labour Shortages?

The SWP was premised on the notion that guestworkers could help to resolve ongoing agricultural labour shortages. Time and again, the global expansion of guestworker schemes in this century demonstrated that migrant labour was regularly and increasingly in demand to fill labour deficits in more developed countries and regions. Few OECD countries are without such schemes. In Australia shortages have varied according to problems of seasonality (where some crops have very short harvest times), or the lack of Working Holiday Makers (WHMs), grey nomads (retirees) and migrant seasonal workers.

Prior to the establishment of the SWP (Chap. 4), and virtually every time issues of labour shortages have arisen thereafter, it was widely argued that unemployed Australians should move into seasonal horticultural jobs. Such arguments ignored the very practical reasons that this did not occur: many of these jobs were located in remote areas (distant from the urban unemployed), horticultural wages were low and conditions were tough. In 2020, Peter Davidson (2020) pointed out that a third of Australians on unemployment benefits were aged 45 or over, a quarter had disabilities and 10 per cent were caring for children. Fruit picking was simply not a good 'fit' for many of these people, and nor were they welcome in the industry. Responding to the suggestion that unemployed Australians might take up horticultural jobs, a key informant who had spent years working in horticulture in Victoria told Kirstie in 2020:

> Never has it been the case that unemployed people come [and work in horticultural jobs]. And industry don't want them. Industry do not want long-term unemployed working. I myself used to hire and fire in the private

> sector and I remember going to my manager and saying 'I'm not going to hire them. We're not a babysitting service for people who don't want to work'. And so we went to backpackers. And that continues to be the case. Industry just don't want people who don't want to work.

At the same time, most Australians did not want to work in an industry known for poor working conditions, inadequate remuneration and long hours, and where social life was limited. At the other end of the age spectrum, although the industry advertises for grey nomads, few were engaged in agricultural work, again because of the poor working conditions and because most did not need supplementary incomes. One union official observed of the unemployed:

> If you're a young Australian person in Bundaberg who's unemployed and you see all these horror stories in the area about how backpackers and seasonal workers are treated and then you go and say to an employer 'I'm willing to work but I'm not going to put up with shit. I want to be paid the award', well that's what I suspect they mean when they say they're [Australians] not suitable and they're lazy. They're not lazy. They just want to be paid right and they want to get decent conditions in their workplace. (Quoted in Howe et al., 2019, p. 65)

In turn, farmers often perceived long-term Australian residents as unreliable and lacking in motivation: some degree of mutual antipathy. This lack of motivation was attributed to welfare dependency, intergenerational unemployment and substance abuse. By contrast, guestworkers and 'new' Australians who had recently migrated were perceived by industry as being more likely to work hard due to the need to fulfil visa requirements, and their lack of local commitments. One harvest labour services official commented

> There's three things [capabilities and characteristics] that workers need. One is reliability, the second one is reliability and the third one is reliability … The reason why [growers] don't like Australian workers, because they are overwhelmingly unreliable. (Quoted in Howe et al., 2019, p. 62)

Backpackers (WHMs), who were granted a second-year visa extension if they undertook 88 days farm work, were often employed in the horticultural industry and faced similarly difficult working conditions. The necessity to 'do their time' meant that WHMs were vulnerable to exploitation, while many growers found them less productive than guestworkers, due to inexperience, and worker turnover could be high (Orton, 2020). Regular media stories about the poor conditions faced by WHMs only served to make the industry even less attractive to Australian workers. Underpaid wages, overcrowded inadequate accommodation and sexual harassment were all issues experienced by migrant workers. In the financial year 2019–2020 the Fair Work Ombudsman (FWO) received 1647 formal complaints related to migrant workers. About a third of these complaints involved backpackers on WHM visas, despite their temporary status often discouraging them from complaining. Most complaints (about 86 per cent) were dealt with via education and dispute resolution. In theory, according to the FWO, avoiding formal disputes 'helps maintain productive and cooperative employment relationships and enables the faster recovery of unpaid wages' (quoted in Uibu, 2020). In practice this meant that most employers received little more than verbal admonishment, although the same FWO spokesperson reported that conditions were 'improving' on farms that had experienced problems in the past. Bad employment experiences were not uniform across the industry, however publicised problems did nothing to make seasonal labour attractive to local workers.

Australians may have been perceived as unreliable, but Iain Campbell et al. (2019) argue that what employers really wanted was workers who were vulnerable and compliant, and who would not complain about unpleasant working conditions—so migrant workers best fitted the bill. Debate has rarely relented from arguments around whether there were 'real' shortages or whether appropriate unemployed people could be induced to work with the right incentives. Either way, SWP workers filled an apparent gap in the seasonal workforce, but one that was not evenly distributed around the country.

Formal surveys have indicated that growers are only occasionally unable to recruit adequate labour in normal circumstances (Campbell, 2019; Dufty et al., 2019; Hay & Howes, 2012). In fact, Joanna Howe and others (2019) argue that technically there was no shortage of horticultural labour but what amounted to a range of labour supply challenges that affected some regions more than others. These challenges were partly due to a dual labour market system; 'compliant' growers who followed government rules and regulations may have been at a disadvantage and found it more difficult to secure labour supply than employers who were willing to pay workers cash and turn a blind eye to visa regulations. However, Howe et al. (2019) also found that it was difficult to generalise the nature of labour supply shortages and the challenges faced by growers were nuanced; in some regions there was a relative oversupply of labour, while in others, the labour supply was highly insecure. One way or another, whether a function of lazy Australians or exploitative conditions, there were at least some labour shortages. Moreover, through this century, demographic shifts that had intensified the rural-urban migration of youth, a cost-price squeeze emanating from the duopoly of Coles' and Woolworths' supermarket power that had added to supply chain pressures, a growing reliance on labour hire contractors and the weakened presence of trade unions, increased the vulnerability of all temporary workers, and especially those from non-English speaking countries: a particular 'layered vulnerability' (Underhill & Rimmer, 2016; van den Broek et al., 2021) that exacerbated the widespread disinterest in seasonal employment.

Despite all of this, at no time in this century was the agricultural industry ever satisfied with labour supply. Media stories detailing Australia's ongoing seasonal labour issues have remained common. In 2018 John Said, the chief executive of Fresh Select, a Victorian agricultural company, explained:

> We don't have a labour issue, we just don't have labour [...] No one [among Australians] wants to work this caper. We have tried everything. We are some of recruiters' best friends. We are forever advertising, it is simply impossible. (Quoted in Thompson, 2018)

Put another way, as a Queensland horticulturalist stated: "The industry is not the repository for the long-term unemployed and we want willing workers not those with no choice' (ibid). Stories abounded of growers being forced to regularly 'leave money on the ground because there's no one to harvest their crops, or to shy away from even planting them in the first place' (Thompson, 2018). Headlines, such as 'Fruit rots on the vine for want of seasonal workers' (*The Australian*, 21 June 2008), and 'Rotten call on fruit pick plan' (*The Land*, 18 October 2018), made the most of such shortages. Dramatic headlines multiplied during the COVID-19 year of 2020, alongside many crude numerical estimates (but often inflated by agriculturalists anxious for government support). In 2020 much-quoted estimates of shortages ranged from 26,000 to approximately 70,000 short-term vacancies in the national fresh produce industry (FP, 4 September 2020). Even assuming the lower figure to be inflated, that number is more than double the annual total of SWP workers (but it also reflected the lack of the usually more numerous WHMs).

To meet ongoing labour shortages, in 2017 the National Farmers Federation (NFF) called for a distinct agricultural visa with two streams: 'A short term (six to twelve month) stream to cater for a seasonal or low skilled workforce [who may return in subsequent years], and a longer term stream to cater for a skilled workforce who may move between employers, industries, or regions or employer groups' (quoted in Dornan et al., 2018). While the SWP program would seem to be oriented to the need for short term workers, the NFF rejected it as 'great but ultimately it is a foreign aid program' whose primary purpose was to contribute to economic development of Pacific Island Countries (PICs) (ibid). That was reiterated in another industry source that argued that the principal purposes of both the WHM and SWP schemes were 'cultural exchange and humanitarian aid' (Kotsios, 2020). While the SWP was in practice almost the converse of this, and the NFF did not indicate how it might have failed to benefit Australian farmers, the push for a scheme that gave priority to Australian farmers was evident, and was supported by the National Party (within the Coalition government).

After a Parliamentary Standing Committee report 'Growing Australia' (Australian Government, 2020), and with national borders closed due to the COVID-19 pandemic, the possibility of an agricultural visa was revived with the Committee making a similar recommendation. The suggestion for such a visa arrived during the pandemic, a minority report argued against it and nothing eventuated until mid-2021 when the Australian government announced the establishment of an agricultural visa for ASEAN countries, encompassing a region of the world that, at the time, was struggling with ongoing and widespread COVID-19 outbreaks. Its introduction was unusually divisive with the Minister for Agriculture, David Littleproud, noting that it would provide a pathway to permanence for migrants from ASEAN: 'We want Asians to become permanent and be part of our regional communities, make up the fabric of our country towns, not just pass by' whereas PIC workers would remain temporary: 'we don't want to take away that precious human capital. They need their men and women there to build their nations' (ABC Landline, 26 September 2021). Concerns were raised that such a visa would jeopardise the strategically important SWP, that the Melanesian states could supply more workers, and the less-regulated nature of the proposed visa would further entrench worker exploitation within the industry (Howe, 2021; Howes, 2021; Howes & Curtain, 2021). Other similar proposals to expand the pool of available workers have included a Pacific Integration Visa, analogous to New Zealand's Pacific Access Category, that would enable short-term migrants, who might initially be directed to regional areas, to gradually acquire permanence (Howes, 2019). For most of the twentieth century, the agricultural industry, notably the NFF and its parliamentary support in the National Party, was absolutely sure of a problematic labour shortage, present well before the arrival of COVID-19 that precipitated new demands. The solution, they argued, was increased migration.

A Crisis and Clues from COVID-19

While Australia largely escaped the early health effects of the COVID-19 pandemic, that was made possible by a rapid and ruthless closure of international borders in March 2020. The impact of this closure intensified and demonstrated the reality of horticultural labour shortages within the country. The downturn in Australian urban economies and the national (and later frequent state) border closures meant both a rapid rise in urban unemployment—especially amongst the loosely informal sector and gig economy that hired many temporary migrant workers—allied to the impossibility of bringing in seasonal workers from the PICs.

Many backpackers and seasonal workers had 'escaped' to their home countries where they could. That brought an immediate horticultural worker shortage and a continued reluctance by Australian workers to respond to these labour needs, as mobility was discouraged, feared and associated with the spread of the virus. Later in the year especially, as harvest times became imminent, the shortage resulted in most states, followed by Australia itself, establishing incentives to generate agricultural labour.

The return of seasonal workers to the PICs before the coronavirus crisis, and their non-replacement (despite all the PICs, other than PNG and Fiji, being almost entirely COVID-19 free through 2020), and the similar absence of backpackers (although a few did shift from the hospitality sector), meant that by mid-2020 farmers were assuming that a labour crisis was imminent for horticulture and the orchard industries. With worsening health circumstances in Victoria in July, and the closing of the border between Victoria and New South Wales (NSW), across which seasonal workers had frequently moved, the situation deteriorated. The then Australian Minister of Agriculture, David Littleproud, suggested:

> Even when our social security payment for the dole was $550 a fortnight, we couldn't get people off the couch to go and pick fruit. There's a real aversion from the Australian workforce to go and pick fruit… It's a very hard incentive to get someone off the couch to travel a couple of thousand kilometres to pick fruit for six weeks, then turn around and come back. (ABC Rural, 30 July 2020)

Some farms had developed isolation and quarantine plans for Pacific workers in the hope that border policy might change, while the collapse of PIC economies (centred on the disappearance of the tourist industry) meant that the PICs were anxious for schemes to resume.

New South Wales set up a state website—'Help Harvest NSW'—as a one-stop shop for those looking for workers or work in rural NSW. The NSW Minister for Agriculture, Adam Marshall, said:

> Farmers are desperate to find contractors and employees willing to work to make the most of the bumper harvest we're about to have. Due to

> COVID-19 international travel restrictions, there simply isn't the number of farm workers the industry needs at the moment. I want to see more Aussies climb off the couch and get out into the regions where there's plenty of good-paying work on until the end of the year. (Quoted in Fresh Plaza, 27 August 2020)

Nothing was heard of the scheme thereafter. A parallel Western Australian scheme, *Work and Wander out Yonder*, was launched to inspire 18–30-year-old Western Australians 'to embrace the opportunity to have a working holiday in regional WA'. This scheme too largely disappeared without trace. In 2021 Queensland, lacking 9000 full-time workers in the main harvest period for strawberries, launched a campaign to attract winter harvest workers offering them a chance to win up to A$100,000 in cash prizes if they worked on a strawberry farm over winter (FP, 7 May 2021). That had no greater success.

The looming labour crisis produced familiar intemperate reactions amongst conservative commentators, evident in the preoccupation with couches: 'Young healthy Australians who live with their parents and receive a taxpayer–funded Jobseeker allowance should be conscripted to work on farms. They should be tested for COVID-19 and sent to picking areas for three months' (Gleeson, 2020b). By contrast, and by chance on the same day, the former Labour Party leader, Bill Shorten, now the Shadow Minister for Government Services, observed that disability and aged care workers 'do the work of angels, [but] are effectively treated like fruit pickers'. A well-known unattractiveness seemed to be built into the system. Even around Bundaberg, with an unemployment rate of about 11 per cent, one of the highest in the country, it proved impossible to attract young people away from the steady welfare lifeline of 'Jobseeker', introduced during COVID-19, while rental housing was almost non-existent (McDonald, 2021). Welfare provision, through a temporary Jobseeker system, may have discouraged some workers but there was no indication of interest in Western Australia, New South Wales or elsewhere. As one journalist observed of the poorly subscribed, administratively burdensome incentives on offer:

> There is no incentive to work if the incentives themselves are no certainty to be awarded. By definition, it's not really an incentive, but a gamble. The federal government is asking "lazy" Australians to take a financial and personal bet, asking them to relocate three hours away from family and friends, outlay $200 a week upfront on accommodation, some more still on fuel, placing their trust in farmers and employment services they have never heard of or met. These are individuals most likely on a low income or no income at all, individuals that are able to afford this sort of gamble least of anyone. (McGlone, 2021)

At the national level, and a little later, Deputy Prime Minister Michael McCormack was hopeful in seeking to encourage and offer relocation grants to people on Youth Allowance (and thus aged between 16 and 24) to work in regional areas:

> Have a go. Come to regional Australia. Bring your mobile, have that Instagram moment up a ladder, picking fruit, blue sky in the background, wonderful country breeze, wonderful friends around. You'll find more friends, you might find the love of your life out in regional Australia. It could change the way you live. (Quoted in *The Australian*, 1 October 2020)

McCormack's message, alongside the relocation grants, was described by the Australian Workers Union as a 'brain fart'. He too had no success. By the end of the year the relocation grants had attracted just 355 applicants across the country (*Sydney Morning Herald*, 28 December 2020). At the same time, the Premier of Victoria said that the state urgently required between 15,000 and 20,000 workers. A smaller number were quarantined in Tasmania but no transport and quarantine scheme could provide that many. Queensland simultaneously identified a shortfall of 7000 workers (ABC Rural, 4 November 2020). A National Lost Crop Register was set up in December 2020, endorsed by the NFF, to monitor crop losses since the start of COVID-19, with farmers recording the value of crops that had rotted and/or been ploughed back into the ground due to lack of workers. Early in 2021, after eight weeks of operation, overall losses were given as A$45 million. Again, these were estimates; some may well have been inflated, just as were estimates of labour shortages, but many farmers had little incentive to use the Register. Unquestionably unfilled shortages, that Australians alone could not or would not satisfy even during a time of hardship, existed at some cost to farmers.

The need for PIC workers was evident in other ways. The Australian government allowed Pacific islanders on labour mobility schemes, mainly the SWP, to stay in Australia during the pandemic and keep working, by granting 12-month visa extensions. While some Pacific workers were able to return home—and were not replaced—border and quarantine restrictions in several PICs, such as Kiribati, meant returning home was impossible for many. In May 2020, roughly 8300 SWP workers remained in Australia. State border restrictions were also eased specifically to enable agricultural workers, including SWP workers (in most cases), to cross where stability was otherwise rigidly enforced. While few arrivals were allowed into Australia, and even Australian citizens in other countries found it difficult to get flights home, seasonal workers acquired special 'privileges'. The first effective resumption of flights was not until September 2020, when a charter flight of about 160 mango pickers arrived in the Northern Territory (NT) from Vanuatu, 'literally saving the mango harvest in NT, valued at over A$128 million' (Gosling, 2020) and more flights came in October, with the number of workers doubling. From then onwards numbers increased in all states and between September and mid-December 2020 over 1000 Pacific and Timorese workers arrived in Australia—with more to follow—from most of the PICS and to every state, including Tasmania and Western Australia, those with the most rigid borders, that Australian citizens had extreme difficulty in crossing. Elaborate schemes had to be developed to quarantine the new arrivals at some cost to industry, but farmers agreed the expense was 'worth it'.

While some international travel into and out of Australia occurred in the year after COVID-19 had flared, it was rare and extremely strictly regulated so that the migration of Pacific islanders for employment was quite exceptional, and distinctly outside usual regulations. It emphasised just how crucial that employment was to the agricultural sector—the first real arrival of workers during the pandemic—despite clamours from business groups for the migration of skilled workers. It was accompanied by greater pressure from PICs for the creation of a regional air corridor, or 'bubble', between Australia, New Zealand and the PICs that would allow these exceptional solutions and the overall SWP program to continue and even to expand, as seasonal labour from other countries dried up. The return of SWP workers, most on specially chartered flights, also emphasised that seasonal labour migration was the antithesis of any 'new normal' since one of arguably the most precarious and insecure forms of international employment was reviving in lieu of any significant changes in this or any other economic system. Indeed much the same process was simultaneously taking place in Germany, the United States and New Zealand, where migrant workers were welcome—if segregated—and flown in at a time when many local people were unemployed (Bedford, 2020; Griffiths, 2020). COVID-19 conclusively demonstrated that migrant seasonal workers were essential and even irreplaceable.

The Geography of Seasonal Work in Australia

Australia has long experienced agricultural labour shortages, but these have never been evenly distributed. Labour supply issues were and are both spatial—relating to the dispersed nature of Australian horticulture—and temporal due to the different harvesting and labour requirements of crops (Campbell, 2019). Overseas workers were particularly important to the horticultural industry where they accounted for 40 per cent and 35 per cent of the workforce on vegetable and fruit and nut farms, respectively. Other farm types, including dairying and broadacre (such as wheat and canola), were less reliant on migrant workers (where they made up less than 3 per cent of the workforce), being highly mechanised or without real seasonal peaks. Agricultural industries with seasonal peaks faced difficulties. Regions with the most secure labour supply usually had a permanent local workforce; plentiful and reliable access to WHMs (e.g. Orange in New South Wales, and Gingin and Binningyup in Western Australia); a population of recently settled migrants from the Global South (e.g. Virginia in South Australia); regularly employed SWP workers (e.g. Katherine in the Northern Territory and Mundubbera in Queensland); or, they relied upon a combination of these labour sources (e.g. Lockyer Valley in Queensland). In such regions, there were generally few illegal workers and growers tended to be experienced and innovative in developing informal and formal networks with agents and accommodation providers. By contrast, growers who had trouble in securing labour were, not surprisingly, often located either in more remote regions or were ineligible to hire guestworkers (either WHMs or SWP workers) (Howe et al., 2019) due to not meeting the government requirements (see below). Peri-urban fruit and vegetable growers especially could also face difficulties securing labour as they had to compete with jobs available in nearby industries and services. Other factors such as farm type and size also played a role in farmer's recruitment experiences. Fruit, vegetable and nut farms that required workers for seasonal positions had a better chance of filling these vacancies as the work did not require special skill sets.

Reflecting the geography of labour shortages, Pacific seasonal workers were employed in all Australian states and the Northern Territory in both peri-urban and remote locations, but employment was greater in some areas. Between 1 July 2012 and 12 February 2020, 38 per cent of all SWP placements were in Queensland; Victoria was a distant second accounting for 18 per cent of all jobs, followed by New South Wales (13 per cent) and South Australia (12 per cent) (Curtain & Howes, 2020). Temperate Tasmania was the last state to be involved in the SWP (Figs. 7.1 and 7.2).

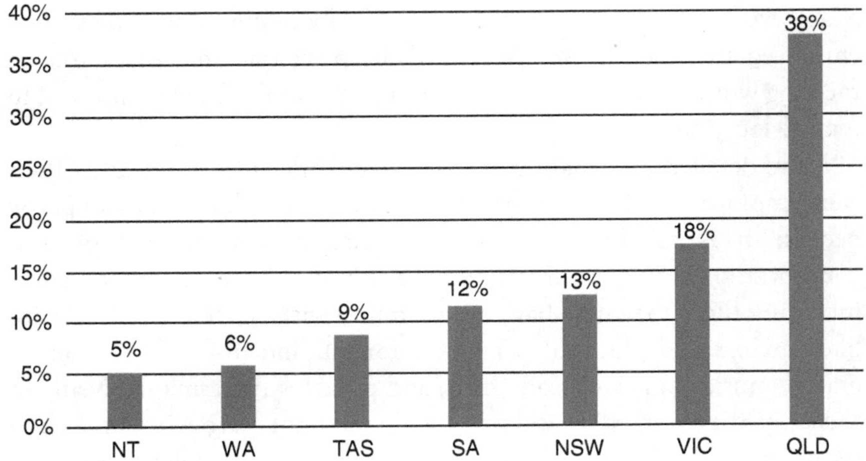

Fig. 7.1 Distribution by state/territory of SWP positions, July 2012 to February 2020. (Source: Curtain & Howes, 2020, p. 13)

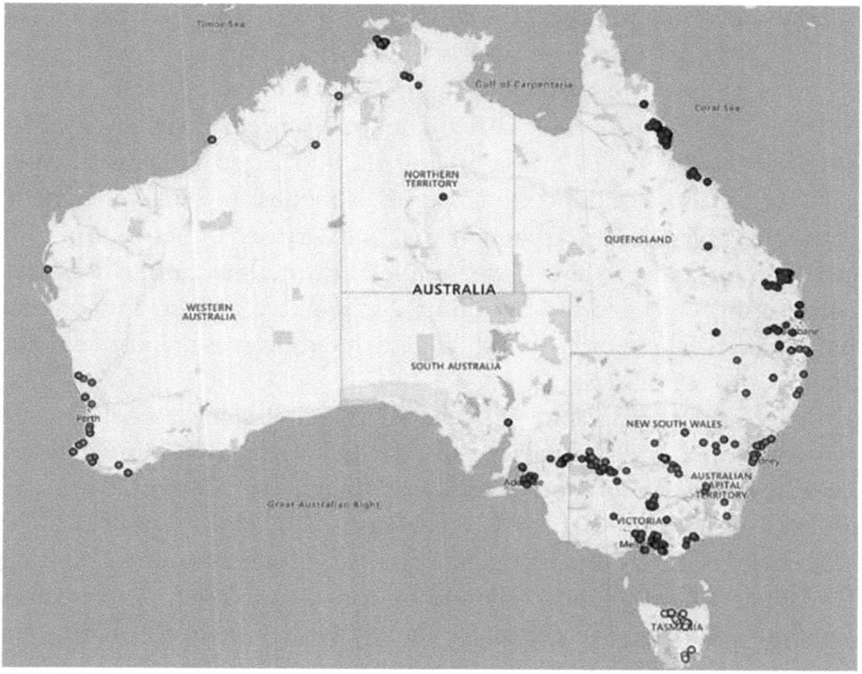

Fig. 7.2 SWP recruitment locations in Australia, 2018–2019. (Source: DESE)

Many workplaces were very remote, and labour hire companies, who employed the majority of SWP workers, played an important role in meeting workers upon their arrival in Australia to facilitate their travel to remote locations.

More detailed information about where workers from different PICs were employed, or the crops they were working on, is largely absent partly because it is so fluid. The Riverina and Bundaberg had obvious concentrations. SWP workers were involved with a variety of crops including those they may have been familiar with, such as tropical crops like bananas and pineapples, and the foreign and unfamiliar—such as grapes, mushrooms and stone fruit, and potatoes in Tasmania. Many of the crops they worked on did not—and could not—grow in home environments. In addition to discovering new crops, seasonal workers experienced new climates. While workers employed in Queensland and the

Fig. 7.3 I-Kiribati SWP workers pick strawberries in a giant glasshouse. (Source: DFAT via Flickr)

Northern Territory, and those working in glasshouses (Fig. 7.3), experienced similar climates to those at home, workers in Tasmania and Victoria particularly struggled with colder, drier climates. I-Kiribati workers were welcomed in New Zealand as the only ones said to be capable of enduring full days at high temperatures in glasshouses. No equivalent existed in Australia. SWP workers' experiences in Australia were nothing if not varied. It is impossible to capture the diversity of employment and related experiences, from remote to peri-urban, from family farms to huge estates, offering quite different work regimes and familiar and unfamiliar crops.

Employers' Perspectives

To recruit Pacific seasonal workers, organisations (contractors and growers) needed to become an approved employer (AE) by applying through the Department of Education, Skills and Employment (DESE). Prospective AEs needed to have been in operation for at least five years, and have a record of compliance with immigration and workplace relations requirements. As well as having a valid Australian Business Number (ABN) and meeting the requirements of being an 'organisation' under Australia's migration regulations, employers needed to be 'committed to employing Australian job seekers first', have had 'good' workplace practices and understand their obligations (such as the need to provide pastoral care) as a seasonal employer. Once approved, AEs were permitted to recruit only a small number of SWP workers in their first season so that they could 'adjust to the obligations and responsibilities under the program' (DESE, 2021). Recruitment caps were reviewed after each placement to ensure requirements were being met.

Estimates of the proportion of SWP workers employed in different activities at any one time varied. According to the Australian Bureau of Agricultural and Resource Economics and Sciences (ABARES), Australians made up the majority of the agricultural workforce, but during peak periods more than a third of seasonal jobs were filled by overseas workers. Most 'overseas' workers were WHMs, and SWP workers were employed by only about 4 per cent of horticultural farms and less than 0.5 per cent of broadacre farms in 2017–2018 (Dufty et al., 2019). A few large employers, many being labour hire firms, employed most of the seasonal workers (see above) (Curtain & Howes, 2020).

In the early years of the SWP, participation was limited, partly because many growers had never heard of the programme. Others were confused about the costs and responsibilities they would incur to become an AE, including subsidised airfares, and the need to provide pastoral care and accommodation, all usually beyond their scope for other workers. Most employers did not use the SWP because they had no need for extra workers, and certain crops, such as those requiring year-long harvesting, were not seen as compatible with temporary SWP employment. Others perceived SWP workers as being less productive than other workers (Howe et al., 2019). Negative racial stereotyping has persisted in some places. In 2020, Kirstie was told by one key informant:

> But on the ground, the intel on the ground is, this is how ridiculous it is [...] 'Pacific Islanders have fat fingers so they can't pick the grapes properly. They're not as nimble.' And I was highly offended and I said: 'That's rubbish'.

It is impossible to know how widespread such sentiments were when the SWP started, but they seem to have diminished. By contrast, many employers eventually perceived migrant workers—including those recruited through the SWP—as more productive, compliant and reliable than their Australian counterparts. One employer posted a tribute to a departing worker on Facebook in 2021 demonstrating the kinds of connections the SWP could facilitate:

> Today we say good bye (just for a short while hopefully) to another great worker. Cant [sic] say thank you enough Joe for your help, friendship, kindness and patience you have shown me and others in the nearly 2 years of being here with us. Gonna be very hard without u.... so hurry back . hope u have safe travels my friend. And thanks again for everything.

Many employers, such as this one, valued their SWP workers and formed relationships that went beyond the labour power the workers provided.

Despite formal government requirements, employment arrangements varied. 'Under the table' or 'cash in hand' payments, and other illegal practices, have long been a feature of the Australian horticultural industry. The prevalence of highly flexible but illegal labour practices, and the availability of WHMs, initially provided a deterrent to participating in a highly regulated scheme such as the SWP (Doyle & Howes, 2015). New Zealand dealt with analogous issues through a crackdown on illegal labour and similar practices before introducing the Recognised Seasonal Employer (RSE) scheme. Australian regulation was slower off the mark, so that employers had less incentive to abandon illegal practices and continued to face bureaucratic barriers to participation for several years (Howe et al., 2019). In 2020, a key informant in rural Victoria told Kirstie:

> What I hear on the ground is the Pacific seasonal workers are not really that valued. So a lot of the talk around town is that they're not that good, you know we prefer the Malaysians. We know why, because they're [the Malaysians] all getting, paid under the award rate. The Pacific islanders are seen as expensive, they have all these conditions, their housing has to be this and that. So for whatever reason, for those reasons they're not seen as a valued proposition.

The same informant explained that most labour contractors continued to employ visa overstayers and other illegal workers because this was exactly what some growers preferred: 'Even the reputable contractors are saying "Yeah, we carry some illegals because some customers want that"'. Some employers had no intention of abiding by labour laws and preferred the black market.

Although many of the up-front costs associated with employing SWP workers, such as contributions to airfares, were eventually recouped, some farmers experienced short-term cash flow problems as a result of SWP-related costs. Other costs such as providing (and sometimes building) suitable accommodation, and having vehicles available for workers to use, provided further challenges (Howe et al., 2019). Employers were often limited in the number of SWP workers they could employ by the available accommodation options nearby. That situation never changed. Some farmers eventually needed to employ a pastoral care worker to assist the workers in access to churches, sporting teams or doctors' appointments (Zhao et al., 2018). In other instances, local residents unofficially volunteered in this pastoral care role in an unpaid capacity (Chap. 8). Such circumstances deterred small farmers especially.

Eventually, in 2018 the Australian government announced various reforms to the SWP process to make it more streamlined and more attractive to employers. Changes included the option of multi-year visas (with a maximum validity of three years), and a reduction in employer contributions to airfares (A$500 was reduced to A$300). Workers were granted the option of changing employers under certain conditions when they risked becoming trapped in exploitative or 'untenable' situations with employers (Curtain & Howes, 2018). The ability to seek alternate employment would have been invaluable to the seasonal workers we spoke to in rural Victoria in early 2012. These workers described their employer as a 'strict' man who swore at them and changed their pay rates (from hourly to piece rates) unfairly. Two workers were sent home at short notice with little explanation from the farmer, and those still in Australia felt their contractor merely dismissed their complaints. The workers had received offers of employment from other local farmers, who were aware of their employer's reputation and commiserated with their situation, but the restrictive visa conditions of the time prevented that

(Petrou & Connell, 2018). The situation eventually escalated, and the workers walked off the job four months early.

Other reforms followed. In May 2019, a Regional Pilot was launched in the Sunraysia, Goulburn/Murray and Riverina regions of Victoria and NSW that allowed SWP workers to move between farms without prior approval from DESE. The three-year trial was intended to meet the needs of smaller farmers with short harvest seasons, who would be unable to meet the minimum work requirements, averaging 30 hours a week, for the four to six months required to participate in the SWP. Nominated AEs coordinated the workers' moves between farms and assisted other employers in meeting workplace health and safety requirements. The results from the trial are yet to be formally assessed, however Kirstie was able to speak with a community service provider in the Sunraysia region in mid-2020 who indicated that while the trial was beneficial for workers, it could create issues for accommodation providers

> Yeah, there's the pilot that's been run in the area that's been quite good in terms of keeping people moving. Not so good for accommodation providers though because they bank on the fact that there are gonna be 30 ni-Vans staying at the facility and then all of a sudden, on the farm, something happens, a rain event happens and they uproot them and move them to another place and they have to leave that accommodation for [another] region, and then that person, that accommodation provider is out of pocket. And the program then suffers vocationally because of that. People start talking locally, you know it's not worth engaging in the Pacific Seasonal Worker Programme 'cause they'll just pull out from underneath you and leave you hanging with an accommodation facility with no tenants and no income.

Such issues highlighted how many different stakeholders were reliant upon the scheme for incomes, the considerable problems of coordination, complex demand structures, unpredictable events (such as drought or floods) and the necessity of taking a holistic approach to assessing the impacts of the SWP.

Recruiting Pacific guestworkers involved more effort than was associated with 'traditional' labour sources, but benefits existed for employers who were willing to participate. Many employers involved in the SWP felt they would be unable to harvest their crops without the programme. Pacific workers were often described by employers as more productive, motivated and reliable than their WHM counterparts due to superior physical strength, experience in physical labour, motivation to earn money to support families at home and familiarity with working in hot or humid conditions (Zhao et al., 2018). Similar perceptions were reported amongst New Zealand's RSE employers (Research New Zealand, 2018). Over time some experienced employers regarded seasonal workers as a core part of their seasonal staff, thus ensuring that workers were available at peak times. One farmer, an apple grower, explained:

> We became interested in the scheme because we didn't want staff churn and wanted reliability at peak harvest. In getting Pacific workers it's a guarantee that you're going to have the workers on-farm, on the day you need them. (Quoted in Zhao et al., 2018, p. 17)

Isaac, a long-term Pacific islander migrant in rural Victoria, echoed these sentiments to Kirstie in 2020:

> We're probably more productive given the opportunity than the backpackers are. We're not here to party and hang out. The feedback from some of the farmers is that the backpackers, six or eight of them will disappear as soon as there's a cheap flight special to Bali on Jetstar, you know? All of a sudden you lose eight good workers because they've made enough cash to go. Pacific islanders won't do that, they're here for the long haul.

Recruiting the same workers again was additionally beneficial to employers, as return workers required less induction and could eventually undertake more skilled tasks such as forklift driving, specialised picking and installing irrigation. Moreover, they were 15 per cent more productive than first-time workers (Zhao et al., 2018). It made sense to value potential return workers; as one employer told Christine Brickenstein:

> We invest a lot in these workers so we want them to return. If you want people back you look after them. You need to show that you actually care

> about them as people not just hands. You also need to build relationships with their communities back home. (2014, p. 248)

However, productivity gains were finite and tended to plateau and sometimes even decline after four or more seasons (Bedford & Bedford, 2017), indicating the importance of continuing to recruit new workers into the scheme. Despite the initial economic and other costs, employers increasingly benefitted from the use of Pacific guestworkers even as some continued to be deterred by the regulated nature of the scheme or 'easier' and cheaper options available through WHMs and illegal labour.

On the Job

For Pacific islanders, horticultural work in Australia was often a mix of the familiar and the foreign: crops that were known and many that were not, production at scales unheard of in PICs, and long workdays dictated by the clock. The ni-Vanuatu workers who Kirstie met in rural Victoria in 2012 were employed on a stone fruit farm. They shared a bag of plums with her and laughed about how, when they first arrived, they would peel the fruit before eating it. Now they knew that the skin of this once unfamiliar fruit could be eaten, they found their earlier naivety amusing. Another group of ni-Vanuatu workers who were employed in peri-urban Adelaide in 2020 worked with tomatoes, a more familiar crop. Unlike in Vanuatu however, where crops are grown in the open, these workers tended tomatoes in giant greenhouses. The scale of planting too was vastly different from the subsistence gardens that most ni-Vanuatu plant in rural areas, but at least the crop was recognisable (Fig. 7.4).

Tasks undertaken by SWP workers spanned the spectrum of what is usually classified as 'low skilled' farm work, and included planting crops, weeding, picking fruit, grading and other work in the packing sheds, forklift truck driving, spraying, trimming, pruning and generally tending the crops, and dealing with machinery. In some workplaces, SWP workers were assigned a single task for the duration of their contract, while in others, the work changed seasonally according to employer needs.

Fig. 7.4 Tongan SWP workers pruning grapes, a crop that does not exist in the PICs. (Source: Facebook)

> Lucy was in her late thirties and lived on Santo, in one of Vanuatu's two urban areas. She was an experienced seasonal worker, having worked in New Zealand through the RSE scheme for five years. In 2019 Lucy decided that the hourly pay rate in New Zealand was too low, and thought she might try working in Australia—where the award wage was higher—instead. The warmer climate was an added bonus. In New Zealand Lucy had picked apples. In Australia she was employed picking tomatoes, a job she would perform for the entire six months of her contract. Lucy much preferred tomatoes to apples and explained 'Working with tomatoes is good. It's much easier than working with applies [...] Apples are hard because the picking baskets weigh more than 25 kg. And then you have to go up and down the ladder. But with tomatoes you can just walk around, and the picking bucket is small.' The work may have been monotonous, but compared to her previous experience, Lucy was pleased to be working on a much less physically demanding crop.

Tomatoes may have been less physically demanding than apples, but this did not mean the work was easy. Jenny, another ni-Vanuatu woman employed on tomatoes, described her workday:

You're in your own world, you just totally forget what's happening around you, because your mind is just on the tomatoes, and you don't realise that the time is going and suddenly it's smoko time, break time. And then after 15 minutes you go back and when you're in there you just totally forget everything. For me, I like it that way. Because I concentrate more on my picking rather than to think of anything else. Most of the time, I sing a song in my head. I'm into the classics—the '70s, the '80s—like Phil Collins, Bryan Adams, the Bee Gees. Pussycat. You know that song "Mississippi"? I'll remember you, yes that's it, that's the song that I got in my head today. (Quoted in Dao et al., 2020)

Of her job, Jenny explained:

I am a picker, my job is just to pick tomatoes everyday. There's five varieties of tomatoes in the glasshouse I'm working in now. All the pickers have to pick six to eight tonnes of tomatoes, in a day. It's very hot, very hot in the glasshouse. Really really, very very very hot. I wear boots. Long sleeves, gloves, and there's hairnets and facemasks. One glasshouse is the size of half of a football field. We plant the tomatoes high off the ground, on top of kind of a slab and we've got all these drippers for the water to run into the plant.

I pick ripe ones only. No green shoulders. That means that we pick tomatoes that are not green on top, it's red or orange. We put them in a bucket hanging around our neck, then we put them inside a box, on top of the trolley. It's 40 boxes in a trolley, and we have to push. In a day, we have to pick two different varieties of tomatoes and on Saturday we do truss cutting—that's where we cut the stem from the tomato plants. When we get home, we're just worn out. It's only Sunday that I don't go to work. (Quoted in Dao et al., 2020)

Unlike Lucy and Jenny, Matthias worked on a stone fruit farm.

Matthias was in his twenties and lived in Port Vila. He was working in Australia for the first time, and was employed on a farm in rural Victoria that grew stone fruit. The women in his group worked inside the packing shed, while the men stayed outside picking fruit. The men also performed other tasks, such as thinning and pruning the trees, according to the time of year. At peak picking time, the men were very busy and often worked

> seven-day weeks in the hot sun, but when the weather was above 37 degrees, they finished work at lunchtime. Matthias enjoyed work on the farm because, unlike in Vanuatu, he could earn a lot of money. However, he found it frustrating that they were paid a piece rate. Matthias and his fellow ni-Vanuatu colleagues could pick quickly, and thought they were faster than many of the other workers. But sometimes, when all the fruit in one place did not ripen at the same time, they wasted a lot of time walking around looking for fruit that was the right size and colour to fill their bins. If the men did not pick enough fruit, sometimes they got in trouble with their supervisor, and it was hard to earn enough money.

Matthias' concerns over the disadvantages of piece rates were repeated many times over, and were often attributed to growers knowing they would consequently pay less. Such issues were widespread; Cavanough and Wherrett (2020, p. 23) describe how, when questioned about why their piece rates were less than the horticultural award, one labour contractor responded: 'it is piece rates…the less you pick the less we pay'. Where fruit was not plentiful that significantly reduced earnings.

The organisation of work on Matthias' farm was typical of the Australian horticultural division of labour; tasks tended to be assigned by gender, with men working in the fields on the more physically demanding tasks, while women most often worked in the packing sheds (Fig. 7.5). There were exceptions however, and more 'delicate' work such as picking berries, mushrooms or tomatoes was often given to women (Fig. 7.6). Indoor tasks such as packhouse work, were perceived by workers (and employers) as being 'easier' than outdoor work, although this was not necessarily the case. Packhouses generally operated on shifts, and women assigned to the day shift often struggled to get out of bed to take over from the night shift workers at a time of morning they described as 'half-night' (still dark). Night shifts are almost entirely unknown in PICs, totally so in rural areas, so night work was highly unusual. Rachel Smith (2016, p. 89) undertook a short period of packhouse work alongside RSE workers and described the conditions

> For those in the packhouses, their work rate was dictated by the speed of the conveyor belts. I was to experience the relentless pace of packhouse

Fig. 7.5 Tongan workers in the packing shed. (Source: Facebook)

Fig. 7.6 Berries were a crop that women commonly worked on. (Source: DFAT via Flickr)

> work when I visited Appleseed and was allowed to work alongside the Li-Lamenu [ni-Vanuatu from Lamen Bay] women. It proved difficult to have the time or attention required to have a conversation during work, because if one looked away for a matter of seconds there was a danger of palettes [sic] of apples falling off the conveyor. Later, when I visited the night shift women that worked in the packhouse at their accommodation, one mentioned they did not like working with machines and 'technology', and summed up their experience in saying they felt the machines were in control and the "world was upside-down". The workers have to keep-up with the machine, rather then it making work easier. Packhouse workers at one company noted their employers tended to slow down the conveyor belts when there were inspectors present, but they were normally set to a high speed, and he felt this put them in danger.

Victoria Stead (2021a, p. 9) found almost identical conditions in rural Victoria:

> Packing sheds are hot, often oppressively so. They are noisy, and workers are on their feet, watching, sorting, and packing the fruit that keeps rolling past, relentlessly, for hours at a time. Most farmers and employers will readily admit that it's tough work. When the ni-Vanuatu women worked on the line, they frequently did so with the farmers whose fruit they were packing standing, just behind them, watching them work. The farmers—or the shed supervisors and quality control officers who otherwise act as the intermediaries between the workers and farmers—exert pressure for speed, but also on grading, pushing for that ideal balance that sees as much of their crops packed into the most profitable Grade 1 or Grade 2 boxes, without risking the shipments being rejected on the basis of there being too many defects. Like all aspects of seasonal work, it's a constant succession of minor calculations that weigh cost, time, profit, and risk against each other.

This hypervigilance, and threats of replacement, were perceived as symptoms of a deep lack of trust and reciprocity; as one worker told Rachel Smith: 'It's like they don't trust us. That's how it seems to us … even though we came for three years now' (quoted in Smith, 2016, p. 89). Being indoors, packhouse work may have kept workers 'safe' from the elements, but it was quite unfamiliar to workers, came with its own risks and was by no means a 'soft' option: women too worked hard in sometimes dangerous conditions. Yet ni-Vanuatu perceptions could be quite different. Many young ni-Vanuatu women working in the packing sheds still regarded themselves as very lucky, since, as one said, it was 'the easiest work they had ever done', much easier than cooking, cleaning, taking care of one's children or even working in a tourist shop (Cummings, 2013).

Gendered divisions of labour reflected Australian norms around 'appropriate' work and were not necessarily consistent with the gendered division of labour in PICs. In rural Vanuatu, as elsewhere in most of Melanesia, women tend to perform much of the daily labour in subsistence gardens (and are involved in marketing) while men are involved more sporadically when land is being cleared and prepared or with some key cash crops. By contrast, in Polynesia and in Kiribati, men are the principal agriculturalists (and fishers) with women playing a more domestic role (Chap. 2). Despite Australian work roles often being different from those at home, some workers believed these work assignments were appropriate. Harry, a ni-Vanuatu man, was in his fifth year of RSE participation and explained:

> Most of the boys, they were picking fruits and they were faster than the women. So, I think farmers, they won't waste their time employing a lady not able to catch up with the boys. So, that is where the opportunity comes to employ more men. Not because of discrimination, but because of the work that they see in the field. I think the work in the field for men is ok, but for women, maybe a bit hard for the women [...] but in the field, the men are suitable because in what type of weather [sic], whether it's raining or fairly hot sun, the boys can take up the work. (Quoted in Villafuerte, 2018, p. 159)

Some workers disliked being typecast into gender roles and wished they could try other things.

> Mary was 33 years old and lived in Port Vila, Vanuatu. In 2012, she was working in Australia for the first time. Mary enjoyed working in the packing shed, but disliked the supervisor who she felt favoured the permanent workers. After several months, and with little warning, Mary was moved to picking fruit outside with the men. She was initially unsure

> about this move, but Mary soon decided she enjoyed picking. Mary was determined to prove that she could work just as well as the men, and was proud when she realised she could pick six bins worth of fruit in a day. The men had told her she could earn more money working outside. Other women in the group were more reluctant to join the men, as they believed they would not be able to pick as fast as them, and would slow the men down—an important consideration when workers are paid piece rates. By contrast, Mary was keen to learn everything about farm work that she could. She explained: 'I signed a contract to come and work on the farm, and work on the farm isn't just packing fruit. They told us we'd be doing farm work, and I want to learn everything about farm work. Even today, I just asked my boss if I could work on the tractors. I want to learn how to drive a tractor.'

Regardless of whether work was performed indoors or outdoors, horticultural work could be physically taxing and monotonous, as various stories demonstrate. Hours could vary, but working days were often long and started in the very early morning. The ni-Vanuatu workers we met in rural South Australia were collected from their accommodation by bus at 6 a.m. every morning. Each day they were allowed a 15-minute 'smoko' break at 9:30 a.m. and a half-hour lunch break at 12:30 p.m. During their lunch break, the workers were allowed to eat some of the tomatoes that grew at the farm, but they could not take them home. As Tommy explained: 'We're 500 workers, if we took the tomatoes, there would be none left'. The days could feel repetitive: whatever task the workers were assigned (picking, sorting etc.), they did for the entire six months of their stay. The bus collected them from work again at 4:30 p.m. and took them back to their accommodation, a 20-minute drive away. By the time they returned home, the workers had been away for almost 12 hours.

RSE workers, who mainly picked kiwi fruits and performed outdoor chores (such as cleaning, pruning and picking), described similarly long workdays, with variations according to seasonal demands

> We work from 7 am to 8.30 am. Then we break for 30 minutes; then we start again from 9 am to 12.30 pm, then we break for another 30 minutes. Then we start work from 1 pm to 3 pm and break for 30 minutes. Then we work from 3.30 pm to 4.30 pm and we finish for the day.

> We start at 7 am and finish at 7 pm and sometimes 5 pm on a normal day with a flat rate of $12.75 an hour. (Quoted in Kumar, 2012, p. 58)

Such long hours came as a shock to some. One ni-Vanuatu RSE worker who was employed pruning kiwifruit vines observed

> Something I experienced over there is when I work and I work and I'm tired, but I still have to work. Because their idea is that if you don't work, you can't have money. So, 'time is money'. But in Vanuatu, you work and you work and you feel you are really tired, so you stop and rest, because you feel you are tired. But over there, you must work to complete your 'hours'. When it is time, you can finish, but if not, you must work. (Quoted in Smith, 2016, p. 85)

For seasonal workers in Australia and elsewhere, the days were long and working conditions could be difficult. The work itself was sometimes dangerous, Australian environments could be harsh and variable, ranging from extreme heat and humidity to cold and frost: the 'sunburnt country of droughts and flooding rains' could pose problems. Moreover, widespread practices of cash in hand and under the table work meant guestworkers could be immersed in a culture where workplace rights had sometimes been habitually ignored, and where what was often no more than subtle exploitation could be widespread. The following sections highlight some of the issues workers faced in terms of health, safety and workplace rights while employed in Australia, and how they sought to deal with them.

Economic Gains

While excitement and new experiences motivated some workers, at the core of the SWP scheme was the unparalleled ability to earn a considerable income in a short time. Incomes, savings and remittances were enormous compared with local earning opportunities; Solomon Islander SWP workers claimed that they could earn the equivalent of two weeks' worth of domestic wages in a single day in Australia (Dun et al., 2022). On average however weekly SWP earnings (net of taxes) were 4.3 times what workers could earn at home (World Bank, 2017). Wage differences were a common theme in any extended conversations with SWP workers. Mary, a 33-year-old ni-Vanuatu worker originally from Malakula Island, lived in Port Vila where she earned a small and sporadic income through the informal economy, sewing dresses for others. In 2012, she explained

> In Vanuatu, you can't earn money like you do here. In a single day, you can earn the same as the weekly salary for a *haosgel* [domestic helper] in Vanuatu. So I thought I should just come and make some money in Australia.

Eight years later, Freddy echoed these themes

> But in Vanuatu … the minimum wages, the salary isn't really good enough to cover all of the expenses you have for your family. It's just enough for food. So that's why I decided to come here so that I can manage my finances and with the money that I earn here, when I go back I'll be able to renovate my house and look after my family.

Such sentiments were widespread; roughly 86 per cent of SWP workers surveyed by the World Bank (2017) cited the ability to earn a higher wage as 'very important' in their decision to participate in the scheme. Quite simply, as one older Tannese put it: 'The only thing about seasonal work is money, everyone needed money and that is one of the easiest and fastest ways for school dropouts to earn some money' (quoted in Kaiapam, 2021, p. 54). Even when opportunity costs relating to lost household productivity and wages were considered, the economic benefits remained enormous. Thus, in the early days of the PSWPS, participating in the scheme raised annual household incomes in rural Tonga by about 39 per cent (A$2600) (Gibson & McKenzie, 2011), a percentage increase that was likely to have been even higher in Vanuatu, where incomes were lower.

Repeat workers, who were already familiar with the work and skills required, tended to be more productive than first-time workers and thus earned more than new, inexperienced workers. Amounts earned varied and while some workers returned home with as little as A$4170 in remittances and savings after a six-month contract, others earned as much as A$10,470 over the same period. Workers from the more 'established' PICs, notably the 'first movers' of Vanuatu and Tonga, alongside Samoa, were able to earn, remit and save rather more than those from PICs that had subsequently participated (Table 7.1). However, earnings were also influenced by such factors as seasonal variation in crops, the crops themselves (from strawberries and broccolini to citruses), the type of work (picking, packing, pruning), contract lengths, pay rates (piece work vs hourly rates) and lost 'downtime' when work was unavailable.

Female workers tended to earn slightly less than male workers, mainly due to the gendered division of horticultural jobs; men tended to be paid piece rates for picking jobs, which was at least 15 per cent more per hour than the minimum hourly rate (A$24.80 per hour in 2021). By contrast, women tended to work in packing sheds and were paid the hourly award rate; provided piece rates were paid correctly, men then earned more than women for the same number of hours. In 2015–2016, men had net

Table 7.1 Earnings, costs, remittances and savings in Australia (A$) 217

	Samoa	Timor-Leste	Tonga	Vanuatu	Other PICs	Pacific average
Total earnings net of taxes in 6 months (per worker)	18,902	17,474	19,942	19,318	14,274	18,252
Pre-departure expenses	1199	1822	1438	1542	1602	1474
Ongoing weekly expenses in Australia						
Accommodation	121	121	121	121	121	121
Food	43	43	43	43	43	43
Transport	33	33	33	33	33	33
Telephone	24	20	24	22	31	23
Health insurance	52	52	52	52	52	52
Airfare deduction	29	46	36	34	47	34
Total expenses for 6 months	7826	8190	8034	7930	8502	7956
Net earnings less total expenses	9851	7462	10,470	9846	417	882
Remittances and savings						
Household members	3997	2665	1459	683	2419	2029
Non-household members	0	0	924	141	76	141
Savings	5854	4797	8087	9022	1675	6653

Source: World Bank (2017, p. 35)

earnings of A$8886 compared with women's A$7208 (World Bank, 2017). Nonetheless, hourly rates, which provided certainty around earnings, were generally preferred to piece rates which 'compel workers to pick at a furious rate in order to ensure a profitable return' (Stead, 2021a, p. 9). Regardless of such distinctions, workers were almost uniformly returning home with much more than they could have earned on the domestic market over a similar period. Indeed, while the average net earnings of A$8822 (remittances and savings) was a small sum in Australia for six months' work (and may account for the reluctance of local workers to take up agricultural employment), it represented far more than what could be earned in rural areas of PICs. For workers, it was a real economic 'win'.

Even as workers were able to earn far more than they would at home, they often encountered issues when wages, payslips or working hours failed to meet expectations. Since almost all work was on field crops, rather than in climate-controlled greenhouses, there was always uncertainty over what the end of the growing season would look like, and how that would be influenced by drought or flood. Harvest labour needs could be difficult to predict. The weather was the simplest embodiment of uncertainty for employment and earnings, rather than fluctuating prices or any absent work ethic. Ni-Vanuatu workers in the Adelaide Plains reported that the only significant problem they faced in Australia was the lack of hours. In early 2020, some were getting only a few days' work per week, and were concerned over being able to meet their regular weekly costs such as accommodation and health insurance. They understood the seasonality of the work, but were frustrated at not working the number of hours specified in their contracts. Donald explained to Kirstie:

> Sometimes we finish at lunchtime, 12 o'clock, so we're not meeting our contracted hours. Sometimes we have two or three days off, so we're not happy [...] It's because sometimes the work drops. But we signed a contract, because the [Vanuatu] Department of Labour told us we had signed a contract to work 30 hours per week, we must work 30 hours. Even if the work drops, we need to work 30 hours because that's in the contract we signed. But what's happened is some of us aren't working the full hours.

Tina added

> This week I only worked 2 days. I don't know about next week.... With just two days work you can't pay housing, transport and medical insurance.

When Kirstie asked Donald if decreased working hours would make it difficult for the workers to meet their savings goals, he responded

> That's a good question but it depends on your hours. If the hours are good you can make some money but if the hours aren't good.... it just depends on the hourly rate. If you work lots of hours you can make a lot of money, but if the hours are small, you only earn a little bit.

Freddy agreed

> I want to talk a little bit about what the team leader mentioned about work. Now our hours aren't like they were before. We have lots of days off, all of our working days, but still we have to pay the accommodation, they never dropped [that cost].... Do you understand what I'm telling you? The [accommodation] fee is A$135, and it's the same price.

These were common concerns. In 2012, some of the ni-Vanuatu workers we met in rural Victoria had faced similar issues. The men were busy working seven days a week in the fields, however the women, who were employed in the packing sheds, had little work. Vanessa was 28 years old and usually lived in Port Vila. She was working in Australia for the first time, and observed

> I've heard there isn't as much work this year, so we're wondering if we will be able to meet the financial goals that we had in mind. We won't be able to achieve what we wanted. We thought we'd be able to save a certain amount in three or four months, but it isn't happening like that. (Quoted in Petrou & Connell, 2018, p. 73)

The women were given the option of doing some picking with the men, but since there were only a few of them and they slowed down the men, who worked in teams trying to pick five bins of plums a day, they did not feel comfortable joining in. Workers felt frustrated and, in some instances, stressed over their financial situation. Lack of work and unpredictable hours was a recurring, but perhaps unavoidable, issue (Blanco, 2009). This was particularly true for workers who incurred large out-of-pocket expenses to travel to Australia. In 2020, Isaac, a Pacific islander living in rural Victoria, described how Timorese workers, who incurred a debt of A$6–8000 to participate in the SWP, were under constant stress whenever their hours dropped for fear of not being able to repay their participation costs: the higher the cost of participation, the greater the pressure to earn.

It is hardly surprising that there was considerable variation as to whether workers felt they had earned as much as expected. The World Bank (2017) found that workers generally expected to earn an average of about A$694 net per week, in line with the then horticultural award wage. However, actual earnings ranged from A$549 to A$767 net per week; naturally, workers were dissatisfied when earnings fell, although some had grand expectations based on stories of heroic previous workers (e.g. Connell & Hammond, 2009; LMAP, 2017; Rockell, 2015). Ultimately, half the ni-Vanuatu SWP workers interviewed by the Labour Mobility Assistance Program (LMAP) (2017) felt they had earned as much as they expected. Those who did not (29 per cent of women and 39 per cent of men) attributed lower earnings to working fewer hours or having to pay more deductions (for accommodation, transport and so forth) than expected. Expectations seemed to become more realistic as familiarity with the schemes grew.

Remittances

Workers travelled to Australia to achieve domestic economic goals and so much of what workers earned in Australia went home. Fees to remit money from Australia (and New Zealand) are amongst the highest in the world. Consequently, many workers preferred to remit only occasionally and carry most of their savings home at the end of their contract. Some workers who favoured this approach withdrew large amounts of cash from the airport ATM on arriving home. Others carried their earnings home in cash, packed securely in their luggage. This approach circumvented remittance fees, but it was risky. Lusa, a long-term ni-Vanuatu migrant living in New Zealand, told Kirstie how cash sometimes disappeared when sticky-fingered customs officials 'checked' workers' baggage upon their return to Vanuatu

> As they were coming through Customs in Vanuatu, I've heard people missing their luggage and those luggages have had money that they've actually worked really hard for in the vineyards. So there's $8000 that are missing, 8000 New Zealand dollars, and that's a lot money for seasonal workers who have saved so hard to actually come back home to Vanuatu [...] it's those who have worked in the customs on the frontlines in Vanuatu who are continuously opening bags and so they have to check through luggages as they come through.

Carrying money home in cash was a false economy when it disappeared at the border.

Recognising the high remittance costs in the region, the Australian and New Zealand governments sought to facilitate access to information about remittance fees through the website Send Money Pacific. This allowed workers to compare remittance services offered by different banks and money transfer operators, and included information about fees, transfer speed, network coverage and exchange rates. In theory, this allowed workers to access the 'best' deal, but the website relied upon workers being familiar with internet applications. Moreover, when SWP workers were employed in small, rural places, there were limited or no choices available regarding banking facilities. Most workers could only access the local bank, money transfer operator or Western Union branch. Knowledge about Send Money Pacific was limited; only 9 per cent of workers surveyed by the World Bank (2017) had heard of the website and just 6 per cent had utilised it. As a result, most workers continued to remit via Western Union, very familiar in the region but one of the most expensive money transfer operators.

Where workers were married (or partnered), most remittances ordinarily went to the home partner; single workers remitted to their parents, and occasionally grandparents, but retained significant sums for their own use. Despite substantial fees, remittances made a massive

Table 7.2 Remittance inflows from labour mobility schemes, Vanuatu (vatu)

Remittance source	2016	2017	2018 (forecast)
RSE	1.9 billion (A$23.6 million)	1.9 billion (A$23.6 million)	2 billion (A$24.9 million)
SWP	0.8 billion (A$9.9 million)	1.1 billion (A$13.7 million)	1.4 billion (A$17.4 million)
Total remittances (inflow)	2.7 billion (A$33.6 million)	2 billion (A$37.4 million)	3.4 billion (A$42.3 million)

Source: Reserve Bank of Vanuatu

contribution to recipient households and contributed significantly to GDP. In Tonga, for example, estimated aggregate net earnings from the SWP steadily increased from A$11.7 million in 2012–2013 to A$36.5 million in 2018–2019. This exceeded both the amount received in Australian aid (A$28.9 million) and exports to Australia (A$2.3 million) in 2018–2019 (Howes & Orton, 2020). In Vanuatu too, earnings from temporary labour migration were significant (Table 7.2). LMAP's (2017) sample survey of returned ni-Vanuatu workers in 2017 estimated that their savings (including remittances) averaged A$8529, similar to those calculated by the World Bank (Table 7.1) and loosely comparable with earnings from the RSE (Underhill-Sem & Marsters, 2017). For the 2149 ni-Vanuatu SWP workers in that year, this would have meant a total of more than A$18 million. Comparable data for other countries were unavailable, but given their much lower participation rates, total remittances were probably proportionately lower.

While some workers kept an eye on the exchange rate and tried to time their remittances with favourable rates, this was usually impossible. The ni-Vanuatu workers Kirstie met on the Adelaide Plains reported that they remitted whenever they were asked for money by family back home, although how often these requests came was unclear. Mobile telephony had certainly increased their frequency. Yet many ni-Vanuatu RSE workers tended to remit only for specific purposes such as school fees or a family event (Bedford et al., 2020). In South Australia, SWP workers spoke of their frustration at sending remittances when the exchange rate was bad, so that their Australian dollars did not go as far as they had hoped. Lucy, an experienced seasonal worker who had been employed in New Zealand and Australia, explained:

> When the currency rate is bad, the money we send isn't enough. Like sometimes, because the rate isn't the same all the time. Sometimes the rate is 76. I think the highest while we've been here has been 76, 75, 74, 73. So when you send A$100 and the rate is 73, your family only get 7300 vatu.

Frustrations about the exchange rate were a common theme on ni-Vanuatu Facebook pages, as these 2019 posts illustrate:

> Good evening members of SWP, I've got some concerns that could help all of our workers overseas [...] When you come to work in New Zealand there are already big deductions that they take out of your pay for tax, rent, transport to work, medical insurance, work clothes and food. My first question is about the exchange rate. Who makes the rate go up and down? Why can't it be stable for a whole year? All of the banks, Western Union, Moneygram, and the others, they don't work for the money we make, but when they change our money from dollars to vatu, it's as if it's the end of the world. I know they're all just doing their business, but please don't rip off another man's sweat. We have the National Bank of Vanuatu. Why don't they set up an agent from the bank to every place where there are RSE workers, and the agent can facilitate sending money and banking savings? I hope my post brings some good ideas that can help everyone in the future.
>
>> Yeah... Absolutely you make a good point... Our government should be serious about it... It's about respect. NZ and Aussie workers brought more money to our country than anyone else... Please look after our citizens... The system must change... I really support what you're saying.
>>
>> True. Especially when you come home and you think you'll wait until the rate is good to change your money, but then someone in your family dies when you're not expecting it, and you don't have a choice, you need money, so you must change it right away. It's a real waste of money.

Once workers returned home, it was not as simple as waiting for the right time to exchange money. Many of the ni-Vanuatu workers employed on the Adelaide Plains came from rural areas, and had limited or no access to banking facilities in their home villages, so needed to withdraw cash in Port Vila before returning home: they had to accept whatever exchange rate was available. The alternative was to wait in Port Vila, spending money on food and accommodation and whittling away their savings while waiting for the exchange rate to improve—an option that was unpredictable and unacceptable. Lucy described this predicament:

> Most of us live out on the islands, so when we get to Port Vila, we have to change our money before we go back to the island. If the rate isn't good and you want to wait, but if you stay in Port Vila, everything costs money. So, no matter if you stay there, afterwards you have to exchange your money and find a way to go back to the island. Because back on the island, life isn't really expensive. You can eat everything from the garden. But if you stay in town, for everything, you must pay pay pay.

Toby echoed these sentiments and explained that sometimes waiting for a better exchange rate was not an option:

> Sometimes when we've been working here and we go back and want to change our money, the rate is really low [...] You can wait [for a better rate], but if you need to eat or something, you just have to change it.

Whatever option they chose, workers felt they were losing some of their hard-earned wages and were at the mercy of the banks and money transfer operators.

Across the PICs, the frequency of SWP workers' remittances varied. Very few ni-Vanuatu workers remitted regularly. Workers from Samoa and Tonga tended to remit more than once a month while workers from Vanuatu often only remitted once during their entire contract (World Bank, 2017). This too related to the lack of banking facilities in outer islands, and the difficulties families could face in accessing remittances. According to Rochelle Bailey (pers comm. October 2020), while guest-workers did not always remit during their absence, return workers generally left their wives or family with a lump sum of money earned during previous seasons for their support while the worker was away. Most SWP workers remitted to household members, sending an average A$2029 to their own household and A$141 to non-household members (Table 7.1). Households generally used this money to meet immediate needs including food and clothing, but it could also be used to purchase items such as communication devices and electrical cooking equipment (World Bank, 2018a; see Chap. 9). Differences in remittance practices reflected workers' age and gender; in New Zealand, older and married RSE workers tended to remit more frequently and in larger amounts (Bedford & Bedford, 2017). It is likely that the same was true of workers in Australia. Female SWP workers from Vanuatu and Tonga tended to remit smaller sums than men, mainly because they earned less (World Bank, 2018a). However much remittances and savings varied by gender or age, by familiarity with employment or duration of stay, they underpinned, provided and justified the rationale for participation in the SWP.

Accommodation

Many employers found providing suitable accommodation for guest-workers to be challenging. Short-term accommodation in rural and remote Australia was often limited or inappropriate for housing seasonal workers. According to Australian government guidelines, employers needed to provide accommodation that was suitable, reasonably priced and safe. In addition, men and women needed to have access to separate bathrooms. The interpretation of what represented 'suitable' accommodation however varied wildly. LMAP's (2017) survey of a sample of a hundred returned workers found that more than 80 per cent rated their accommodation as good or quite good. Yet, on average, workers shared their bedrooms with 4 people, bathrooms with 8 people and kitchens with 26 people.

Often, employers used existing accommodation options such as caravan parks, but relying upon pre-existing infrastructure could be problematic when living conditions were cramped and costs were high. In 2012 Kirstie visited a group of ni-Vanuatu seasonal workers housed in a caravan park in rural Victoria. Although the park's location allowed them to walk to and from work, it was 5 km from the nearest town centre and transport to supermarkets, banks and other services was scarce. The units where the workers lived were small, cramped and expensive; four adults were accommodated in a space normally used for two adults and two children. Some workers even had to share beds.

> Richard was in his second year of SWP participation. He lived on the island of Tanna, Vanuatu with his wife and three children. According to Richard, the bed sharing arrangements in Victoria were totally inappropriate: At home, I wouldn't sleep in the same bed as one of my brothers like this. I would sleep with my wife, yes, but to share a bed with a brother, no. And to do it for six months, oh no, it's not good'. (Quoted in Petrou & Connell, 2018, p. 72)

The cabin amenities were basic at best. Cooking facilities were limited to an electric stovetop, a microwave and a kettle. Each unit contained a single bar fridge, although workers also had access to an extra fridge in the park's function room. Living conditions were crowded and barely catered to the occupants: the workers felt they could not relax in such conditions, something they sorely needed after long days of physical labour. From the workers' perspective, and while they appreciated their proximity to the workplace, the accommodation was far from 'suitable'.

Almost a decade later, Victoria Stead (2021a, p. 6) described ni-Vanuatu being housed in virtually identical accommodation in rural Victoria:

> The group lived at a caravan park in a tiny town 20-odd kilometres from the centre of Shepparton that boasts a pub, a post office, a takeaway food kiosk, a swimming pool, and a football field. The eighteen women shared four cabins, with the two men of the group in a fifth. Inside, they slept on bunk beds in cramped spaces, with sheets hung up to give at least some privacy, and a tiny kitchen area in each cabin where they prepared food in small convection ovens they purchased in town. The cabins were barely insulated, stiflingly hot in summer and bitterly cold in winter.

Overcrowding and inadequate facilities were common issues, particularly when employers needed to provide accommodation for large groups of workers. The ni-Vanuatu workers we met over the years often identified overcrowding as the key problem of their accommodation. Over the 2019–2020 summer season, Kirstie met some members of a group of several hundred ni-Vanuatu workers employed in peri-urban

Adelaide. The size of the group meant they needed to be accommodated across three different caravan parks and a renovated ex-hospital. At one caravan park, eight or nine men shared a dorm room. They described the accommodation as being too crowded, and felt they had little opportunity to relax, sleep properly or even store the goods they planned to take home to Vanuatu. Tom shared a room with eight other men and grimaced:

> It's not good. Because when you work and you're tired, you come back and you want to have a good rest. But because there are lots of us, there's always people talking or making noise, and you can't have a good rest, you get woken up.

Not only were the rooms crowded, but cooking facilities were limited; roughly thirty-five workers shared a single stove with four gas burners, although they also had some rice cookers they could use, which was handy provided they wanted rice! As a result of the limited cooking facilities, some workers found themselves dining late at night as they waited their turn. Donald was in his first year of SWP participation and complained:

> So when we come back from work in the afternoon, it's very hard for us to cook early. Sometimes it reaches 9 o'clock, 10 o'clock and the boys are just finishing cooking.

Those who had previous experience could compare their accommodation to other locations. Tom had worked in the SWP once before, and although he was employed on the same farm, he had been accommodated at a different caravan park:

> The year before last year I slept at [a different] caravan park and when I stayed there I thought it was good, there were only four people in one room, four people in another room, and there was lots of space. If you bought a suitcase, you could bring it back to the room, you could bring your bags and put the things that you'd paid for in your room and they were safe. But here, no. Here, you bring your suitcase and put it on the side and it squishes you on the side. And some people are just using them as pillows.

Housing inadequacies, over years and between sites, could be simply the luck of the draw and depend on the number of workers to be accommodated, but accommodation was invariably overcrowded. That was not unique to the SWP; similar concerns had long been raised by RSE workers, about tightly packed accommodation, and the difficulties in getting a good night's sleep. George confided:

> There are now seven men in my bedroom. It's too much, all snoring and farting and we can't open the window because it's too cold. I bought myself a mattress, I'm moving to the lounge. (Quoted in Bailey, 2014, p. 162)

In this instance, accommodation issues persisted for three years, until workers were finally moved to new lodgings.

Beyond caravan parks, existing infrastructure was sometimes modified—by employers or investors—to provide accommodation but not always adequately. In Tasmania, in 2020, 20 Tongan workers were found living in 'squatter like conditions' in a converted primary school, owned by the local mayor (Cooper, 2020). When workers' living conditions came to light, local residents were furious at the workers' longstanding mistreatment. Another Tasmanian mayor, Peter Freshney, explained: 'The flaw, I believe, in the policy, is the approved employer makes the declaration that the properties are compliant, but there are no checks and balances in the process where local government can confirm that is the case' (quoted in Sullivan & Hynninen, 2020). A group of four Tongans at Childers (Queensland) were allocated a single room for which they each paid A$219 per week. A similar group were put in a caravan in Western Australia and had each to pay A$120 a week (Holani, 2017). Some local councils, such as Leeton Shire in NSW, established their own policies and recommendations around seasonal workers to prevent such issues occurring. These efforts were patchy, remained reliant on the interests, initiative and assertiveness of local councils and required ongoing monitoring.

Overcrowding and inadequate accommodation have plagued the scheme. The Tonga Australia Seasonal Workers Association (TASWA) reported in 2016 that at one site in NSW, thirteen workers shared a house with one toilet and bathroom: 'In mornings when they get up and prepare to go to work, and desperate to use the toilet, inconveniently, some had to use a bucket of water as toilets and then poured it outside' (TASWA, 2016, p. 15). TASWA reported overcrowded conditions, with workers packed in 'like sardines', in more than 13 places in NSW alone, and conditions were attributed to employers either recruiting too many workers without access to adequate accommodation or trying to cut costs, or both. Most Tongan workers were staying in properties owned by contractors or their close family members and TASWA calculated that at one motel where five workers were accommodated in a single room, they were paying a total A$1050 per week. TASWA reasonably concluded that labour hire contractors were making most of their income through accommodation.

Overcharging for accommodation proved a common issue and was still flagrant in 2020. In Tasmania, a group of almost 70 Tongan workers were found to be each paying more than A$100 a week for 'slum like' housing. Media reports indicated the accommodation provider was receiving around A$7000 a week for a property that, on the domestic rental market, would make around A$600 a week and accommodate a single nuclear family. Daniel Walton, the president of the Australian Worker's Union, commented:

> What kind of employer allows their workers to live in slum-like conditions? […] These are people who were too afraid to speak up because they were scared of losing their jobs. Employers like Costa are taking advantage of vulnerable people for profit and they deserve to face the full brunt of the law in this matter. Unfortunately, this is not a one off. (AWU, 7 February 2020)

One key informant, an Australian Pacific islander resident in rural Victoria, observed to Kirstie that issues around accommodation usually occurred when there were too many middlemen trying to make some easy money out of the scheme from workers unable to complain.

Conversely, if infrequently, not all employers were enamoured with the attitude of the workers. As one employer said of a group of workers from Tanna:

> I don't want them back ... They left the place in a pig-sty; the way they left the place was just obscene, the way they lived, and the level of cleanliness or lack thereof was just vile... It's going to be a number of years before they are going to cope culturally in our environment. I think it's too different and at the moment we are asking them to make too much of a shift. Even for them to go from Tanna to Vila is a significant shift, so to come from Tanna to here is too much. They are too frustrating to work with. (Quoted in Williams, 2010, pp. 89, 83)

But such blatant and mean-spirited racism was probably rare.

Not all SWP workers experienced issues with accommodation, and only the worst experiences tended to make the news. Good practices could also be newsworthy. In Koo Wee Rup (Victoria), Joe Vizzari, the manager of Vizzari Farms which produced asparagus and broccoli, was so impressed with his ni-Vanuatu employees' work ethic that he invested in 'quality accommodation' for 150 workers. The accommodation and amenities included a kitchen that supplied workers with two meals a day, a cleaning service and a bus service. Joe observed: 'It is pretty involved, it's almost like another business, but it works for me. I've always got people, on standby, ready to go to work whenever it's required according to the weather' (ABC Landline 2 May 2015). Providing suitable accommodation was possible and could benefit both workers and employers. Few employers made that commitment.

Many guestworkers were housed in remote locations, where choices were few and monitoring and enforcement of regulations difficult. Nonetheless, overcrowded and inadequate housing at extortionate cost was only partly a function of isolation and limited accommodation options. Poor accommodation was also the outcome of crude and niggardly assumptions that Pacific islander 'guestworkers' were temporary and did not require or appreciate more than the most basic accommodation and facilities—even as some repeat workers spent more time in Australia than at home. Problems with accommodation were not unique to the SWP program and have faced various migrant horticultural workers in Australia (Barry, 2020). However, SWP workers were less able to complain and were also generally not in a position to seek new accommodation. As Mary explained to us in 2012:

> I'd like it if there were just two of us in one room, me and a friend. But when there are three or four of us, no, it's not good. But we just have to stay, because where else would we live?

In recognition of these issues, and after several particularly damning media stories, the Department of Foreign Affairs and Trade (DFAT) commissioned an audit of accommodation sites in 2020. The audit found overcrowding was an issue at some sites (although most sites were not visited in person). Recommendations included the development of an accommodation cost calculator; price benchmarking against available rental accommodation in the area; and ensuring AEs remedy any overcrowding issues that arise (BellchambersBarrett, 2020). But such obvious conclusions depended on accommodation being available and practices being monitored. Accommodation issues within Pacific guestworkers schemes were longstanding, and were linked to accommodation shortages, a longstanding rental crisis in much of regional Australia (especially the Riverina), geography, poor oversight and limited options for workers to move to new lodgings. As we discuss below, poor accommodation was not merely unpleasant, and demeaning to the status of the workers, but could have real implications for workers' health, while the lack of adequate accommodation limited flexibility of employment.

Getting Around?

Alongside accommodation, employers needed to provide transport for SWP workers so that they could easily get to workplaces. Some employers provided free transport, some asked a nominal fee, while others charged much more than was 'reasonable'. One group of SWP workers in Queensland paid A$75.60 a week in transportation costs, an extortionate amount for a 15–20-minute commute to work, and a sum that eventually made a substantial dent in their earnings (Bailey, 2019). The ni-Vanuatu workers Kirstie met in rural South Australia were similarly charged A$59 per week each for a short (roughly 20-minute) bus trip between their workplace and accommodation. Overcharging seemed most common where employers hired vehicles through a third-party company (Bailey,

2019). Again, this was not unique to the SWP; 'inflated deductions' for transport, accommodation and equipment were relatively common in the Australian horticultural industry (Howe et al., 2019). Nonetheless, SWP workers often had little idea of what could be considered a 'reasonable' cost for transport, despite the costs being vastly in excess of those in their own countries.

Beyond travelling to work, seasonal workers needed access to goods and services in nearby towns. Employers sometimes provided little assistance and expected workers to find their own solutions. The ni-Vanuatu workers we met in rural Victoria were unsure whether they were allowed to walk the 5 km from their accommodation to the town centre, and so instead relied upon expensive taxis (A$40 return) or lifts from friends so that they could perform essential tasks such as their weekly grocery shopping, sending remittances and other errands (Petrou & Connell, 2018). In South Australia, one group of ni-Vanuatu workers initially used cars provided by their employer, but after several accidents, these were replaced by a mini-bus with a designated ni-Vanuatu driver. This allowed the workers to travel to the local township of Gawler, where supermarkets and other services were located, with relative ease. At another caravan park in South Australia, ni-Vanuatu workers could easily walk to a nearby train station and travel to Adelaide, giving them much greater flexibility. Access to towns was variable, and relied upon workplace locations, existing infrastructure and the whims of employers.

In some extreme instances, employers actively discouraged SWP workers from travelling even to nearby town centres. In Robinvale (Victoria), contractors and employers used invented 'rules' to 'discourage' and 'stop' Tongan SWP workers from visiting Mildura, the regional centre, at night. One employer provided transport to Mildura so that workers could shop, but did not allow them to travel to Mildura for other reasons, which he believed met his pastoral care responsibilities. While employers had a duty of care, travel was managed to ensure that workers could not abscond, despite being 'contrary to basic human rights principles to enable freedom of movement *within* national borders' (Tazreiter et al., 2016, p. 116). Other employers were more relaxed, and allowed workers to travel to Melbourne to visit friends and family when they had time off—although they kept a record of these movements and required contact phone numbers. One employer explained

> We're not there to prevent them from going anywhere, to stop them, as long as we know where they are in case something does happen or if we need to get in touch with them, and I mean, they could be away somewhere and something could happen back ... we just like to know where they are and what they're doing and make sure that their host employer is aware that they're going to be missing and it's not going to interfere with their work and they've just gone away to visit friends, and when do you expect to be back? ... It's not a control freak thing, it's just a matter of knowing where they are in case we need to get in touch with them [...] It's not a matter of trust, I mean, it's just a matter of us knowing where they are at all times so if something happens. ... They're here under your care. (Quoted in Tazreiter et al., 2016, pp. 114–115)

Most employers were more relaxed about travel, recognising the need for shopping and some components of a social life. Minimally, carefully monitoring workers in and out of working hours was paternalistic, and was widely resented.

While there were certainly nuances specific to the SWP—such as the perceived 'need' for constant surveillance and restricted movement to ensure employees were cared for—transport issues were not unique to the programme and represented a wider issue within the Australian horticultural sector. Experiences of localised immobility were also a feature of seasonal worker programmes. In the Canadian SAWP, parallel employer-enforced, state-sanctioned immobility allowed employers to 'fix' workers in place for their own benefit; if workers were always on the farm, they were always available to work the long, highly 'flexible' hours required in the horticultural industry. Just as for the Tongan workers, discourses of care were used to justify curtailing workers' localised mobility and disallowing access to bicycles and other forms of transport required to access local services (Reid-Musson, 2017). The power imbalances evident in worker-employer relationships that facilitate mobility restrictions are explored in greater detail in Chap. 8.

Health Costs

An ongoing issue in meeting horticultural labour needs was the nature of the work itself. Seasonal horticultural labour conditions could be harsh and gruelling, even occasionally dangerous, with health risks associated with long hours of physical labour. The 2006 parliamentary enquiry noted

> seasonal workers are at the bottom of the employment food-chain in terms of wages, conditions and skills. This contributes to the unreliable nature of seasonal work, and the unreliability of seasonal labour supply. Growers stressed that fruit-picking is hard, hot and dirty work requiring long hours exposed to the elements. There is no escaping the fact that harvesting crops—picking fruit, pruning trees, working in a packing shed—is physically demanding work. During the summer months when many crops are ripe for harvest, workers routinely spend eight to ten hours a day, six days a week, in the field. Much of this time is spent up a ladder, bending over or on hands and knees picking fruit and thinning bushes, one row at a time. The occupational hazards of picking fruit are ever-present, especially in the mango industry. The committee was told in detail how sap from a mango cutting, which squirts in all directions, can inflict serious burns when it makes contact with exposed skin. (Australia, 2006, p. 14)

In addition to the physical grind, unfamiliar foods, new working conditions and alien climates could take a toll on workers' welfare. In New Zealand, the government regularly collected data on RSE workers' health and, according to its 2019 survey, approximately 10 per cent of workers arrived in New Zealand in poor health with issues such as dental problems (53 per cent), boils (50 per cent) and skin diseases (26 per cent) (Research New Zealand 2019). No comparable data were available for the SWP, but much the same was probably true, thus some workers were already at risk of minor health issues upon arrival in Australia, although none were likely to pose an immediate problem.

Most information on workers' health in Australia remains anecdotal and it was only when tragedies occurred that problems were widely reported. In Australia, at its worst, there were 15 SWP worker deaths between 2009 and 2019. The causes varied from car crashes to pre-existing but undetected health problems such as heart conditions. Workplace incidents were not the cause. However, unsuitable accommodation, long hours with inadequate rest and difficult working conditions may have sometimes been contributing factors, emanating from socio-economic disadvantage. In 2017, a Tongan SWP worker allegedly died from a combination of 'overwork, lack of food and medical care'. Falepaini Maile, president of TASWA, collected the worker's belongings and observed: 'The saddest thing was when the police and I and others went to where he lived [. .] He had a plastic container for all his belongings and we found a couple of Maggi noodles' (quoted in Locke, 2017). After the death of another Tongan worker in late 2019, Alison Rahill, the executive officer of the Archdiocese of Sydney's anti-slavery taskforce, argued that investigation into SWP deaths considered only the physical condition of the dead man.

> There doesn't seem to be any investigation into the hours that have been worked, whether there's enough fresh water to drink during the day, the temperatures, the living conditions, is it overcrowded at night, are people able to sleep? (ABC Mildura, Swan Hill Facebook page, 30 December 2019)

The Vanuatu High Commissioner, Samson Fare, raised similar concerns about fatigue management in 2020

> It's not just the long working hours ... When you look at their accommodation, where they live, you would have 250 of them living in one place. And because there are too many of them, there are not enough rooms in the kitchen for cooking. Sometimes they have to stay up until 10 or 11 pm at night to cook. (ABC News Radio, 20 January 2021)

Bad nutrition also played a role in poor health outcomes. For some seasonal workers lack of information about Australian foods was a problem. In all-male contexts unfamiliarity with diet planning and cooking was not unusual. More frequently nutritional problems were the outcome of attempts to save money by cutting down on food quantity and quality (Blanco, 2009; Bailey, 2014), irregular access to distant shops and fresh food, bulk buying of processed foods (such as rice and sugar) and the lack of familiar foods. Referring to the situation of Samoan workers, the Samoa First Union (Union Aid Abroad APHEDA, 2020, p. 12) observed:

> For many of [our workers] they have never needed to manage money before and so have no skills. Often this results in the boys sending back all of the money and then not having enough money to buy healthy food. Sometimes they get skin infections like boils and are told by the manager that it is because they only eat noodles and need better food. I think this is true. They also can't afford the medicine to heal the infections. These workers need to know how to manage their money and how to buy food that is nutritious and not just filling and easy. These are life skills that mostly these young men do not have. They also cannot easily refuse their family if they are told to send back all of the money.

Similar concerns have been raised amongst ni-Vanuatu social media users. A Facebook post on a popular seasonal workers page advised in 2018

> It's not all about earning dollars, but it's the lack of knowledge as well to be considered. More than 80 percent of workers are still consuming unhealthy food daily. Breakfast—bread/butter/tea, Lunch—bread/butter/juice and dinner—big white amount [sic], rice/chicken. We should have breakfast like a King, lunch like a Prince and dinner like a College Kid. The question is "The lifespan of the SWP/RSE workers for the next 30 years".

Like most aspects of long-term engagement in seasonal worker programmes, the question went unanswered.

SWP workers paid compulsory health insurance fees, so that should they fall ill or suffer an injury while in Australia, they were—in theory—covered. However, various barriers existed to accessing health care. Some farms were remote from health facilities, and workers were often reluctant to take time off to visit doctors. Another Facebook post warned in 2019

> Health: there are no clinics on the farm or doctors, so when you're working to meet your goals, think about your health as well. Make sure you don't lose working days because you're sick and have to go and visit the hospital for treatment. You must eat well and rest well.

While some SWP workers clearly understood the potential problems around health and nutrition, acting on this guidance proved difficult when peak season labour requirements meant long hours and few days off to rest and recover.

Unfamiliar climatic conditions could also take a toll. Workers in Queensland encountered similar environments to those at home, but in other states workers could face extreme heat, frost and even bushfires (Fig. 7.7). Speaking about Tongan SWP workers employed in Robinvale in 2013, one informant observed:

> I mean it gets really hot here, like, not every day is 45 but you can have two weeks of 40 plus and they are still out there and that's why they like them, because they will keep going in that heat. They can [get medical help] but you've got to go through—jump through hoops to get medical help for them. They haven't got a Medicare card. … [The] hospital treat them. They are pretty good. … But in terms of going to the doctor and that sort of thing, if you can't pay, that's sort of less extreme where you just need health support, you don't [get it]. (Quoted in Tazreiter et al., 2016, p. 118)

Fig. 7.7 SWP work involved long hours in the sun. (Source: DFAT via Flickr)

Challenging climates, coupled with complicated requirements to access medical care, could result in poor health outcomes.

Health problems could be aggravated where workers were far from medical facilities, and when health issues did occur, workers were often worried about the expense of attending medical appointments—a fear that was not unfounded—and time lost from work. Even insurance was not cheap. In 2015, Tongan workers employed in Queensland paid between A$22.70 and A$24.80 a week to cover medical insurance, but many had no health insurance cards. One group of workers were given 'health insurance letters' from their labour contractor, which were, not surprisingly, declined by medical providers. Despite weekly deductions from their payslips, the workers, without genuine health insurance, were forced to pay large out-of-pocket expenses for medical attention (Holani, 2017). TASWA similarly found that many Tongan seasonal workers were provided with no information about health insurance, despite being charged weekly deductions.

> A group working at [Farm X] in Mundubbera recruited by [XXX], had rashes and itchy skin because chemical spraying took place while TSWs [Tongan Seasonal Workers] were at work picking fruits. No Health Insurance benefit during this skin burns and rashes period. When TSWs addressed it, they were told that the chemicals won't affect them and they can even drink it and won't be poisoned. (TASWA, 2016, p. 20)

The ni-Vanuatu workers who Kirstie met in rural Victoria in 2012 told similar stories of expensive health care and not wanting to miss work to attend medical appointments. Such sentiments were shared by Tongan seasonal workers in Robinvale, as recounted by a local Pacific islander migrant:

> I know some of the seasonal workers, they came here, they sick … so if anyone is sick at work, I can come and pick them up and bring them to the doctors, but they—a lot of time they don't—they just push me away because they don't want to encourage them to have a sick day. (Quoted in Tazreiter et al., 2016, p. 118)

In peri-urban Adelaide, ni-Vanuatu workers were equally frustrated and confused by the Australian health system, believing their insurance worked only for hospital visits and 'serious' health problems, and not for regular GP visits. Cultural issues were also at play with workers confused as to why they needed to pay for medicines when there were often free in Vanuatu. Financial barriers, coupled with concern about taking time off work, sometimes led workers to put off seeking medical help—meaning minor conditions could sometimes develop into something more serious, as was the case with Mark, a ni-Vanuatu worker:

> During the 2011/2012 season, Mark did not seek treatment for his boils... he said he was afraid of having to pay high hospital fees and due to 'shame' in taking time from off [sic] work. Although Mark did not directly use the term shame with me, other ni-Vanuatu who shared a room with him did, explaining that by not attending work and earning money, Mark felt shame because he was disappointing his employer, his fellow workers, his family and community. Additionally, he imagined that his wages would be used to cover his illness ... He said other workers told him not to tell his pastoral care providers and that he had to work in the painful condition that he was in. This led him to suffer from more serious complications due to delays in treatment. I was informed of other cases where this occurred for ni-Vanuatu. The main concern here was the medical costs and the time off work. Mark did not factor in that he was covered by his medical insurance or his entitlement to sick leave. (Bailey, 2014, pp. 160–161)

A number of factors—many relating to misunderstandings about financial needs and obligations—could provide barriers to seeking health care when needed.

Poor or inadequate health checks in home countries sometimes prevented SWP workers from accessing health insurance in Australia when they were deemed to have pre-existing conditions (some unknown and undiagnosed at home, where rural health facilities can be rudimentary). A Tongan worker diagnosed with tongue cancer while in Australia was initially sent back to his accommodation because his health insurance did not cover treatment for a prior condition, but was eventually allowed to remain in hospital until his death (Tazreiter et al., 2016). Such extreme circumstances and medical conditions were fortunately rare.

Australian citizens who provided pastoral care or support to SWP workers related similar medical system woes. In mid-2020, Kirstie spoke with Terry, a Tongan-Australian with relatives working in rural Victoria through the SWP. Terry expressed his frustration at how workers needed to pay out-of-pocket expenses to access health care and follow a confusing process of filling out forms to get their money back from the insurance company.

> Because they come work like an Australian, why they can't give them [Medicare]? That's the main thing to give them, healthcare and Medicare [...] They deduct A$20 every week and then it's a long process to go and pay on their own and bring a receipt and give to MADEC and MADEC [gives it] to the insurance company to bring back the money. Because they still paying A$20 every week from their pay [...] Why can't they give them healthcare? Because they take the 15 percent off the tax. For me I work here, [Australian citizen] seasonal worker is only 13 percent. They take taxes higher than for Australians.

Elsa, a Samoan woman living in New Zealand who Kirstie spoke with in 2020 had a similar perspective:

> Yeah well sorry, but you guys suck at looking after your workers [...] what you just told me [about workers not attending medical appointments because of the cost] is true. And the [insurance] form. And some of them don't speak really good English and they just screw up one bit of the form, and they are sitting waiting. And some choose to not even go through because they don't understand and they don't have the pastoral [support at the appointment ...]. I'm sure the Australian program is fantastic but just the support for people in those circumstances [is lacking].

Such health and insurance issues are not unique to the SWP and the RSE and are common to guestworker schemes more generally. In North America, Mexican migrant agricultural workers in the United States and SAWP workers in Canada have also experienced health problems after (sometimes long-term) exposure to agricultural chemicals, and suffered injuries after using machinery without adequate training (Holmes, 2013; ILO, 2012). Such injuries sometimes led to workers losing their jobs and the health insurance that came with it, and some then being left in limbo; they could not return home because the health facilities there were inadequate for their needs, but they no longer had health insurance or employment in Canada so could not access adequate health services there either (McLaughlin & Hennebry, 2013). Guestworker concerns and experiences in Australia have been much like those elsewhere, and might well have been predicted. In their submission to the Australian government, Union Aid Abroad APHEDA noted their

> serious concerns about the lack of sick leave and health insurance for workers that are resulting in workers losing pay for work related injuries as well as having to pay for medical costs. This would appear to be at odds with the spirit of attending to the issue of pastoral care that is often spoken of by Pacific leaders in relation to Australia's duty in relation to Pacific workers. (Union Aid Abroad APHEDA, 2020, p. 4)

Despite this, new climatic and working conditions offered some degree of novelty; workers often posted pictures of themselves on social media—usually to illustrate their working conditions (notably new machinery and crops), their accommodation and their fellow workers. Prominent in many such pictures were the layers of clothing, gloves and hats, unfamiliar in the PICs but essential in Australia. Occasionally, real extremes generated distinctive photo opportunities for social media (Fig. 7.8). In New Zealand, where the climate is colder than most parts of Australia, ni-Vanuatu workers suffered from such health complaints as aching bodies, muscles seizing up, nosebleeds and coughing blood. Moreover, some felt their employer treated them as if they were 'just muscles', had little concern for their welfare and dismissed their health concerns as trivial (Smith, 2016, 2019). Yet not all concerns have been trivial; SWP workers have suffered from 'bleeding episodes' as a result of working with chemicals, a practice rare in the PICs (Brookes, 2018).

At best, workers were confused about health care and needed to go through a complicated process that made little sense to them in order to obtain refunds for accessing health care and similar services. Unfamiliar systems, and trying to save money—and hence avoid sick days—compounded these issues and could lead to more serious health problems

Fig. 7.8 A ni-Vanuatu seasonal worker experiences a new climate. (Source: Facebook)

when workers deferred treatment. More accessible, streamlined systems, improved pastoral care around medical issues and useful data on SWP workers' health issues would all have helped. Healthy workers are content and productive workers, and health care is an important part of pastoral care.

Work and Welfare

For a decade, Pacific seasonal workers have been an important part of the Australian horticultural labour force, meeting labour shortages that are both geographic and temporal so differing over time and place: picking mangoes in the tropical Northern Territory is a world away from indoor mushroom picking and packing in peri-urban Sydney. Employers were usually keen to employ SWP (and other migrant) workers, believing them to be more serious and harder workers than the domestic equivalent—as long as participating in the SWP was not bogged down in bureaucracy. The type of labour that SWP workers provided (compliant and always available) was particularly desirable, since it was flexible according to growers' needs, if not necessarily so for workers' needs. It is easy enough to describe the SWP scheme in simple terms. It existed to fill a labour gap that could not easily be filled within Australia, so migrant workers were very well paid by PIC standards; they worked hard for a short time and returned to their home islands with a substantial sum of money (Chap. 9). Since workers were temporary, unsure of their rights and the costs of various things in Australia (from health insurance to accommodation and transport fares), they could be, and sometimes were, easily exploited by unscrupulous recruiters and employers. Dealing with different climates, unusual food and accommodation, demanding work regimes and long hours, in a multicultural society, were quite challenging enough.

While the most egregious examples of exploitation provided media fodder, inadequate pastoral care was not uncommon, if rarely at critical levels, despite various attempts at regulation. As guestworkers moved into new places, and new cohorts arrived, the same old problems recurred. As one ni-Vanuatu man said: 'More than anything else, what the company wants is to find men that will shut up and work. They prefer people from the islands who don't know English or even Bislama, because they can't complain' (quoted in Cummings, 2013, p. 394). Yet many employers did want workers with English skills who might even take on responsibilities. Here, as in other spheres of life in Australia, there was a wide spectrum of possibilities, and quantification was impossible.

Horticultural labour was physically demanding, but decent working and living conditions could be elusive for at least some, particularly in a poorly regulated industry. As we expand on in the next chapter, cultural differences ensured that many problems and complaints were ignored, and workers often felt unable to complain, sometimes lacking the knowledge and resources to do so, while also fearful of losing their job. Not only have problems of accommodation, wages and conditions of employment been continuous, they share very obvious similarities with circumstances in parallel schemes, notably the neighbouring RSE. In an age of laissez-faire governance, where regulation is weak this seems unlikely to change. In regional Australia such problems are never entirely out of sight—even where the bulk of employers and contractors provided fair treatment and an adequate context—and partly accounted for Australians being uninterested in taking up such jobs. In the next chapter we build upon these themes to examine the social costs of participation in guestworker schemes, but also the manner in which social contexts and contact could mitigate and counter problems, and how workers addressed them.

Bibliography

Australia. (2006). *Perspectives on the future of the harvest labour force.* Senate Standing Committee on Employment, Workplace Relations and Education.

Australian Government. (2020). *Growing Australia.* House of Representatives Standing Committee on Agriculture and Water Resources.

Bailey, R. (2014). *Working the Vines: Seasonal migration, money, and development in New Zealand and Ambrym, Vanuatu.* PhD thesis, University of Otago, Dunedin.

Bailey, R. (2019). *Limiting possible exploitation in transportation services for seasonal workers.* ANU InBrief 2019/12.

Barry, K. (2020). Momentarily immobile: Backpacking, farm work, and hostels in Bundaberg, Australia. *Geographical Research, 59*(1), 46–55.

Bedford, C. (2020, November 24). *New Zealand's seasonal labour shortage: Why local workers aren't enough.* ANU DevPolicyBlog.

Bedford, C., Bedford, R. & Nunns, H. (2020). *RSE Impact Study: Pacific stream report.* Bedford Consulting.

Bedford, R., & Bedford, C. (2017). *RSE earnings and remittance surveys: Samoans, Tongans and Ni-Vanuatu employed in Hawke's Bay, the Bay of Plenty and Marlborough*. Ministry of Business, Innovation and Employment.

BellchambersBarrett. (2020, December 9). *Pacific labour mobility accommodation review: Key findings and recommendations*. BellchambersBarrett.

Blanco, C. (2009). *Temporary migration, remittances and development: Early impressions of the Australian Pacific seasonal worker pilot scheme*. Honours thesis, University of Sydney, School of Geosciences.

Brickenstein, C. (2014). *Foreign Seasonal Workers in Agriculture; a comparison of regulations and practices in the Pacific and in Europe*. PhD thesis, Macquarie University, Sydney.

Brookes, S. (2018, May 23). Agri Labour Australia suspended from government scheme amid underpayment, unsafe work claims. *ABC News*. https://www.abc.net.au/news/2018-05-23/government-suspends-labour-hire-firm-amid-underpayment-claims/9788742.

Campbell, I. (2019). Harvest labour markets in Australia: Alleged labour shortages and employer demand for temporary migrant workers. *Journal of Australian Political Economy, 84*, 46–88.

Campbell, I., Tranfaglia, M., Tham, J., & Boese, M. (2019). Precarious work and the reluctance to complain: Italian temporary migrant workers in Australia. *Labour and Industry, 29*(1), 98–117.

Cavanough, E., & Wherrett, C. (2020). *Blue Harvest. Wage theft & other labour infringements in the NSW Mid-North coasts's 2019/20 berry harvest*. The McKell Institute.

Connell, J., & Hammond, J. (2009). *Iso Iso*: The first ni-Vanuatu guest workers in New Zealand. *Pacific Economic Bulletin, 24*(1), 83–95.

Cooper, E. (2020, December 2). 'Squatter-like' property owned by Burnie Mayor Steve Kons used to house Tongan workers. *ABC News*. https://www.abc.net.au/news/2020-02-12/burnie-seasonal-worker-accommodation-concern-from-union/11959286.

Cummings, M. (2013). Imagining transnational futures in Vanuatu. In A. Quayson & G. Daswani (Eds.), *A companion to diaspora and transnationalism* (pp. 381–396). Hoboken, New Jersey.

Curtain, R., & Howes, S. (2018, July 13). *Making the SWP more employer friendly: A work in progress*. ANU DevPolicyBlog.

Curtain, R., & Howes, S. (2020). *Governance of the seasonal worker programme in Australia and sending countries*. ANU Development Policy Centre.

Dao, A., Green, M., & Kass, T. (2020, May 16). We feed you. *The Saturday Paper.* https://www.thesaturdaypaper.com.au/news/politics/2020/05/16/we-feed-you/15895512009836?fbclid=IwAR1X5r411aDQfb397ODGaQey8a3YtJ_hdIBAGnPLm0d4w5GXp9hE1XjDvn4.

Davidson, P. (2020, September 22). 'If JobSeeker was cut, the unemployed would be picking fruit'? Why that's not true. *The Conversation.*https://theconversation.com/if-jobseeker-was-cut-the-unemployed-would-be-picking-fruit-why-thats-not-true-145951

DESE. (2021). Department of Education, Skills and Employment (dese.gov.au).

Dornan, M., Howes, S., & Curtain, R. (2018, September 13). *Is a new visa for agricultural workers needed?* ANU DevPolicyBlog.

Doyle, J., & Howes, S. (2015). *Australia's Seasonal Worker Program: Demand-side constraints and suggested reforms.* World Bank.

Dufty, N., Martin, P., & Zhao, S. (2019). *Demand for farm workers: ABARES farm surveys 2018.* ABARES.

Dun, O., McMichael, C., McNamara, K., & Farbotko, C. (2022). Investing in home: Development outcomes and climate change adaptation for seasonal workers living between Solomon Islands and Australia. *Migration and Development, 11*(3), 852–875.

Gibson, J., & McKenzie, D. (2011). Australia's PSWPS: Development impacts in the first two years. *Asia Pacific Viewpoint, 52*(3), 361–370.

Gleeson, P. (2020b, August 16). Pickers can't be choosers. *Sunday Telegraph*, 106.

Gosling, L. (2020, September 24). Australia needs a new Pacific compact. ANU DevPolicyBlog.

Hay, D., & Howes, S. (2012). *Australia's Pacific seasonal worker Pilot Scheme: Why has take-up been so low?* ANU Development Policy Centre Discussion Paper No. 17.

Holani, A. (2017). *Enhancing the development impact of labour mobility in Pacific Island countries: The role of the seasonal worker program and the PACER plus.* PhD thesis, University of Adelaide.

Holmes, S. (2013). *Fresh fruit, broken bodies: Migrant farmworkers in the United States.* University of California Press.

Howe, J. (2021, November 5). Pay ruling puts an end to slave labour on farms. *Sydney Moring Herald.*

Howe, J., Clibborn, S., Reilly, A., van den Broek, D., & Wright, C. (2019). *Towards a durable future: Tackling labour challenges in the Australian Horticulture Industry*. University of Adelaide.

Howes, S. (2019, December 12). *Time for a permanent Australian step-up in Pacific labour mobility*. ANU DevPolicyBlog.

Howes, S. (2021, July 23). *Foreign workers on Australian farms: Not if but how*. ANU DevPolicyBlog.

Howes, S., & Curtain, R. (2021, June 17). ASEAN ag visa makes no sense. ANU DevPolicyBlog.

Howes, S., & Orton, B. (2020, January 21). For Tonga, Australian labour mobility more important than aid and trade combined. ANU DevPolicyBlog.

International Labour Organisation. (2012). *Pacific seasonal migration to New Zealand and Australia: Literature review*. ILO.

Kaiapam, S. (2021). Impacts of labour migration on migrants and their families: A case study of Vanuatu. MA thesis, University of the South Pacific.

Kotsios, N. (2020, November). Picking up the pieces. *AgJournal*, 6–7.

Kumar, R. (2012). *Migration and development: The development impacts of short-term labour migration of Ni-Vanuatu workers to New Zealand*. MA thesis, University of the South Pacific.

LMAP. (2017). *LMAP Vanuatu reintegration and tracer study report for management*. LMAP.

Locke, S. (2017, November 1). How labour hire villains make millions from Australia's 'modern slavery' market. *Australia: ABC News*. https://www.abc.net.au/news/rural/2017-11-01/exploitation-or-slavery-of-tongans-malaysians-in-horticulture/9102676.

McDonald, H. (2021, May 28). Bitter Harvest. *Inside Story*.

McGlone, T. (2021, April 11). Seasonal work on Australia's farms: 'No one wants to do this sort of work'. *Guardian*.

McLaughlin, J., & Hennebry, J. (2013). Pathways to precarity: Structural vulnerabilities and lived consequences for migrant farmworkers in Canada. In L. Goldring & P. Landolt (Eds.), *Producing and negotiating non-citizenship: Precarious legal status in Canada* (pp. 175–194). University of Toronto Press.

Orton, B. (2020, August 19). *Working holiday makers not a long-term solution for horticulture*. ANU DevPolicyBlog.

Petrou, K., & Connell, J. (2018). 'We don't feel free at all': Temporary ni-Vanuatu workers in the Riverina, Australia. *Rural Society, 27*(1), 66–79.

Reid-Musson, E. (2017). Grown close to home™: Migrant farmworker (Im) mobilities and unfreedom on Canadian family farms. *Annals of the American Association of Geographers, 107*(3), 716–730.

Research New Zealand. (2018). *Recognised Seasonal Employers Survey 2018—Working Report*. Research New Zealand.

Rockell, D. G. (2015). *Pacific Island labour programmes in New Zealand: An aid to Pacific Island development?: A critical lens on the recognised seasonal employer policy*. PhD thesis, Massey University, Palmerston North.

Smith, R. (2016). *The goal of the good house: Seasonal work and seeking a good life in Lamen and Lamen Bay, Epi, Vanuatu.* Unpublished PhD Thesis, University of Manchester, Manchester.

Smith, R. (2019). Be our guest/worker: Reciprocal dependency and expressions of hospitality in ni-Vanuatu overseas labour migration. *Journal of the Royal Anthropological Institute, 25*(1), 1–19.

Stead, V. (2021a). Precarity's reach: Intersections of history, life, and labour in the Australian horticultural industry. *Journal of the Royal Anthropological Institute, 27*(2), 303–320.

Sullivan, K., & Hynninen, E. (2020, February 14). Seasonal worker program faces fresh scrutiny after 70 workers found living in one house. *Australia: ABC News.* https://www.abc.net.au/news/2020-02-14/seasonal-worker-program-under-scrutiny-after-70-people-in-house/11960818.

TASWA. (2016). Issues and concerns by Tongan seasonal workers regarding exploitation, Sydney (mimeo).

Tazreiter, C., Weber, L., Pickering, S., Segrave, M., & McKernan, H. (2016). *Fluid Security in the Asia Pacific* (pp. 103–127). Macmillan.

Thompson, B. (2018, March 23). Australians avoiding farm work despite abundant jobs, award rates. *Australian Financial Review.*https://www.afr.com/companies/agriculture/australians-avoiding-farm-work-despite-abundant-jobs-award-rates-20180323-h0xv9f

Uibu, K. (2020, September 15). There are no human rights here. *ABC News.* https://www.abc.net.au/news/2020-09-15/backpacker-farm-workers-speak-of-wage-exploitation/12545294?nw=0.

Underhill, E., & Rimmer, M. (2016). Layered vulnerability: Temporary migrants in Australian horticulture. *Journal of Industrial Relations, 58*(5), 608–626.

Underhill-Sem, Y., & Marsters, E. (2017). *Labour mobility in the Pacific: A systematic literature review of development impacts.* New Zealand Institute for Pacific Research.

Union Aid Abroad APHEDA. (2020). Inquiry into strengthening Australia's relationships with countries in the Pacific region: Submission 20. Submission to Joint Standing Committee on Foreign Affairs, Defence and Trade—Foreign Affairs and Aid Sub-Committee: Canberra.

van den Broek, D., Wright, C., Howe, J., & Reilly, A. (2021). Pro-market governance, migration status and worker vulnerability: The case of Australian horticulture. *Economic and Industrial Democracy, 42*(4), 1305–1325.

Villafuerte, L. (2018). *Seasons of change: Ni-Vanuatu and the Recognised Seasonal Employers (RSE) scheme*. PhD thesis, University of Auckland.

Williams, A. (2010). *The experiences of the recognised seasonal employer scheme and Ni-Vanuatu temporary workers in Te Puke, New Zealand.* Master of Planning thesis, University of Otago.

World Bank. (2017). *Maximizing the development impacts from temporary migration*. World Bank.

World Bank. (2018a). *The social impacts of seasonal migration: Lessons from Australia's Seasonal Worker Programme for Pacific Islanders*. World Bank.

Zhao, S., Binks, B., Kruger, H., Xia, C., & Stenekes, N. (2018). *What difference does labour choice make to farm productivity and profitability in the Australian horticulture industry? A comparison between seasonal workers and working holiday makers*. ABARES.

8

Social Worlds

Having examined the physical realities of seasonal employment in Australia—earning a livelihood; coping with transport, food and accommodation; dealing with an unfamiliar climate and work regimes—we now turn to the workers' social experiences of the Seasonal Worker Programme (SWP). Despite the focus on employment, and the long hours, many workers enjoyed their time in Australia, encountered no serious problems during their stay, sought to return and had stories to recount back home. Moreover, within Australian communities, Pacific seasonal workers reinvigorated small towns, took part in local sporting events and church services (often singing in choirs), contributed to local economies and made firm friends along the way. At the best of times, the SWP offered an opportunity for social interactions, cultural exchanges and invaluable life experiences. Yet, for some workers, remote workplaces, long and sometimes unpredictable hours, poor food and accommodation, boredom and downtime, along with low or unpredictable wages, resulted in social isolation and few positive experiences. All too often these negative experiences seemed to outweigh the pleasurable moments.

For many SWP workers, speaking out when things went wrong was difficult; culturally, most Pacific islanders, especially Polynesians, are

K. Petrou, J. Connell, *Pacific Islands Guestworkers in Australia*,
https://doi.org/10.1007/978-981-19-5387-3_8

taught not to question authority and seniority. This was not the only deterrent since, for workers to return to Australia, they needed to be invited back by employers, a power imbalance that meant workers were often unwilling to risk being labelled troublemakers. Pressure from home country governments to behave responsibly reinforced this reluctance; after ten Samoan workers were removed from the Recognised Seasonal Employer (RSE) scheme in 2019 for 'bad behaviour' (extramarital affairs, theft and smoking marijuana), the then Samoan Prime Minister, Tuilaepa Sailele Malielegaoi, warned: 'This action serves as a reminder and a deterrent to seasonal workers that are disobedient and show any undesirable behaviour which will compromise more jobs for Samoans, that they will be weeded out without hesitation' (RNZ, 25 June 2019). That it was the Prime Minister himself who issued the warning emphasised the significance of seasonal employment for the Pacific Island Countries (PICs). Nonetheless, despite such warnings to 'toe the line', through their own ingenuity and agency, many workers devised means of maximising benefits from the system, and coping with difficult circumstances, partly through social media, whose usage grew in parallel with the SWP scheme. The tensions between what are sometimes social costs—either in Australia or at home—and economic gains provide the focus of this chapter, and extend into the following one.

A Spectrum of Experiences

The complexities of workers' experiences were considerable. Some lived on isolated farms in small monocultural groups from a single island; others were in multicultural workplaces—perhaps indoors in packing sheds or working together in the fields—and with accommodation on the fringes of cities. Some were young, first-time seasonal workers, uncertainly finding their way; others were seasoned veterans. Some were shy, poorly able to communicate in English; others were confident, garrulous and anxious to experience a different world, whether through urban excursions, football teams or church groups. Tongans particularly have been formidable participants in both rugby teams and church choirs. Other workers merely sought to accumulate income and were constantly in touch with home, while some were simply too exhausted to think beyond their next workday. A small part of that diversity has been well summarised in one account of a large group of workers from different PICs resident in a converted Tasmanian hotel:

> It's something between a workplace, with all its office politics, and a school or college camp with all its drama. There are romances, power plays, deep friendships and bitter rivalries. Several women, including senior group leaders, say they've had to push back against some sexism and harassment from male colleagues. Older women with a few seasons under their belt are over the drama—they've seen affairs form and fail around them, seen fresh young egos blow in on their first season and get caught up in a party scene. The emotions are amplified by deep homesickness. They're not here for fun, they just want to make their money and go home. (Davidson & Earl, 2020)

Reaching conclusions on the formation and consequences of the social worlds that seasonal workers created and passed through in their short stays in Australia is impossible, but certain characteristics were common to the lives of most.

Seasonal workers travelled to Australia to work, and this was often almost all they did. Many were literally and metaphorically distanced from local places and people by geography, language and culture. Limited pastoral care and the nature of horticultural labour—which was physically exhausting and could require long, unpredictable work hours—further exacerbated these issues. Workers spent most of their time working and living, on or close to the farms, and were essentially under constant supervision, thereby creating a space of intentional or incidental segregation from the wider Australian community.

Connection and Isolation

The geography of Australian horticulture meant that SWP workers generally lived and worked in isolated locations with limited transport options. Consequently, making connections with local communities could be difficult, and SWP workers' social worlds could be confined and introspective. However, one advantage of workers being recruited from the same locality was that they shared kinship ties, language, religious affiliation and other interests that provided social support. Some workers from Tanna and Ambrym had their own string bands, sources of solace and even occasional income generation. But these intra-group connections were not always enough. In rural Victoria one ni-Vanuatu worker, Matthias, was working in the scheme for the first time in 2012. He described his feelings of boredom and isolation when he could not easily leave his accommodation:

> There isn't anywhere to go here. When you come back from work, your head is full of thoughts, but you can't go anywhere to think. When you finish work, you just have to go home … It would be better if there was somewhere else that we could go. (Quoted in Petrou & Connell, 2018, p. 72)

Geographical and social isolation could be further compounded by long working hours and physically demanding employment. Even when transport was available, seasonal workers were often too tired to travel to town centres to meet and socialise with local people (Blanco, 2009; Petrou & Connell, 2018). This could leave workers feeling a lack of connection to their host communities. Victoria Stead (2019, p. 152) describes the 'sadness' of ni-Vanuatu women in rural Victoria who experienced an 'almost complete lack of interaction' with locals. Working long, monotonous hours with little time off could result in quite extreme social isolation.

> At the packing shed, the ni-Vanuatu group sorted and packed boxes of apples, pears, pomegranates, and plums from local farms. Lined up alongside a conveyor belt … the group of twenty women worked up to six days a week, depending on the demand at any given time. Every day except Sundays, they travelled back and forth between the caravan park and the packing shed in two white minibuses. On Sundays, they used the vehicles to travel to church services in two neighbouring towns, and then to the main town, where they would do their weekly supermarket shopping and spend an hour or two browsing second-hand shops, before returning to the

> caravan park ahead of another week of shifts. This was the pattern repeated almost without fail every week. With the exception of the people in the churches they attended, the manager at the post office where they sent back remittances, the bosses and other employees at the packing shed, Tracey at the labour hire agency, and me, they had very little contact with the wider community (Stead, 2021, pp. 310–311)

For Pacific islanders, whose culture assumes and emphasises the importance of connections to others, work and life in Australia could be depressing, yet it was contextual. In 2020, when Kirstie asked ni-Vanuatu workers in peri-urban Adelaide if they were involved in any local sporting clubs or churches, they explained that they only had one day off a week and had to use that time to do grocery shopping and any other errands to see them through the week. The workers generally travelled into town as a group and descended on the local supermarket to buy food and see to other needs. Donald was a team leader for the group and explained,

> Here, we only have Sundays off and that's the time that we worship, do devotion, some people want to go shopping to get some food. Because we work the full week and sometimes we finish late and we're tired, we don't have time [to run errands]. So Sunday for us is a special time, so after worship we go shopping … we manage it by buying food for the full week to last us until next Sunday […] That's how we manage ourselves here.

Karl similarly explained that after a long week of work, their day off was an important opportunity to rest: 'Our day off is our time to rest and sleep, so we just rest and sleep'. Playing in local sports teams or making other regular commitments was difficult, and sometimes unappealing, in some cases made even more difficult by shyness, unfamiliarity or language problems. Social interactions with local communities were often irregular, and unpredictable working hours made it difficult to plan ahead.

Here too, these issues were not unique to the SWP and were partly due to the nature of horticultural work. Analogous feelings of social isolation have been described for RSE workers in New Zealand (Bailey, 2014) and Seasonal Agricultural Worker Program (SAWP) workers in Canada (Basok, 2002; Reid-Musson, 2017). In both schemes, just as for the SWP, long working hours and the locations of workplaces meant social interactions outside the workplace were often limited at best. New Zealand's geography enabled most workers to be located physically closer to townships, but tiredness, scarce free time and unfamiliarity still resulted in limited opportunities for socialising. In 2020, Emma, a Samoan migrant living permanently in New Zealand, told Kirstie:

> And you know, it's very strict, RSE for Samoa is no alcohol, pretty much no sex, no alcohol, no life. You just work and home and prayers and work and that's it for 6 months or however long.

Life as a seasonal worker could be arduous, repetitive and monotonous.

Not all seasonal workers experienced feelings of social isolation. Kirstie spoke to a pastoral carer about a group of a dozen ni-Vanuatu men who were employed by a broccoli and lettuce grower in Victoria in 2020. Rosita, a retired minister and member of the Fijian diaspora who, along with her Australian husband, Carl (also a minister), provided informal pastoral care. She described how the employer, Clive, 'a very kind man', looked after the workers, expected the men to work hard for him and treated them well in return, providing the workers with their own vehicle. Rosita laughed as she explained: 'These [workers] are looked after so well they're telling [Clive], the farmer, not to tell anyone else to come here because they want to keep coming'. The workers' accommodation and workplace were located such that they could easily access the local community—having a vehicle at their disposal helped—and they would often drop in on Rosita and Carl for a cup of tea and a chat, or to bring broccoli and other crops to share with them. In this case, geographical proximity, a generous minister and a caring employer enabled the workers to create strong connections with members of the local community.

Most PICs imposed behavioural norms on workers that could also hinder opportunities to socialise with locals. Many countries discouraged their workers from drinking alcohol or—an even worse transgression—smoking marijuana, while employed in Australia. Vanuatu has long promoted its workers as 'alcohol-free', and ni-Vanuatu workers needed to agree to abstain from alcohol while employed overseas. This 'alcohol-free brand' was intended to prevent 'bad' or risky behaviour, such as drink driving, and associated reputational issues, but also provided a competitive edge for ni-Vanuatu workers who were 'guaranteed' (in theory at least) to be on their best behaviour. Overall, the ni-Vanuatu workers we spoke with believed the sentiment behind the 'alcohol-free' brand was 'good', but acknowledged that there were two sides to the issue.

Twenty-five-year-old Alice was working in the SWP for the first time. She thought the alcohol-free brand was 'really good' for the scheme as it helped everyone to focus on earning money. She had heard rumours of ni-Vanuatu going to work in New Zealand before the alcohol-free brand was introduced. These workers drank alcohol 'all the time', and weren't able to save any money. Her husband, Keith, who was in his second year of guestwork, agreed with this sentiment. They had heard of RSE workers dying as a result of drunken misadventures and thought it was good that the same would not be able to happen in Australia. Samuel was also in his second year of SWP participation. During his first contract the workers had been allowed to drink alcohol. Reflecting on these contrasting experiences, Samuel remarked: 'There is a good side and a bad side [to the alcohol free policy]. We work seven days a week, and sometimes you need something to help you relax, right? […] But yes, it's good in some ways, but not in others …' Netty was participating in the SWP for the first time and thought overall the policy was good. She suspected some of 'the boys' had probably been drinking, but doubted the boss would make them go home if they were good workers.

Nonetheless, the alcohol-free policy received much criticism and could sometimes lead to tensions within groups of workers when some consumed alcohol and others were unsure of how to react, when reporting it could lead to blacklisting or other sanctions.

Edna was 37 years old and participating in the SWP for the first time. Originally from Ambrym, she lived on Santo with her husband and children, where she was employed as a housekeeper in a hotel. Edna had decided to come to Australia so that she could earn money to build a house and invest in a small business. She found the alcohol-free policy stressful and claimed most of the workers in her group had been drinking. Edna herself had abstained, along with two other workers, because her church did not allow alcohol consumption. Edna thought it was wrong that the others were breaking the conditions of their contract, but when she reprimanded them, they

> were cross with her, and as a result, she felt she could not speak to them about it again. Edna felt extremely worried about the situation, and although she wanted to report it to the bosses, she felt that it was not her place to do so. Instead, she called her husband at midnight to confide in him over the phone. She explained, 'I'm facing a hard time [...] I couldn't keep it to myself, so I told them off [for drinking], but they were all cross with me'.

Other workers intentionally used accusations of alcohol consumption as a means of blacklisting co-workers whom they disliked, and creating space for friends or family to participate in the scheme (Bailey, 2018b). In a context where competition for places in the scheme was high, ni-Vanuatu workers had little choice but to comply (or be seen to comply) with the alcohol-free policy.

Displaced from home, families and communities, not everything went smoothly for SWP workers, hence their desire to stay with 'good' employers such as Clive. Alcohol abuse was a constant concern but seemed to be most problematic in the earliest years. Savings consequently suffered. Workers could feel disconnected from local communities, and while they sometimes offered valuable opportunities for connection, usually through churches, often distant pastoral care could not always contend with the range of problems that could occur.

Socio-economic Impacts on Host Communities

Employers obviously benefitted financially from participation in the SWP, but the scheme had wider economic benefits for local communities, as workers paid for accommodation or spent wages locally. The social impact was more complex. Local and national media reporting on the scheme ran the gamut of attitudes to migrant workers, complicated by the unfamiliarity of people in many rural areas with Pacific islanders (rather more so in Australia than in New Zealand). In some instances, the unannounced arrival of large, visible groups of migrant workers took communities by surprise, and relationships started off on the wrong foot. When a group of SWP workers arrived in South Australia for the first time in 2016, the local newspaper reported,

> Many Mallala residents were left wondering as to what brought a sudden influx of immigrants to the town at the start of October. Confused residents took to facebook [sic] to ask questions and express bewilderment. Unfortunately due to a lack of information surrounding the issue, some misunderstandings and incorrect assumptions had begun to circulate. (Centenera, 2016)

Versions of 'incorrect assumptions' were not unusual in such circumstances, and this article noted there was some concern over migrant workers taking local jobs—an all-too-familiar stereotype around the supposed perils of migrant workers. However, the article went on to report that 'Initial negative reactions are simmering out [sic] as residents are coming to understand that nothing is to be feared from the new residents'. Poor planning and lack of communication could lead to unnecessary worries and make even basic acceptance more difficult.

Initial trepidation about seasonal workers was fairly common, especially where small towns were largely monocultural, outsiders unusual and Pacific islanders totally unfamiliar. A thrift store worker in rural Victoria explained her evolving reaction to a group of ni-Vanuatu workers:

> They come into the store in large groups, wearing hoodies and backpacks, so we watched them carefully. Now we know they are good people and that's just the way they are. They spend a lot of money here. (Quoted in Bailey, 2019a)

Often such attitudes were patronising, sometimes racist. Others in rural Victoria were nervous about SWP workers congregating at supermarkets. Fears over 'African gangs' in the region, along with people's inability to distinguish Pacific islanders from Africans, had contributed to one community's initial concern about the workers. Local residents feared criminality and violence, familiar tropes of an underlying racism. Jenny, a ni-Vanuatu worker in South Australia, described how she had continued to experience racism in a town that was beginning to feel like a second home:

> I'm getting used to it now, because it's four seasons, four years I've come back. I'm getting used to it, because I'm always here at Mallala. I feel good about it because it's my second home. But on the other side, um… to be honest, the other side, it's like… some whitefellas, they don't think that we belong here, we're not supposed to be working here. The job we're doing, it's theirs. Here in town, my first season, they told us straight in the face: "Go back to your place". And they just laugh at us. (Quoted in Dao et al., 2020)

By contrast, the ni-Vanuatu workers we met in rural Victoria in 2012 were welcomed into the community by a local council member who arranged for them to attend various community events including the council Christmas celebrations and the funeral of a respected Indigenous elder. The workers felt immediately welcomed by the community, and the community were not left guessing who they were, where they had come from and what they were doing in a small rural town. The workers described how locals would often say hello to them when they ventured into town, had even invited the workers to barbeques at their homes, and were generally kind and friendly people.

Social wellbeing and community connections probably contributed to economic involvement. On average, SWP workers spent about A$8000 over a six-month period on recurring expenses such as accommodation, food, transport, phone credit, health insurance and other deductions (World Bank, 2017). Guestworker spending contributed directly to local economies and thrift stores were particularly popular with workers who sometimes belatedly discovered the need to purchase things like warm clothes for the duration of their stay in Australia or household goods to take home with them at the end of their stay. In some instances, second-hand and charity shops even provided support for workers while they were in Australia. In Rosebud, Victoria, a church-run op shop periodically (about once a year) closed to the public and invited a small group of ni-Vanuatu workers to spend a Saturday there and freely take whatever they needed. At least one more nearby charity store did the same: small gestures of reciprocity where the workers were welcomed into and supported by the local community could make a big difference.

In addition to their contribution by spending at local businesses, guest-workers generated significant value for their employers. This was emphasised by the accidental natural 'experiment' of COVID-19, when SWP workers and working holiday makers (WHMs) were in limited supply during the 2020–2021 harvest season. Despite the complication of assessing work undertaken by WHMs, and probably boosted estimates from an industry anxious for national support, growers' estimates of lost crops suggest the substantial value of seasonal employment. The Northern Territory mango industry alone, producing just over half the nation's mangoes, estimated a potential loss of around A$100–128 million, based on gross values, with the wider NT horticultural sector losing A$251 million with an additional indirect impact on related economic activities (such as transport) estimated at A$128 million. In a similar vein the NSW tomato industry estimated a loss of A$50 million while the strawberry industry lost A$22 million (PacNews, 3 September 2020; 25 November 2020; Fresh Plaza, 9 January 2021). In two months some 55 growers from five separate states and territories reported losses on the National Lost Crop Register for a wide array of crops including berries, tomatoes, carrots, citrus, bananas, pumpkins, chillies and leafy green vegetables, in excess of A$38 million. These estimates imply that producers were unable to find any other labour, even at higher wages, to minimise losses; anecdotal information and repeated references to crops rotting suggest that alternative short-term labour sources were probably rare. While it is impossible to extrapolate from such fragmented data and back of the envelope estimates, any real measure of the economic contribution of seasonal workers, the data strongly suggest that the annual value of seasonal employment is significant and far exceeds the value of workers' savings and remittances (Chap. 7). Bluntly, if crudely, the economic benefits to Australian agricultural producers from the SWP have been huge.

Beyond economics, SWP workers sometimes made invaluable contributions to their host communities, notably during natural disasters, demonstrating a deep connection to and a reciprocal care for these communities. Over the 2019–2020 summer, a period that became known as Australia's 'black summer', bushfires affected 24 million hectares of land and destroyed thousands of houses and livelihoods. Some impacted areas included farms and communities where SWP workers were employed. In northern New South Wales, Colin Foyster, an avocado farmer, recounted how SWP workers from Papua New Guinea (PNG) 'stood shoulder to shoulder' with him and other community members as they fought the fires. Together, they helped Foyster save 7000 trees as well as homes and irrigation infrastructure (Fig. 8.1). Another group of workers from Vanuatu, who were evacuated from their farm, became heavily involved in church fundraising efforts to assist locals impacted by the fires. For their part, employers have also helped out when workers' home communities have been impacted by natural disasters. After Cyclone Pam caused devastation to much of Vanuatu in 2015, some approved employers (AEs) organised shipping containers of goods such as building materials,

Fig. 8.1 Workers from PNG helped fight bushfires. (Source: Colin Foyster)

solar panels and water sanitation pumps to help their workers' communities rebuild from the disaster (Bailey & Ng Shiu, 2016). The contributions that SWP workers made to Australian communities—and vice versa—went far beyond the economics of agriculture.

Churches, Sport and Diaspora Connections

Inevitably, many workers had positive interactions with local people. In South Australia ni-Vanuatu workers spoke about how they enjoyed chatting to a man who worked at the local post office; he had visited Vanuatu, and the workers felt he 'understood' their culture. Merely experiencing a brief sympathetic ear was invaluable. Otherwise, social interactions could be difficult—Australian small towns can be conservative and outsiders not easily absorbed—but much depended on the location of workplaces and accommodation, and being able to fit social activities around long, sometimes unpredictable, work hours. Social connections were also assisted by the return of workers to increasingly familiar places.

Chance encounters with locals sometimes led to social invitations; a group of Australian Baha'is living in Gawler (South Australia) invited some ni-Vanuatu Baha'is to lunch, but when the same group of Baha'i workers were invited to socialise with the Australian Baha'is again, they had to cancel at the last minute due to work commitments, and contact between the two groups eventually waned. In Robinvale, where a large Tongan diaspora existed alongside the temporary Tongan SWP workers, church provided 'a space of common ground' for these two groups to mingle and the SWP workers were 'warmly welcomed and very much a part of the community' (Tazreiter et al., 2016, p. 110). Christianity and the church are central to most Pacific islander cultures, so that it is not surprising that churches provided an important point of connection. Of all Australian organisations it was the churches who were most involved in seeking to develop linkages with workers, providing some pastoral care and inviting them to services and other forms of fellowship.

In New Zealand too, RSE workers were similarly welcomed by churches that considered ni-Vanuatu workers an 'asset' to their church community. Local church newsletters welcomed the RSE workers and sought donations, such as warm clothes and musical instruments, to help make the workers' stay more comfortable. Nonetheless, and as in Australia, geography could make it difficult to attend church services located at some distance from accommodation.

Churches were in some respects surrogate trade unions and less 'threatening' to employers. Yet the outreach and accessibility of churches varied. In peri-urban Adelaide, ni-Vanuatu workers tended to hold their own worship sessions at their accommodation sites. These included video sermons that they would watch together every Sunday, and visits from a resident Pacific islander priest who rotated across the various sites where the workers lived. In Swan Hill, Victoria, ni-Vanuatu workers similarly chose to hold their own worship services on Sunday nights. By contrast, those in Rosebud, Victoria, attended the local church, a choice made easier by access to their own vehicle and proximity to town. Where workers did attend local churches, religion facilitated community connections. As Mary reported for her group of workers from Vanuatu,

> We went to the Church of Christ and they welcomed us, and to the Presbyterian Church. We were in the daily newspaper, so when we walk around in town, everyone says 'Hey, I saw you in the newspaper! Hello, what's your name?'. So now when we go to the shop, they all smile. Especially at the supermarket, we go there a lot for our shopping.

In New Zealand, ni-Vanuatu workers forged friendships through church attendance which, in some cases, led to workers subsequently hosting their New Zealander friends in Vanuatu (Bailey, 2014). Churches also filled gaps in official pastoral care programmes, through, for example, providing a sympathetic ear when workers faced challenges and coordinating clothes drives during the winter months to ensure workers had enough warm clothing. In 2019, a group of local churches worked together with the Pacific Islands Council of South Australia (PICSA) to collect and distribute winter clothes to 500 ni-Vanuatu workers employed on the Adelaide Plains. Local church and community groups frequently provided such practical support to workers.

In some cases, usually when employed close to towns, SWP workers found themselves working alongside Pacific islanders who lived in

Australia permanently. Forging connections with these communities helped the workers settle into their new environment. At the same time, the growing use of SWP workers in the horticultural industry, and an increasing reliance on contractors to recruit this labour, could negatively affect Pacific diaspora populations who worked in these same jobs. In the Sunraysia Mallee region (Victoria), the local Pacific population, many of whom were Tongans, relied on personal relationships with farmers to secure seasonal work. Where farmers changed their crops or switched to using contractors to recruit workers instead, the Pacific diaspora could lose employment to SWP workers (Nishitani & Lee, 2017) Nonetheless, long-term resident Pacific populations were generally eager to welcome guestworkers to Australia and could provide an important source of care and other assistance.

Sport provided another opportunity for workers to connect with local communities, although regular participation in sport was constrained by geography and working hours. Workers organised their own informal touch rugby and football games. Some employers however could provide additional, irritating barriers, with one employer in New Zealand discouraging his workers from participating in sport lest they injure themselves and become unable to work (Bailey, 2019a). Other employers, however, saw the benefit of providing recreational opportunities for their employees. In January 2020, VegetablesWA organised an Australia Day soccer tournament for the seasonal workers in the region. One of the organisers explained: 'These workers will be in the country anywhere between six months and three years, so it's important that we do our part to make them feel welcome and appreciated' (quoted in Wheeler, 2020). Most employers broadly agreed with this sentiment, and photos of seasonal workers playing in short sporting tournaments were a regular feature on seasonal worker Facebook groups (Fig. 8.2). Women were often notably absent from such photos, but this may have been due to their lower participation rates in the scheme. Once again, the ability of SWP workers to benefit from attendance at more or less familiar churches, participate in sporting activities and receive assistance and support from established Pacific islander communities was a function of location and distance. Those in more remote places benefited least.

Fig. 8.2 Ni-Vanuatu workers participate in a soccer tournament. (Source: Facebook)

Keeping in Touch with Home

An absolute constant was that workers missed home, amplified by many having been selected as productive members of young families. However, keeping in touch with home was expensive; a mobile phone call to Tonga cost A$0.90 per minute in 2020, while to Vanuatu the cost was approximately A$1.95 per minute. Enthusiasm for personal contact could bring costs of A$500 or more over the duration of their stay. Nonetheless, maintaining contact with family at home was important in mitigating negative impacts of physical separation (Chap. 9). In 2015, most SWP workers surveyed by the World Bank (2017) communicated with home at least monthly, mainly by phone (96 per cent), with a minority relying upon Skype (3 per cent) or email (4 per cent). Reliance on phones was attributed to costs, slow internet speeds and poor connectivity in home countries, but also to a digital divide between Australia and the PICs, and between urban and rural areas. Fortunately, telephone reception was generally good, but high costs meant that many workers relied on receiving rather than making calls, as was the case for the ni-Vanuatu workers we met in 2012. Edna, whose husband and children lived on Santo, explained,

> It is very expensive for me to call them. I always have to pay A$30 when I call, A$20 … You can't call, you just have to text instead. But I prefer for them to just call me every time for 200 or 300 vatu [A$3.60]. I can talk to them for five minutes, so that's good. It's cheaper. So I speak to them every day […] I've only called them once and it cost A$30. But otherwise, every day they call me. Yes. So I don't feel too lonely because every day I hear the voices of my children and my husband, so in one way I'm still able to be close to them […] If I lost contact with them, I don't think I would cope.

Our discussions with SWP workers, and research on their use of social media, suggested that reliance on the internet as a means of communication had increased significantly by 2021. Improved network coverage, including the availability of 3G and 4G networks, and the relative affordability of smartphones meant that the internet had become much more accessible within the PICs. Youth-dominated populations have been key in driving the rapid uptake of mobile phone technology, and thus the PICs largely 'leap frogged' computer-based internet access. In Vanuatu, for example, by 2010, 95 per cent of urban households and 82 per cent of rural households had mobile phones (VNSO, 2012), and mobile phones were the primary communication device used by rural and urban residents alike. The internet was at least as significant as traditional forms of news media—despite access to these media remaining concentrated in urban areas (Vanuatu National Statistics Office, 2017). Social media—largely accessed via smartphones—had become an important platform for discussing and sharing news, current affairs and political activism (Finau et al., 2014). Facebook emerged as the largest online social network in the PICs and figures from 2012 indicated that the Pacific already accounted for 700,000 users, with 16–34-year-olds the most common users (Cave, 2012; Hobbis, 2019). While more recent figures are unavailable, ongoing infrastructural updates and improvements have meant that the growth in Facebook, and other social media, has continued rapidly.

Mobile phones and social media made it much easier to keep in contact. Through regular phone and internet communication, including posting photographs and videos, some workers managed to maintain an active role in household and community decision-making. One RSE worker in New Zealand stated that if his daughter had trouble sleeping, his wife would call him and he would talk to her so that she could go to sleep (Cameron, 2011). Some seasonal workers preferred to rely almost entirely on social media, but many used both. Ni-Vanuatu workers on the Adelaide Plains told Kirstie that calling home was too expensive, and most relied on Facebook messenger to stay in touch with friends and family in Vanuatu. Donald explained, 'We just use our mobile phones [to message family]. It's one of the cheapest ways to talk because if you call it's very expensive for credit'. Beyond their primary international communication role, smartphones had become as popular amongst seasonal workers as amongst the Australian population as a whole and were used for all manner of things, including keeping in contact within Australia, for entertainment and to access information that was difficult to retrieve at home where internet speeds were slower. Simon, a Solomon Islander SWP worker, explained,

> I learn a lot of things. I always downloaded new things. One thing is how to make brick, how to make flowerpots, out of cement. I downloaded it from here, and I go back home and I told my wife, and she's doing it. And there are many people interested in it, and they paid for my wife … [to make] flowerpots. (Quoted in Dun et al., 2022, p. 13)

Timorese were similarly adept at keeping in touch by smartphones and skype, exchanging photographs, building networks with resident kin in Australia, participating in multiple Facebook groups and building something of an online 'quasi-family' (Wu, 2019). Beyond contacting home, smartphones enabled a cheap form of entertainment, especially welcome after hard physical labour. By 2021, social media had also opened up conversations with a national (and sometimes international) audience (Petrou & Connell, 2020). Seasonal workers from different PICs, working in different areas of Australia and New Zealand, increasingly relied on Facebook groups as a means for sharing information and experiences, asking questions, navigating the system and celebrating successes. We expand upon how SWP workers have employed social media to navigate the SWP later in this chapter.

Labour Rights and Power Imbalances

Like guestworker schemes around the world, SWP employment was characterised by power imbalances and structural precarity. Employment through the scheme was, by definition, temporary, and job security was non-existent. Workers could experience income insecurity if work dried up, and they had little control over employment conditions. While workers were originally tied to a single employer, by 2021, new provisions enabled them to change employers in limited circumstances such as when an employment situation became 'untenable' (although what this might mean was undefined). It was not clear how often this option had been taken up (although it was probably rare), or even if workers were aware that it existed. Indeed, SWP workers generally felt unable to leave their employer when their labour rights were not protected, or hours of employment were not what they expected.

The ni-Vanuatu workers we met in Victoria in 2012 were upset about their conditions, but unwilling to act on this.

> Joshua was 39 years old and came from rural Santo where his livelihood revolved around subsistence agriculture. It was his first time working in Australia, and he and his fellow workers had experienced various problems around pay rates, poor pastoral care, crowded accommodation, fluctuating hours and an unkind employer: 'We've come, and we're here now. They can treat us however they want, and we just have to put up with it'.

These sentiments were echoed by his colleagues. Edna explained,

> So we just have to be humble, and put up with it. We have to keep working until it is time to go home, and then we'll be able to breathe easily again. That's just how it is. (Quoted in Petrou & Connell, 2018, p. 75)

Workers' ability to complain was further constrained by limited English skills, being unsure about what was appropriate and to whom they might complain to.

Since many SWP workers were reliant upon being invited back to work in subsequent seasons, they had no wish to jeopardise future opportunities, either for themselves or their home communities. Alice, who lived in Port Vila, was a member of the same group and explained,

> We've come and we haven't come to just hang out and do nothing and then come back next year. We actually have to prove ourselves that we can do whatever work they want us to do.

Working hard—whatever the circumstances—was understood as a requirement for future employment. Some of the men from the same group of workers briefly went on strike when their pay rates were switched from hourly to piece rates, a move they believed was contrary to the contracts they had signed. However, the strike fizzled out when they decided they would rather ensure they could return to Australia the next year than risk being labelled troublemakers. It is hardly surprisingly that workers who did not plan to work for the same employer again were often the most willing to fight issues such as underpayment and poor working conditions (Howe et al., 2019; Smith, 2016). Moreover, a cultural aversion to challenging authority, typified by Tongan workers in Robinvale, meant that workers were more likely to abscond than speak out over workplace problems. One Robinvale employer explained:

> INTERVIEWEE C We learned, culturally, they would never complain.
> INTERVIEWEE A Never complain, never even asked outright for something.
> INTERVIEWEE C And so we asked them a direct question and they responded to it.

Interviewer They wouldn't complain to MADEC who's their official employer?

INTERVIEWEE C No … culturally, they never ever will (Quoted in Tazreiter et al., 2016, p. 117)

The SWP scheme—like guestworker schemes elsewhere—put the onus on workers to report problems, challenges and other unacceptable or concerning experiences to the relevant authorities. Yet, the process for lodging a complaint to the Fair Work Ombudsman—the government body that deals with such issues—was too abstract to make much sense. Isaac, who had frequent contact with SWP workers employed in rural Victoria, spoke out in 2020:

> And this is one thing that has always stuck with me, the whole you know, they can go to Fair Work Ombudsman's office. I'm like, you are kidding me, are you serious? A lot of Australians don't know how to engage with Fair Work Ombudsman's office. I've been on the phone to them and ripped the person on the phone for being absolutely useless because I was frustrated that they weren't acting on issues that I was reporting to them exactly about Pacific islanders and exploitation. And how, you know, they have to file a report and if it's worth it then we'll follow up, we don't have resources on the ground. You know, this is what they get when they contact the Ombudsman's office. And how is someone from the Pacific who has no understanding of our systems supposed to engage in that? Or trust it? And understand the logical sequence of events that may happen once they make the initial report? It's ridiculous for a bureaucrat just to say Fair Work Ombudsman's there and that will handle it. It's not gonna handle it. It won't. And if I don't trust it, Pacific islanders who are coming into the country for the first time, they're just not going to do that.

Processes for handling complaints existed, but they were slow, challenging to understand and difficult to navigate.

Employers were aware of their power over migrant workers and in a few cases enjoyed lording it over them. In New Zealand, for example, one employer singled out a ni-Vanuatu worker in a group meeting and then went on:

> One man—you—(pointing at him) complained about your pay. You wasted two hours of *my* time, with your stupid complaint … My team have better things to do than listen to a bunch of men complain about things they simply don't seem to be able to understand—no matter how many times we explain it. From now on stupid complaints will mean you have no future with my company—you have been warned. (Quoted in Cameron, 2011, pp. 108–109)

Ed Cameron observed that this was rather like a headmaster addressing pupils at the start of a school year, and certainly not a context for discussing grievances let alone challenging authority. That was not necessarily exceptional. In July 2020 an RSE manager was taped threatening to withhold the allowances of Solomon Islander employees, claiming he would report them to New Zealand Immigration and prevent them from boarding their return flights home. The threats stemmed from his displeasure that they had complained to the Ministry of Business, Innovation and Employment about their employment circumstances (Fonseka, 2020). When we met a group of ni-Vanuatu workers in rural Victoria, they described how a member of their group who had been acting as the team leader was sent home without warning because their boss felt this man was challenging his authority. According to the workers and their pastoral carer, the employer objected to their listening to and taking orders from their team leader, viewing it as a challenge to his own authority. Edna explained that the remaining workers were threatened with similar actions:

> Right now, when we're there [at work], we have to listen to their talk. They're always saying 'You'd better work well; if you don't, you'll be sent back to Vanuatu. You'd better work well, if not, you'll be leaving. Do you want to go back home?'. These are the kinds of things they're always saying to us.

While this appeared an unusual case, with few employers overtly behaving in this way, it was everywhere clear where power lay. Workers often felt they could do little to address these issues and inequalities.

Several workers expressed concern about overt racism in the workplace. In rural Victoria in 2012, Mary and her female ni-Vanuatu co-workers felt they were being victimised by their supervisor who gave them conflicting instructions, belittled them and generally treated them poorly:

> I really like the work, but the way they are treating us on the farm, we have to deal with … I think it is because of the colour of our skin… (Quoted in Petrou & Connell, 2018, p. 74)

While probably an extreme example (and established local residents agreed that this particular farm was not a good place to work), it raised questions about race and racism in small rural places. More generally, some workers felt frustrated at being given conflicting instructions and being constantly surveilled by supervisors. Grace and Anne were two ni-Vanuatu workers employed in a packhouse in Shepparton. They worked at the 'crossfire' of conflicting instructions from packhouse supervisors and farmers, and were under constant threat of being sent home—as another member of their group had been—if they could not work fast enough in the sorting line. Grace was on her third contract at the packhouse and commented, 'It's like they don't trust us. That's what we think, how it seems to us' so that the workers were 'constantly, anxiously alert to the insecurity of their employment […] and fearful of the consequences of being found to have criticized or complained about their conditions' (quoted in Stead, 2021, p. 311). Workers felt that complaining simply was not an option.

Employer-related problems in the workplace were generally attributed by industry and the media to a few employers who were 'bad apples': a not inappropriate metaphor. Speaking about employers who mistreated WHM workers, Tasmanian blueberry and raspberry farmer, Greg McCulloch, observed: 'You get some bad ones — and they are the minority — but we all get tarred with the same brush' (quoted in Uibu, 2020). Consequently, and despite the longstanding and recurring nature of many of the issues discussed here, few attempts were made to address or rectify the power imbalances inherent in the design of guestworker schemes. The relevant bureaucracies assumed that the 'bad apples' were anomalies and would disappear from the system so rarely took action. Moreover, while grievance processes may have been adequate in a formal Australian context, the focus on worker-initiated complaint processes was not consistent with Pacific islander cultural practices.

Blacklisting and Standing Down

Being 'blacklisted'—permanently banned from the scheme for 'bad' behaviour—was an ongoing fear for workers. Home country governments were keen for them to toe the line and 'behave', lest their country gained a reputation for troublesome workers and opportunities to participate in the scheme disappeared. When we aired concerns at a conference in 2012 about the treatment of ni-Vanuatu Pacific Seasonal Worker Pilot Scheme (PSWPS) workers in rural Victoria, Vanuatu's then commissioner for labour accused us of lying, despite our stressing that no 'blame' was attached to the workers. In 2015, when Samoan workers took strike action in Australia, the Samoan Prime Minister both denied that mistreatment had occurred and warned that troublemakers would not be allowed to return to Australia.

> We have a list of their names and the villages they are from. If anything happens again, the Village Councils will be notified so they can make a decision on those who have wasted the opportunity given by government … Causing embarrassment to the country … people might think Samoans are lazy. (*Samoa Observer* quoted in Union Aid Abroad APHEDA, 2017)

Once again the highest level of authority was involved. At the destination, the pressures were little different; the Australian government reinforced these messages in their pre-departure handbook.

> Remember that while you are in Australia you are a representative of your country. It is important that you represent your country and that you do not do anything that could reflect poorly on you, your family or your country. If you behave poorly, you may not be allowed to return to Australia as a seasonal worker in the future. (Australian Government, n.d., p. 3)

From the handbook, it was not clear what 'poor' behaviour might entail. In some respects this vagueness was important as workplace norms could vary, but it also afforded flexibility to authorities. The Vanuatu Department of Labour was somewhat more specific, and their Migrant Worker Disciplinary Policy laid out specific periods of blacklisting for different behavioural transgressions (Table 8.1). Between 2007 and 2018, Vanuatu's Employment Services Unit (ESU) had put 1300 workers on their stand-down list and 106 workers on their blacklist; some were included for criminal reasons, such as violence, alcohol abuse and damage to property, but just as many appear to have been blacklisted to achieve and maintain workers' compliance (Bailey, 2018b). Moral overtones about 'good' behaviour were evident at both ends and hinted at collusion.

The list of infractions that the Vanuatu Department of Labour had compiled, based on past experience, indicated that seasonal workers could cause and encounter problems in Australia (and New Zealand). The workers were not without blemish. Occasionally crimes resulted in criminal cases, but misbehaviour mainly offered variants on 'drunk and disorderly'. Living in all male communities, without traditional leadership and being bored during downtime, could be difficult to negotiate.

Table 8.1 Migrant Labour Disciplinary Policy, Vanuatu Department of Labour

Offence	Period of Stand Down
Deliberate poor productivity Failure to board a flight to the receiving country Other misbehaviour	1–2 years
Repeated difficult to manage misbehaviour Damage to property	3–4 years
Fighting Affairs Sexual assault Abuse of alcohol Use of drugs Being sent home by the employer	5 years
Fraud and falsification Theft Drunk driving Reoffending after stand down and reinstatement Involved with NZ/Australian police/courts	Ban

Source: Employment Services Unit (2019)

Blacklisting 'troublemakers', and their villages (at least in Samoa), occurred periodically over the years, and in 2019, the entire village of Poutasi (Samoa) was banned from participating in the RSE after three workers from the village were 'allegedly involved in alcohol-induced criminal activity' (RNZ, 25 June 2019). This proved an effective means for disciplining workers and keeping them quiet. In 2020, Kirstie spoke with Emma, a long-term Samoan migrant living in New Zealand, about the RSE scheme:

> See the problem is what you guys are gonna hear is not the full story because they [RSE workers] are gagged. And every time an RSE worker speaks, they know that their reputation and their village and their family would be affected. So you're not gonna get honest answers. (quoted in Petrou et al., 2021, p. 304)

Blacklisting and deportation, and the shame and even punishment brought on home communities, were to be avoided at all costs (Fatupaito et al., 2021). Blacklisting did occur for legitimate reasons, and for ni-Vanuatu was consistent with cultural norms around migration more broadly; migration was considered a privilege and those who abused this privilege were sent home, as traditional leaders had sent troublemakers home from Vila in quite recent times (Petrou, 2020). Some workers used blacklisting to their own advantage and made untrue reports of misbehaviour to get others banned from the scheme, so creating space for friends and family members. Blacklisting and stand-down periods sometimes followed clashes between employers and workers. Rachel Smith (2016) describes how several ni-Vanuatu workers were stood down from the RSE scheme for drinking alcohol. The workers did not dispute that they had drunk alcohol, but claimed that the boss had bought it for them. Rather, they believed that they had been singled out for requesting their holiday pay, which totalled roughly NZ$1200 (A$1117) per worker. The boss had tried to get them to agree not to claim this pay, which he would use to cover their flights the next season. When the workers rejected this, their boss reported them to the Vanuatu Department of Labour for drinking alcohol and they were barred from participating in the scheme for five years.

Blacklisting was not necessarily problematic, but in the hands of unscrupulous employers—or co-workers—it could be. Putting up with poor treatment seemed like a minor sacrifice when speaking out could mean jeopardising personal participation and that of a wider community or even country. With plenty of others willing and available to take workers' places in the scheme, the risks from 'misbehaving' were significant.

Underpayment and Elusive Wages

Underpayment of wages was a recurring problem for seasonal workers and a common theme in media coverage. Most SWP workers expected to earn close to A$700 per week, consistent with the horticultural award rate. Those who earned less than this—as happened when bad weather intervened or the crop was disappointing—were naturally dissatisfied (World Bank, 2017). Media stories about underpayment of wages, and concerns and confusion over pay rates and deductions could lead workers to fear they were being exploited—as they sometimes were. In 2012, the ni-Vanuatu workers Kirstie met in rural Victoria believed they had signed contracts promising an hourly rate, and so were annoyed and disappointed when they were switched to a piece rate. Joe, a 24-year-old man from Santo, was working in Australia for the second time. He had a young daughter and had travelled to Australia to 'improve his future'. Joe was extremely frustrated that he and the other men had been shifted from hourly to piece rate payments:

> Our belief was that we had signed a contract when we were in Vanuatu stating that we would be paid an hourly rate, not a piece rate. When I signed the contract, I agreed with this rate, and so I came to work in Australia. Then, when I came, I found the conditions had changed. I don't think this is right… We did strike for a little while, but now we don't want to ruin our reputation in Australia.

The change in pay rates was probably stipulated in the workers' contracts, and thus was legal, but the workers neither understood nor expected this change. Inadequately explained changes to pay rates were a recurring issue. In Mundubbera (Queensland), Tongan workers' contracts stipulated that they would be switched to piece rates after a period, but it was not clear how much the piece rates were, nor what proportion of work would be paid at these (unspecified) rates (Holani, 2017). Several years later, in 2020, pay rates continued to present problems. Isaac, explained of rural Victoria:

> Another thing is that they sell the program on an hourly rate on the horticulture hourly rate award, and then when [the workers] get here they're told about the piece rate. And you know, not many people in Australia can understand the piece rate [let alone seasonal workers]

Workers' fears over exploitation were not unfounded, and illegal deductions from payslips were an ongoing issue. All SWP workers had various deductions made from their wages for such costs as accommodation, transport, medical insurance and airfares or other reimbursable costs that employers contributed to. RSE workers had similar deductions from their wages; however, in New Zealand, these were formalised and clearly stated in workers' contracts. This was not so in Australia and inflated deductions for transport and accommodation frequently occurred (Holani, 2017; Howe et al., 2019). Tongan workers in Mundubbera, for example, were charged potentially illegal amounts for accommodation and some paid fees for health insurance coverage that may not have existed (Holani, 2017). They were not alone. In a case brought before the Fair Work Ombudsman, a labour hire firm underpaid a group of ni-Vanuatu workers by A$77,600. During their contracts, the workers' wages were paid infrequently and leave entitlements were underpaid. The company was fined A$186,000 and the sole director fined A$41,300. However, the underpayments occurred in 2014 and it was not until 2017 that the ruling was made; the workers had to wait three years for the case to make it through the legal system (Union Aid Abroad APHEDA, 2017). By then they had returned to Vanuatu.

SWP workers often put up with such issues in the workplace because, as discussed above, they did not want to be branded troublemakers for speaking out. A Queensland mandarin farmer explained:

> There is no system of checks and balances to see how much the grower pays to the labour hire company and how much money the workers end up getting. You can only see that if workers complain and show you their payslip but of course they never complain because it is such a big opportunity and nobody wants to make trouble. That is a big problem in the scheme, that when things go wrong nobody complains. (Quoted in Brickenstein, 2014, p. 250)

These concerns were well-founded and labour hire companies in particular were likely to be at fault. Employer reprisals were not unheard of, and some unscrupulous employers went so far as to deport—or try to deport—workers who complained legitimately. In December 2015, a group of Fijian workers employed on a casual piece rate were paid as little as A$1.21 per hour rather than then the award rate of A$21.61. The workers had 'agreed' to this rate after their labour hire company confiscated their passports. When they eventually protested, the labour hire company attempted to deport six workers who were identified as the 'ringleaders' of the protest (Union Aid Abroad APHEDA, 2017). Such extreme cases reached the Fair Work Ombudsman, but legal remedies needed external support. Achieving restitution in less egregious and obvious cases was more difficult, the process was drawn out and compensation took time. It is impossible to know how many 'lesser' cases were never remedied.

Sometimes underpayment accompanied other dubious practices and poor workplace conditions. In 2019, after ongoing litigation, 50 ni-Vanuatu workers were awarded an unspecified amount in a settlement from Agri Labour Australia, and the company was removed as an AE. The workers, employed on a farm outside Shepparton (Victoria), had been paid only A$8 an hour and were expected to work in 'shocking' conditions which included an unsafe level of exposure to agricultural chemicals. Tulia Roqara, one of the workers, described what it felt like trying to pick tomatoes straight after they had been sprayed with chemicals.

> "It was really hard to breath it in, especially when we start picking," Ms Roqara said. "I got chest pains and others (workers) got bleeding from the nose and the ears." (Quoted in Schneiders, 2018)

Explaining why she had pursued the case, Tulia said: 'The most important thing, it's not all about the money. [It's] that we have the same rights Australians have' (Schneiders, 2019). That case utilised the Protecting Vulnerable Workers amendments to the Fair Work Act, introduced after a number of wage scandals in the industry and was won with the support of the National Union of Workers. While a few workers were unafraid to speak out, and so could improve conditions for others, structural contexts and the locus of power made this rare.

Underpayments and illegal deductions were not specific to the SWP, but were a wider problem across Australian horticulture and far beyond (Chap. 11). Dishonest pay arrangements, such as 'cash in hand' rates for illegal workers, or piece rates that resulted in salaries far below the minimum wage, were common industry practices (Cavanough & Wherrett, 2020). Speaking about the illegal horticultural workforce and the profits to be made from shady dealings, Tony Natale, a table grape grower, observed: 'The contractors—you see them driving in Mercedes. Is that all from the 5 percent [contractor's fee]? … They'll charge you for the superannuation and what have you. Whether it goes to the poor guy at the other end, I doubt it' (quoted in *The Australian*, 7 October 2008). Removing multiple problems of poor pay rates and illegal deductions from workers' payslips is seemingly insuperable within a decentralised and temporary seasonal worker programme, until broader illegal industry practices are stamped out.

Pastoral Care and Monitoring

Effective pastoral care could be invaluable and could avoid or address many of the problems outlined above. Sometimes all it took was a single person to make a difference. In 2020, the ni-Vanuatu workers we spoke with described their employer appointed pastoral carer as *em i mama blong mifala* (she is like a mother to us). John, who had worked in both the SWP and RSE schemes, explained:

> She is good, she's a really good person. She's concerned about us, right? Like, she's a good person and she's concerned about how we're going at

> work. Sometimes, when she comes to visit us, she asks how we're finding the work. She says 'If there is anything at all, just tell me and I'll talk to them at [the company]'. She's a good person, she's really concerned about us and our work.

Timely, effective, adequately resourced pastoral care could ensure that workers felt heard and cared for, however patronising that might be. However, employers often found providing pastoral care difficult, as it represented an unfamiliar, extra layer of responsibility and coordination (Howe et al., 2019).

The most effective pastoral care was often informal, unofficial and provided by the diaspora. Recognising the difficulties SWP workers faced, some Pacific islander community organisations, like Tonga Australia Seasonal Workers Association (TASWA), were established specifically to assist migrant workers and support their concerns. In other instances, existing organisations took on this role. In addition to holding clothing drives, PICSA welcomed new arrivals with community meals, arranged celebrations around independence days and other significant events, and organised trips to visit other workers stationed around South Australia. Connections between the local Pacific communities and seasonal workers were often unplanned; in South Australia, one group of guestworkers happened to stumble across a long-term Pacific islander migrant in a doctor's waiting room, leading to ongoing connections with the established Pacific Islands community. Similarly, ni-Vanuatu workers in Rosebud coincidentally attended the same church service as Rosita and her husband, a former minister who eventually took on the informal role of pastoral carers for the group. Rosita explained in 2020 how they became involved with the workers:

> But the first lot [of workers we met] was two or three years ago and they just appeared at church. And because we were there we just felt, as Pacific islanders, we welcomed them and that became their church.

In the difficult circumstances of COVID-19, in the Sunraysia Mallee region, local Pacific communities provided food relief for workers who faced periods of reduced hours and earnings. More generally, such local communities were often eager to be involved with and assist seasonal workers in any way they could. Mark, a Papua New Guinean living in New South Wales, lamented that the government did not provide contacts to such community groups when workers arrived in Australia:

> Now if people from home come here and they don't have a home or people to relate to, then there's a missing link there. I for one, I would like to know if there are seasonal workers here [...] not that we [could] give financial support, but we have moral support, cultural sort of engagement with them.

In recognition of the role Pacific islander community organisations had been playing and the important resource they represented, in late 2020, the Australian government had begun a trial to officially involve such communities in providing pastoral care and culturally appropriate support to seasonal workers.

The cultural appropriateness of pastoral care was vital. Seasonal workers interpreted Australian events and practices through their own cultural lens. Not recognising this could lead to avoidable misunderstandings. Isaac, an established Pacific migrant living in rural Victoria, described how a group of Timorese workers misinterpreted the relationship between their own representative and their employer.

> As soon as the trust breaks down in the smallest way then the trust of the whole process starts to be questioned, you know like the Timorese were saying the local federal rep who I know very well, and he's a very nice man, he's straight up and down, he wouldn't have been in the department for as long as he has been if he wasn't. They were suggesting that he was on the take because he was talking to the AE, and perhaps not understanding the way that he should be talking to them [the workers]. He didn't bring them along, he didn't have them onside, so you've got to have a smart operator that can see where the workers are coming from and the power dynamic isn't reinforced in any way… And yeah, so just certain things like that can threaten workers' trust in the scheme. So once that seed has been planted then the whole program and every element of it just comes into question.

Isaac only became aware of that situation after his Timorese colleague alerted him to it. In such contexts, involving local Pacific communities in pastoral care, where feasible, was invaluable.

Pastoral care could fill in some of the gaps left by other monitoring bodies. Limited resources and remote work locations made it difficult for authorities to monitor seasonal workers' experiences or those of the horticultural workforce more generally. According to one study of the New Zealand RSE scheme, in 2017 some 9000 seasonal workers were being monitored by just six labour inspectors and six compliance officers (Lees-Galloway, 2017). Our own conversations with key informants indicate that the Australian system has been similarly understaffed. Speaking about the illegal workforce in rural Victoria, one frustrated Pacific community member told Kirstie about an encounter with the Australian Border Force, the government law enforcement agency responsible for investigating migration status and visa breaches:

> You know, I talked to Border Force a few weeks ago and told them that they need to look at this farm, and I got a phone call saying 'Oh, we work proactively. The farms, we've been up there recently and inspected 70 farms and explained the system and we're really expecting farmers and contractors to run everybody through their books. We're under-resourced, blah-di-blah-di-blah'. Great, so we'll just spread the word out that you can make hay while the sun shines because the department doesn't have any fricking resources to come out here and raid.

Recurrent media articles about the types of abuses both WHMs and SWP workers encounter in the horticultural labour force suggest that any 'working proactively' proved inadequate.

Unions: Fighting for Rights … Sometimes

Unions could play an important role in supporting workers in navigating systems, but some workers have been actively discouraged from joining them. Pre-departure briefings for Timorese workers in 2019 included advice that the National Union of Workers, the most relevant union for agricultural workers, was unhelpful and they should take any complaints to their team leader or the employer (Annie Wu, Pers. Comm, 2019). In 2017, the labour hire company MADEC allegedly pressured ni-Vanuatu

SWP workers into resigning from the National Union of Workers (NUW). MADEC prepared resignation documents for the workers and told them they would not be offered further work if they did not sign them. They also claimed that they would not recruit any more workers from Vanuatu if the workers continued their membership. Such illegal actions, breaching Australian laws over freedom of association, were later described as a 'misunderstanding' by MADEC's CEO (Union Aid Abroad APHEDA, 2017). Other SWP workers never got the chance to join unions, with employers claiming union representatives never 'bothered to turn up' to worker briefings. In contrast, unions reported they were often not invited or provided with insufficient notice or details to attend worker orientation sessions, especially in remote sites. When union representatives did attend, employers sometimes remained in the room during information sessions or left union representatives with too little time to adequately present their information to workers and openly discuss workers' rights (Howe et al., 2019, p. 110), while workers themselves had less time for queries and ultimately less ability to raise concerns.

Lack of familiarity with unions—rare in PICs and where they do exist, notably in Fiji, unions are usually urban based—and instructions from employers not to join, further discouraged or blocked access. As Saina Tomi Setu from Samoa First Union observed,

> Farm workers are not unionised in Samoa as the farms are generally owned by the villages and those who are not in formal employment work without payment. No information is provided to the workers before they leave for Australia and they know nothing of unions but because they are confused by their rights they are looking for help. I have been able to speak to some of our boys in Australia and explained what unions do and they were confused as they often are told by the managers that unions won't help them and they should save their money. However recently some of them have found that the information that they received from the NUW organisers was very useful to help them understand the Australian way (Union Aid Abroad APHEDA, 2020, p. 11)

When workers did contact unions, outcomes were generally positive. Jenny, a ni-Vanuatu worker in South Australia, described her experience of speaking out about underpayment:

It was okay the first season, but we had some rip offs. So what we did was, we called the labour hire bosses and I told them, See, this is what's happening. My friends were a bit scared, because back home, our culture, we respect the bosses. We don't go close to them. But I told the others that if we don't speak up, this thing is going to go on and on and on. I'm like this—I speak up every time something is wrong. Most of us here don't have a very good education background, so that's why I stepped in.

When they came, they told everyone, "Okay, you guys that don't want to join the union come and sign your name." It was a form to say that you're not going to join the union. They told us, "The union is not going to do anything good for you guys. If you join the union you're going to be sent back home." I didn't believe their words. The room was filled with everybody. One boss was reading from the iPad, he kept going on, on, on about issues from the policies and the contract that we signed. After everything he said, "Do you understand?"

Back home in Vanuatu, if you whitefellas come, when you say something, we don't talk but we just say yes. It's kind of a respect. We just nod our heads and nothing else. Because I knew my people, I knew that they don't understand anything. I stood up and I spoke in our dialect and I told them, "Do you people understand what this whitefella is talking about?" And everybody they said, No, Jenny… this, that, this.

I said, "Okay, now you people shut up, I'm going to ask him a question." So I started. I started with our payslips. Then, why are we not supposed to join the union? Why, if we're not working, do we still have to pay for the bus? And I told him, "You want us to do our jobs here, do your job here."

He got pissed off with me. He got very pissed off with me, then he walked out. And everybody went silent and one of the guys, he turned around and told me, "Jenny, see what you did?" I said, "No, I'm talking for the better of the crew. We have to have some privileges here. They're not going to treat us like our forefathers, our grandfathers that were here as slaves, during the blackbirding. No, we're not going to be like that."

The union helped us fight the case. We were supposed to go to court, but we had this outside settlement and we won. And the chief executive, he came down here. Might be he's a city person, but he came down here to Mallala to say, "You're a strong woman." He said sorry. I said, "It's not for me, it's for all of the Ni-Vans working here. You just have to apologise to them." So the chief executive and the general operations manager, they apologised. We all said, "Yes, it's okay" (Quoted in Dao et al., 2020)

Unions could be invaluable in supporting workers when things went wrong. When the ni-Vanuatu workers in Victoria won their case for compensation for sickness from ingesting chemicals and receiving wages below the award rate (see above), it came with union support partly because one of the employees of the union was a ni-Vanuatu man. Such fortuitous conjunctions were rare, and lesser offences slipped by. Their resolution relied upon workers speaking out and knowing who to contact—something that could be difficult when union representatives did not attend arrival briefings, or when employers otherwise 'discouraged' union membership or involvement—and the effort involved, at some cost, rarely seemed worthwhile when it might jeopardise ongoing participation in the SWP.

Negotiating Online

While many workers were uncomfortable about directly confronting problems in the workplace, they had not simply accepted the status quo. Workers, their families and other interested parties employed social media to share information about the seasonal labour schemes, typified by the role of Facebook for ni-Vanuatu workers. In April 2021, 'Seasonal Workers Vanuatu' was the largest ni-Vanuatu Facebook group targeted at seasonal workers with roughly 26,900 members. The group included seasonal workers, agents, their families and others interested in the schemes. Common posts in the group included asking for information about or trying to contact other workers or family members; the context of employment (what does Australia look like? What does commercial agriculture involve?); self-management and organisation (such as warnings about 'bad' behaviour); and discussions about policy and developments in the scheme. Perhaps the most important function of these groups was to allow workers to share tips about navigating the uncertainties of the schemes, such as pay rates, participation fees and access to entitlements like superannuation and holiday pay (Petrou and Connell, 2020). A typical post from 2019 asked:

Any clarification please!! Is it possible that people can sign a contract before doing medical and police clearances?!?

Watch out for con men bro…

Bro, you can do it either way but only if you're going to travel in less than a month, because I believe that the medical clearance is valid for less than 12 months.

You can do it that way, sign all of your documents first, and then when everything is organised, just pay for the medical and police clearance.

Another post from 2019 discussed holiday pay entitlements:

Is it alright for a company to hold back their workers' holiday pay and say it is to pay for airfares and visas next year?

Good question. It is your right to take your holiday pay if you want to, but it can really help with all of the deductions if you don't claim it… But you have every right to take it…

If you're confused about what they've told you, just search on Google for Australia and NZ minimum wages for holiday pay. Have a read, and you'll see that holiday pay must be paid out.

Other posts from around the same time warned about scams:

Someone is going around to ni-Vanuatu SWP workers in Australia, asking them to sign a letter authorising him to withdraw their superannuation money and send it to the VNPF [Vanuatu National Provident Fund]. We don't know much about this person, but we're suspicious about fraudulent activities. He's an Australian. Please, does anyone know anything about this person?

Other discussions centred on wages and employment conditions. Group members believed that Pacific and Australian governments both had a role to play in ensuring the fair treatment of seasonal workers. The following comments below, from 2019, typified conversations around these themes:

Our Government could have laid out a few vital conditions to the Aussies before signing the PACER+, and one amongst many being Australia should

> consider improving the work conditions and pay packages of our fellow countrymen/women engaged in the SWP.
>
> I feel for that lady [who has been underpaid]. Many people are interested in the SWP but you have to choose the right employer. You shouldn't just listen to people saying lots of employers in Australia are good, and then go and sign up with any group to go. If you don't choose the right one, you'll regret it later.
>
> Lots of farms in Australia cheat foreign workers who do seasonal/farm work through temporary work visas. Some German backpacker friends of mine told me that one farm in Australia only paid them $300 for a week's work. So they don't want to work on farms in Australia anymore. There's a lot of worker exploitation still happening on Aussie farms… Just like slavery days. All of the workers must keep exposing them to the point that the Australian government must take legal action against all of the sick employers who do this.

Many posts provided sage advice about procedures to follow and people to contact if workers did encounter problems. A typical post from November 2020 explained:

> There are procedures that you can follow if you're not happy with your working conditions and you want to complain, and there are people who can help you.
>
> 1. Speak to your team leader
> 2. Speak to your supervisor
> 3. Speak to the people at your farm
> 4. Speak to your employer
> 5. Speak to your pastoral carer
> 6. Speak to the Vanuatu High Commissioner in Australia
> 7. Speak to your agent
> 8. Speak to the union or Fair Work [Ombudsman], this is the last step for dealing with your case. Do not abscond as a last resort, because you are just creating problems for yourself.

In addition to these types of warnings and cautions, group members aired concerns over the impacts that seasonal work was having on family life:

> I want to travel, but I've seen some women post because they're concerned about their husbands … But I have a question, do you want to travel for work, or to find a partner? Because now there are lots of broken homes as a result of seasonal work. If you have a problem in your home, it would be good if the agent could screen everyone before they send them to work for the RSE or SWP … Because some people form happy couples overseas, and then when there is a complaint they blame their partner at home. If you can't travel and behave responsibly, then you shouldn't travel. That's just my opinion.

Not only did social media provide an opportunity for virtual conversation, but warnings and advice sometimes enabled real-world outcomes and connections within Australia. Tim, a Tuvaluan living in Australia, described to Kirstie how he had been contacted by some i-Kiribati workers who had experienced problems with their employer. Tim described how, through Facebook, he had been able to help SWP workers connect with people and services to assist them with problems that they faced:

> It just happened that I met this guy on Facebook. He is a seasonal worker in Mildura. And that's when he said 'Oh, can you help with [some problems I'm having at work]?'. I said, because I speak Kiribati as well so, I said 'Look I can direct you to the, probably the right person who you can talk to her and she might be able to help you with your situation'.

Through harnessing both virtual social networks and connections to Pacific islander communities in Australia, Facebook provided an important means for geographically isolated seasonal workers to access culturally appropriate assistance in Australia.

Social media were not merely used for informing and educating present and potential workers, but they were also a means of describing and discussing their social worlds, and explaining what their work and life looked like in Australia, often accompanied by photos of themselves at work or resting and cooking in their temporary homes. Photos and videos of days off and special celebrations or excursions were often shared (Fig. 8.3). Here workers could be much more enthusiastic about work and life and the new experiences that came with being in Australia.

Fig. 8.3 Ni-Vanuatu workers celebrating Vanuatu's Independence Day in Broome, Western Australia. (Source: Facebook)

Social media positively contributed to changing the working and social context of employment, as information was shared and workers became more knowledgeable of their rights and duties. Facebook groups provided seasonal workers with a virtual community—and one much larger than they would be likely to encounter in real life. In this community, seasonal workers shared information and ideas in real time, aired grievances, worked through problems, discussed the benefits and pitfalls of seasonal work and found positive outcomes. Various suggestions were made on how to shift the balance of power in guestworker schemes. Most simply focused on improving the work-life experience by legislating for adequate standards and regulating them, variously seen as the province of local and national governments, of unions and employers. Such a complex 'division of labour' meant that regulation—and easy and quick recourse to mechanisms of redress—falls between multiple cracks. Yet, while social media provided a unique avenue to help workers navigate problems and uncertainty, it could not affect the structural precarity of the scheme itself. Some therefore employed their own methods: a Samoan worker, indignant at the cost of medical insurance from which he had gained nothing, feigned illness to get free medicine to take home (Fatupaito, 2021). Some workers employed creative means to benefit from the system.

Transient Times

For many SWP workers, life in Australia was quite different from that in their home countries. Especially where workers came from remote islands (particularly from coral atolls where comparable agriculture was absent and the highest point was two metres above sea level), substantial adjustments were needed. It is unsurprising that their new lives could be puzzling and precarious, and that while they were continually pressed by participating governments to 'behave', they were less protected by these governments and open to exploitation. While 'bad apple' employers and contractors were ever present, and some concerns never disappeared, social media, informal pastoral care and occasional union involvement helped mitigate the worst problems.

Problems did however regularly emerge from challenging workplace practices, and the means of redress for workers to rectify problems were often at odds with cultural norms. Confrontation was avoided. Employers and contractors did sometimes cut corners, breach guidelines, underpay and mistreat workers, and some dissent and disappointment were probably inevitable. How widespread such problems were is impossible to determine. The worst examples that attracted media attention were not typical; most workers did not face blatant exploitation. Indeed, many workers spoke positively about their work and living experience and exhorted others to understand guestwork as 'an opportunity to make a difference to your family and your lifestyle' (Facebook, 2020). As Alice put it to us quite simply—effectively speaking for many ni-Vanuatu workers: 'It's a good experience to come and work here. I like the work here'. Likewise most employers valued their employees, however short-term they might be, and sought to minimise their adjustment problems; doing so could potentially enhance labour productivity and encourage return. For some employers, workers could become close friends. Speaking of the imminent departure of a group of Timorese workers who had been stuck in Australia for 19 months due to COVID-19, one farm manager said: 'It actually gets a bit emotional putting them on the bus because they're like family to us [...] They call this place their home' (quoted in ABC News, 18 May 2021). A mandarin farmer in Mundubbera claimed: 'I do not ever want to go back to the situation where we were totally reliant on backpackers ... I don't know anyone in this area who has tried using the scheme and not gone back the next year with a request for more workers' (Jenkin, 2015). Another stone fruit, citrus and almond farmer who had been employing Timorese workers for five years reported: 'They have the ability to virtually run this company without needing steady supervision from myself. The Seasonal Worker Programme is a brilliant program and I hope it increases forever on' (quoted in ABC News, 15 August 2021). Such farmers were the primary beneficiaries of the SWP.

Communities too, or at least groups within them, whether sports teams, churches or even charity op shops, sought to assist temporary workers to settle in and recognise the contribution they could and did make. Migrants who found themselves working near established Pacific islander communities were usually particularly fortunate, in being guided on how to navigate life in Australia. However temporary their stay, the benefits could be reciprocal and more than merely economic. Influxes of seasonal workers brought some vibrancy to struggling small towns, and even assisted in times of drought and bush fire. With repeated return migration to some areas, such as the Riverina, the boost that workers brought to rural places, both economically and socially, could be long-lasting and in a small way contributed to a revitalisation of the countryside.

When problems did arise, workers sometimes spoke out and were aided by unions, pastoral carers, local Pacific islander communities and others. A large community of seasonal workers, their families and others interested in the scheme used social media to navigate the promises and pitfalls of seasonal work, to understand their rights, and to debate and discuss the pros and cons of temporary labour migration. Seasonal workers were far from mute subjects and, where they could, were actively and creatively exercising agency to avoid the worst aspects of the scheme and capitalise on the best parts. There were many examples of both. In the end what mattered most to them, was the outcome at home, which we turn to in the next chapter.

Bibliography

Australian Government. (n.d.). *Working and living in Australia: Pre-departure guidebook for seasonal workers*. AGPS.

Bailey, R. (2014). *Working the Vines: Seasonal migration, money, and development in New Zealand and Ambrym, Vanuatu*. PhD thesis, University of Otago, Dunedin.

Bailey, R. (2018b, July 12). *Blacklisting seasonal workers*. ANU DevPolicyBlog.

Bailey, R. (2019a, April 26). *Pacific workers in rural Victoria*. ANU DevPolicyBlog.

Bailey, R., & Ng Shiu, R. (2016). *New responses to natural disasters through Seasonal Labour Mobility Programs*. ANU InBrief 2016/12.

Basok, T. (2002). *Tortillas and tomatoes: Transmigrant Mexican Harvesters in Canada*. McGill-Queen's University Press.

Blanco, C. (2009). *Temporary migration, remittances and development: Early impressions of the Australian Pacific seasonal worker pilot scheme*. Honours thesis, University of Sydney, School of Geosciences.

Brickenstein, C. (2014). *Foreign Seasonal Workers in Agriculture; a comparison of regulations and practices in the Pacific and in Europe*. PhD thesis, Macquarie University, Sydney.

Cameron, E. (2011). *Social development outcomes of participation in the New Zealand Recognised Seasonal Employer (RSE) scheme for ni-Vanuatu seasonal migrant workers*. Master's thesis, Massey University, Palmerston North.

Cavanough, E., & Wherrett, C. (2020). *Blue Harvest. Wage theft & other labour infringements in the NSW Mid-North coasts's 2019/20 berry harvest*. The McKell Institute.

Cave, D. (2012). *Digital Islands. How the Pacific's ICT revolution is transforming the region*. Lowy Institute.

Centenera, N. (2016, December 1). *Mallala hospital becomes seasonal worker home* [Online]. Two Wells, South Australia: Two Wells and Districts Echo. Retrieved from https://www.twowellsecho.com.au/2016/12/01/mallala-hospital-becomes-seasonal-worker-home/

Dao, A., Green, M., & Kass, T. (2020, May 16). We feed you. *The Saturday Paper*. https://www.thesaturdaypaper.com.au/news/politics/2020/05/16/we-feed-you/15895512009836?fbclid=IwAR1X5r411aDQfb397ODGaQey8a3YtJ_hdIBAGnPLm0d4w5GXp9hE1XjDvn4.

Davidson, H., & Earl, C. (2020, October 18). The fruit pickers: Inside Australia's seasonal worker program, *The Guardian*.

Dun, O., McMichael, C., McNamara, K., & Farbotko, C. (2022). Investing in home: Development outcomes and climate change adaptation for seasonal workers living between Solomon Islands and Australia. *Migration and Development, 11*(3), 852–875.

Employment Services Unit. (2019). *Migrant Worker Disciplinary Policy*. Vanuatu Department of Labour.

Fatupaito, A., Utuva, L., Tauave, S., Alofipo, A., Meleisia, M., Schoeffel, P., Arthur, T., & Alexeyeff, K. (2021). Samoa's New Labour Trade. *Journal of Samoan Studies, 11*(1), 51–63.

Finau, G., Kant, R., Logan, S., Prasad, A., Tarai, J., & Cox, J. (2014). Social media and e-democracy in Fiji, Solomon Islands and Vanuatu. 20th Americas Conference on Information Systems, Georgia, mimeo.

Fonseka, D. (2020) *Boss allegedly assaulted seasonal worker* [Online]. New Zealand: Newsroom. Retrieved April 17, 2021.

Hobbis, G. (2019). New Media, New Melanesia? In E. Hirsch & W. Rollason (Eds.), *The Melanesian World* (pp. 546–560). Routledge.

Holani, A. (2017). *Enhancing the development impact of labour mobility in Pacific Island countries: The role of the seasonal worker program and the PACER plus*. PhD thesis, University of Adelaide.

Howe, J., Clibborn, S., Reilly, A., van den Broek, D., & Wright, C. (2019). *Towards a durable future: Tackling labour challenges in the Australian Horticulture Industry*. University of Adelaide.

Jenkin, S. (2015, March 5). *Issues with the Seasonal Workers Program: An employer's perspective*. ANU DevPolicyBlog.

Lees-Galloway, I. (2017). *Government continues to fail migrant workers*. Labour New Zealand.

Nishitani, M., & Lee, H. (2017). Invisible Islanders: Precarious work and Pacific settlers in rural Australia. *Pacific Studies, 40*(3), 430–449.

Petrou, K. (2020). *If everyone returned, the Island Would Sink: Urbanisation and migration in Vanuatu*. Berghahn.

Petrou, K., & Connell, J. (2018). 'We don't feel free at all': Temporary ni-Vanuatu workers in the Riverina, Australia. *Rural Society, 27*(1), 66–79.

Petrou, K., & Connell, J. (2020). Overcoming precarity? Social media, agency and Ni-Vanuatu seasonal workers in Australia. *Journal of Australian Political Economy, 84*, 116–146.

Petrou, K., Dun, O., Farbotko, C., & Kitara, T. (2021). Pacific labour mobility on pause: Consequences of temporary immobility during the pandemic. In Y. Campbell & J. Connell (Eds.), *COVID in the Islands: A comparative perspective on the Caribbean and the Pacific* (pp. 299–319). London: Palgrave Macmillan.

Reid-Musson, E. (2017). Grown close to home™: Migrant farmworker (Im) mobilities and unfreedom on Canadian family farms. *Annals of the American Association of Geographers, 107*(3), 716–730.

Schneiders, B. (2018, May 18). Key farm scheme badly rorted, migrant workers paid $8 an hour. *Sydney Morning Herald.*https://www.smh.com.au/business/workplace/key-farm-scheme-badly-exploited-migrant-workers-paid-8-an-hour-20180518-p4zg53.html

Schneiders, B. (2019, August 1). Exploited Pacific migrant farm workers settle significant case. *Sydney Morning Herald.* https://www.smh.com.au/business/workplace/exploited-pacific-migrant-farm-workers-settle-landmark-case-20190801-p52cvi.html.

Smith, R. (2016). *The goal of the good house: Seasonal work and seeking a good life in Lamen and Lamen Bay, Epi, Vanuatu.* Unpublished PhD Thesis, University of Manchester, Manchester.

Stead, V. (2019). Money trees, development dreams, and colonial legacies in contemporary Pasifika Horticultural Labour. In V. Stead & J. Altman (Eds.), *Labour lines and colonial power: Indigenous and Pacific Islander Labour Mobility in Australia* (pp. 133–158). ANU Press.

Stead, V. (2021). Precarity's reach: Intersections of history, life, and labour in the Australian horticultural industry. *Journal of the Royal Anthropological Institute, 27*(2), 303–320.

Tazreiter, C., Weber, L., Pickering, S., Segrave, M., & McKernan, H. (2016). *Fluid Security in the Asia Pacific* (pp. 103–127). Macmillan.

Uibu, K. (2020, September 15). There are no human rights here. *ABC News.* https://www.abc.net.au/news/2020-09-15/backpacker-farm-workers-speak-of-wage-exploitation/12545294?nw=0.

Union Aid Abroad APHEDA. (2017). *Temporary labour mobility in Australia and the Pacific Islands.* Union Aid Abroad APHEDA.

Union Aid Abroad APHEDA. (2020). Inquiry into strengthening Australia's relationships with countries in the Pacific region: Submission 20. Submission to Joint Standing Committee on Foreign Affairs, Defence and Trade—Foreign Affairs and Aid Sub-Committee: Canberra.

Vanuatu National Statistics Office. (2012). *Vanuatu Household Income and Expenditure Survey 2010.* Vanuatu Government.

Vanuatu National Statistics Office. (2017). *2016 Post-TC Pam Mini-Census Report: Basic Table Volume 1.* Vanuatu Government.

Wheeler, T. (2020, February 5). *South West farm workers take part in Aus Day soccer tournament.* Western Australia: Manjimup-Bridgetown Times. https://www.swtimes.com.au/news/manjimup-bridgetown-times/south-west-farm-workers-take-part-in-aus-day-soccer-tournament-ng-b881447078z?fbclid=IwAR0lJDs0BLSFocKvhI9cTyCeAnH7btO5IfYyJEVs9zRDZH-CLaaZsN8IxHR8.

World Bank. (2017). *Maximizing the development impacts from temporary migration*. World Bank.

Wu, A. (2019). *Timorese participants in the Australian seasonal workers programme: Migration, remittances and development.* PhD thesis, University of Sydney.

9

Home Again

The potential and promise of income gain and development benefits was undoubtedly the key objective of the Seasonal Worker Programme (SWP) for Pacific Island Countries (PICs), the workers, their extended households and sometimes their communities. Pacific guestworkers remitted money or returned home with substantial savings when compared with local income earning opportunities. The incomes generated through the SWP were transformative for various households and villages, with multiple welfare gains, useful new goods, opportunities for small business development and greater involvement in traditional exchange practices. More formally, earnings trickled upwards into the national economy. At the same time workers had the opportunity to 'see the world', experienced other lifestyles and cultures and acquired skills and habits: social remittances that could prove useful. This chapter examines these various impacts and outcomes.

The material rewards from the SWP were not always unproblematic, and were sometimes associated with social tensions. Negotiating the changed social relationships that evolved as a result of incomes and absences could prove difficult. Family separation exerted an emotional toll on households and communities, particularly when it happened repeatedly. The seasonal absence of a large proportion of young working-age males could

K. Petrou, J. Connell, *Pacific Islands Guestworkers in Australia*,
https://doi.org/10.1007/978-981-19-5387-3_9

result in increased vulnerability and even possible food insecurity for participating households and communities. As more urban workers participated in the scheme, concerns were raised about a possible skill and brain drain, and of challenges to equity and regional development. At the same time, social remittances—including the changes in ideas and ideologies that returning migrants brought back—could be both constructive and destructive. Ultimately, whether returned migrants acted as agents of universal positive change is an open question, and it is where we end this chapter. Relatively little research exists into the social impacts of the SWP on home societies, and this chapter necessarily draws on examples from the Recognised Seasonal Employer (RSE) scheme to a greater extent than previous chapters. There is every reason to believe that the impacts of the two schemes in the PICs have been much the same.

New Wealth, Familiar Patterns

From the first tentative initiation of the RSE and SWP schemes, there was never any doubt that the objective of the participants was income generation (Chap. 5). Savings, along with new skills, could be channelled into small businesses, especially stores, and bring welfare gains through improved housing, piped water, education and so forth. Overall, with their large number of participants, Vanuatu and Tonga were the main economic beneficiaries of the SWP. Compared to Vanuatu, new remittance flows were less significant for socio-economic change in Tonga and Samoa, both of which had long been the beneficiaries of substantial remittance flows from large diaspora populations. By contrast, in Vanuatu, where outmigration was minimal and international remittances scarce, the new cash inflows were very significant. They were not entirely without precedent, since many ni-Vanuatu had participated in short-term migration to New Caledonia during the 1960s–1970s, but otherwise contemporary international migration was rare in the Melanesian states, and remittance flows were mainly domestic and small. In Tuvalu, by the 2010s, where some degree of dependence on remittances had characterised much of the past century, the 'new' remittances from seasonal workers exceeded those from seafarers to the extent that they were regarded by a prominent Tuvaluan public servant as 'key to the sustainability of life in Tuvalu' (Falefou, 2017, p. 259). Significant income flows through remittances were therefore a novel phenomenon for participating households in most PICs.

Workers often described seasonal employment as 'improving their living' by facilitating access to housing and superior material living standards, and regarded the earnings potential of the scheme as a 'very important' motivator for participation (Chap. 7). For a finite experience, where the potential earnings were reasonably well-known, most seasonal workers had a 'target' or 'goal' to achieve; investing in land (where that was feasible) or housing, starting a small business and paying school fees were longstanding and virtually ubiquitous goals, repeatedly appearing across all PICs (World Bank, 2017; Bedford, 2013; Blanco, 2009). The workers we met over the years were invariably working towards at least one of these goals. But that was never easy; as the recruiter for one of the largest employers in Mundubbera suggested:

> Most of these workers tend to not have long term plans for the livelihood of their families...many of them don't have plans for life after the programme and that's very risky... They can't do it on their own. They need support and it would be really great if government can provide some sort of career planning or support to help them set up plans for their future. (Quoted in Holani, 2017, p. 132)

Establishing or merely contemplating a new improved life required support and guidance. Most workers simply followed those who had preceded them.

Joshua was 39 years old and came from rural Santo, Vanuatu. He was a semi-subsistence agriculturalist and found it hard to make enough money from copra and livestock to pay school fees for his three children. In 2012, Joshua travelled to Australia for the first time, with the aim of earning money for school fees. He was also organising to send some solar panels home. If he had enough money left over, Joshua thought he might be able to start building a house:

> In Vanuatu, school fees are expensive. We make copra [to earn money]. Sometimes, when the price of copra isn't stable, it goes up and down. When the price goes down, we have to try hard to earn money. Cows are exactly the same. When the market for cows is good, I can make good money, but when it's not good… So that's why I decided to come, I thought I'd try and work [in Australia] because I have to pay school fees. And If I see that I have more money left over, I might use it for something, like I could build myself a house.

Twenty-four-year-old Avok worked as a civil engineer in Vanuatu. His family was originally from Malakula, but lived in the suburbs of Port Vila. He had a two-year-old son, but was no longer in a relationship with his son's mother. Avok also travelled to Australia for the first time in 2012. He remarked that he would normally have been interested in the kinds of things that other workers in his group were buying, including laptops and DVD players, but he had decided that he preferred to save his money and put it towards establishing a small business on land his deceased father had purchased in a relatively affluent area of Port Vila:

> I want to start a business when I go back. A small business, not a big one […]. A small store, or I'd like to build a rental property. I'll start with this first, and I think there might be some other things I can do as well, but this is my first plan.

Edna was 37 years old and lived on Santo, although her family was originally from Ambae. She had worked at a motel for 13 years but resigned so that she could travel to Australia for the first time in 2012. Her husband had also worked at the hotel but resigned when he became busy with church duties, and was making money by cooking and selling *gato* (fried bread). The couple had two children aged four and ten. Edna explained her decision to travel to Australia:

> I decided to come to Australia because I've been building a small house, but I haven't finished it […] Vanuatu is good, but the wages we can earn there are small. So you pay for food, pay the electricity bill, pay the water bill, and you can't save much […]. And if your family has any problems, you use your savings to help them. […] If someone dies, we give [money]

> to help our family. So I thought I if I came here, maybe if I work, I can save some money and... it can help me to pay for some materials to complete my house.

Eight years later, the SWP workers we met in peri-urban Adelaide had almost identical goals; Lucy was in her twenties and unmarried. She had been working at a takeaway store in Port Vila before travelling to Australia. Lucy wanted to buy some land with her new income because her family had been going through a land dispute. Netty was in her forties and had already bought some land, where she planned to build a house for her children and pay their school fees. Adam had already bought land on Santo and wanted to build a rental property that would provide an ongoing income. He also wanted to repair his car, and hoped that these combined investments would provide a secure economic future for him and his family. Workers from the smaller island of Ambae have used their incomes to buy land on much larger Malakula and Santo, where they could then relocate for new employment opportunities, business possibilities and access to high schools (Bailey, 2021). These various goals could usually only be accomplished in the two main urban areas, where there was some possibility of purchasing land, resulting in the rise of a new group of urban residents, who were not customary landowners, but who had chosen to make town their permanent home (Cummings, 2016). In peri-urban Port Vila, increased urbanisation was one consequence of labour migration.

As workers became more familiar with the schemes, made more than one trip, and realised what could be achieved, aspirations became more subtle and more complex.

> Nora was 45 years old and married to a *man-Ifira* [someone whose ancestors come from the island of Ifira], since deceased. Together they had one daughter and two sons. Nora joined the SWP to pay her children's school fees and meet other daily expenses. After her husband died, Nora's life wasn't easy as she had to look after her children. She decided to join the SWP in 2015, but things didn't turn out well in her first season, so she decided to have a break until 2018. In 2019, she returned and signed a contract for six months to work on [Farm X] where she worked with pineapples and bananas. She used her 2018 savings to pay school fees for her

children, aimed to build a permanent family house with her 2019 savings, and intended to return again in 2020 to build an adjoining rental property; she enthused: 'I encourage all single mothers or widows who are in the same situation as me not to give up. Think hard about planning and making something for your family to help them'. (Facebook, 2020)

David had similar goals

David was 27 years old and lived on Malakula. He had participated in the RSE for two seasons and had travelled to northern New Zealand through the RSE program. He was recruited directly by [Company Y] and worked on a kiwi fruit farm. He came back from his second season and has confirmed that he would then be able to complete his house. His next aim was to buy a boat that he could take back to the island. David advised everyone who travels to think really hard about their aims when they are working in Australia and New Zealand under the schemes. He encouraged everyone who hadn't gone yet to save up and participate in the schemes so that they could achieve something at home. 'Because if we just work in Vanuatu, we'll never achieve our goals because of the limited finance. But it's good if you go and you can save up and meet your goals'. (Facebook, 2020)

Despite such investments in housing and businesses, remittances from ni-Vanuatu SWP workers were more likely to be used for basic needs such as housing improvements, school fees, clothing and food, whereas in Tonga and Samoa, where these needs had largely been achieved, business investment was a greater priority. Nonetheless Alisi Holani (2017) still found in 2015 that the most important use of remittances was purchasing food, followed by education fees and utility bills. Any business development was more likely to follow the actual return of the workers. For many workers, the most immediate and obvious goal was to pay off the loans that enabled them to participate in the SWP, to meet the costs of airfares, clothes, passports and visas. Such costs were substantial, household objectives had priority and it normally took RSE workers several seasons before they felt financially stable enough to invest in small businesses (Bailey, 2019a), emphasising why many workers were enthusiastic about returning for additional seasons.

Differences in the gendered use of remittances were evident. Women from Tonga and Vanuatu seemed to spend remittances on immediate family needs including school fees, and customary obligations but also small businesses. Such immediate needs also included repairing damage from cyclones and other hazards. By contrast, men tended to spend more money on hardware supplies, including water tanks and solar panels, and tools, and community and village obligations, but they also spent income on such leisure activities as drinking kava or alcohol (World Bank, 2018a). Transported remittances and 'savings' were often consumed quickly when workers returned home. Rather than holding on to their money, Tongan SWP workers tended to save for a specific goal, and then spend the money (Holani, 2017). Few male SWP workers appeared to dictate to their wives how remittance money should be used, and women in Tonga and Vanuatu often assumed greater responsibility and autonomy in managing household finances and remittances as a result of their husband's SWP participation (World Bank, 2018a). With so many goals and objectives, saving money made little sense to workers, was not a culturally appropriate way to utilise this newfound income and kin put pressure on 'spending' (sharing) money to reinforce kinship ties and relationships (see below).

Seasonal workers had various goals for their earnings, most little different from wider migration contexts and remittance flows, primarily aimed at improving their material standards of living, and putting in place variants of modernity. In Vanuatu as elsewhere, the new income was directed at some form of modern improvement, and matched the aspirations and uses involved in other remittance flows within the Pacific (Connell & Brown, 2005). These goals varied little with nationality, both within the Pacific and amongst seasonal migrant workers much further afield (Binford, 2002, 2003). Indeed these aspirations and eventual achievements were reflected in many of the 'success stories' that were commonly shared on social media, in the *Vanuatu Daily Post*, or the *Samoa Observer*, to help 'boost' the scheme, and encourage others to share these dreams and objectives. The dramatic photograph of vehicles said to be purchased by seasonal workers (though no single trip would make that possible), published in the *Samoa Observer* (Fig. 9.1), with an accompanying laudatory story, symbolised and exemplified how the schemes were regarded. Below we consider these flows and goals in more detail, and how the social impacts of this new wealth have been experienced.

Fig. 9.1 New vehicles purchased by seasonal workers, Savai'i island, Samoa. (Source: RNZ, 8 September 2020)

Remittances in Kind: Technology and Consumer Durables

In addition to money, workers often returned home with consumer goods or sent them in containers. From the start, as with cash flows, goods brought back from seasonal migration mirrored in-kind remittance flows elsewhere, and a distinct spatial and temporal pattern existed. In the early years in Vanuatu, RSE workers from Ambrym returned home with mattresses (Bailey, 2014), whereas workers from Tanna wanted chainsaws (Hammond & Connell, 2009). Workers from some parts of PNG similarly favoured chainsaws (Brickenstein, 2017), which both assisted in housebuilding and enabled faster, more efficient land clearing; there were obvious benefits to obtaining these tools, particularly for the men who generally performed the hard labour associated with these tasks. Owning such items also brought a degree of social prestige, even if, over the longer-term, negative environmental consequences could follow; some approved employers expressed concern at the potential environmental degradation brought about by land clearing with chainsaws:

> Recently we were looking at a satellite picture of one of the islands (in Vanuatu). You can see there are more houses now, but also because there is a limited timber supply as they also buy chainsaws and take down a lot of trees, so it is having an impact on the local environment back there. We also considered where is our responsibility here concerning these impacts? But this just becomes too difficult to answer and must be dealt with in these countries. (Quoted in Brickenstein, 2017, p. 68)

In Lamen Bay (Epi, Vanuatu), construction of new housing and an exponential increase in the number of Western-style showers and flushing toilets led to both 'large-scale erosion', as land was cleared for construction, and put exceptional pressure on local water resources (Craven, 2013). However, the rate of participation in migration schemes in Lamen Bay has been particularly high, and such environmental stresses did not necessarily follow labour migration.

Typically, in 2010, Tannese RSE workers were already going through familiar processes of paying school fees, building and renovating houses, and starting small businesses. To support these projects, they returned home from New Zealand with such goods as 'solar panels, hand tools, power tools, sewing machines, a microwave, toasters, electric kettles, a freezer, second hand beds, duvets, sheets, mattresses, second hand furniture and bicycles' (Cameron, 2011, p. 102). By 2015, chainsaws, TVs and motorbikes were arriving in Kiribati. As other national groups participated, the outcomes were much the same. By 2019, Solomon Islander SWP workers carried similar goods home; 'batteries for solar panels; boat engines; children's clothing, toys and learning materials; clothes; electricity generators; farming chemicals (insecticides, fertilisers); furniture; household utensils and equipment; building materials (e.g. timber, concrete); mobile phones; sanitation infrastructure (e.g. water pipes, toilet); solar panels; stereos; tools (e.g. drills, whipper snippers, brick mould); televisions; vehicles; and water tanks' (Dun et al., 2022, p. 864). Most workers considered it entirely natural to acquire goods that enabled a greater level of personal comfort.

Variants on this multiplicity of durable assets and material possessions appeared throughout the PICs in similar form, depending on what needs had already been met, but then it is hard to envisage what realistic additional alternatives might exist. Some goods had particular priority: solar panels were popular since they could often fit into suitcases and resulted in savings on fuel, which was required to run the generators that powered electric alternatives. They also enabled increasingly ubiquitous mobile phones to be charged. Such material goods offered real welfare and development gains: for example, throughout Melanesia solar panels enabled lighting so that children could study at night.

Almost all of the goods that workers brought back home were bought in Australia (or New Zealand)—where they were usually cheaper—and either carried or shipped home in a container shared between workers (Figs. 9.2 and 9.3), hence that expenditure had no multiplier effect in the

Fig. 9.2 Ni-Vanuatu workers with a container of goods ready to be shipped home. (Source: Facebook)

Fig. 9.3 Containers of RSE purchases are unpacked upon their arrival in Port Vila. (Source: Facebook)

PICs. Similarly, but at a different scale, income spent on school fees went to the national capital. Only a relatively small proportion of savings were spent on purchasing items in home villages, one reason why the small stores we discuss below often failed. Nonetheless, workers employed other means of ensuring money circulated within villages, and sometimes contributed to community projects which maintained a certain degree of equity between households.

Not everything carried home necessarily brought about 'development' in the formal sense of either improved welfare or economic growth; consumer electronics included mobile telephones and laptops while other items such as stereos ('ghetto-blasters') and headphones were also popular. That inevitably meant that early on many complaints were directed to workers 'wasting' money on consumer goods that did not contribute to local communities or families. Such concerns demonstrated the clash that could occur between loosely 'traditional' (money used to reinforce relationships) and 'modern' (money used for individual consumption) values, and a deep ambivalence about the outcomes of expenditure.

Housing

Early seasonal labour migrants had basic objectives: 'fix the school fees ... make a proper toilet' and build a nice kitchen (quoted in Calnitsky, 2016, p. 20). More adequate housing remained a key goal for seasonal workers. In Vanuatu, building a 'good house' (one constructed from bricks and cement) became something of a moral and practical imperative, and an expected outcome from seasonal employment. As a Tannese worker told John in 2019: 'quite simply when you see a good house in the village it is because of work in Australia'. Ni-Vanuatu guest-workers who did not invest in housing were often criticised by other villagers. When Kirstie visited the island of Paama in 2011, a handful of villagers had participated in the RSE, but none had built a house (although one was stockpiling construction materials). To many villagers, this visible lack of investment in housing indicated that these workers had effectively wasted their time. Investing in consumer items (particularly electronics), as most Paamese had done, was considered morally suspect and frivolous

> Of everyone who works in New Zealand these days, not one of them has built a house here ... They go, they come back, and I haven't seen them do anything with their money ... They've just bought things like TVs, generators and solar panels, but they haven't built a house. It's no good. (Duncan, quoted in Petrou, 2020, p. 96)
>
> This man here went to New Zealand for seven months, but he hasn't achieved anything, he hasn't built a good house. (Evan, ibid.)

For Paamese, building a 'good' house was a long-held social norm. During the 1970s many Paamese men travelled to Noumea where they worked on short-term contracts before returning home, and using their savings to build modern houses which both acted as a symbol of success and neutralised their wealth so that a relatively egalitarian socio-economic status was maintained between villagers. The electronics and other flashy consumer items, purchased by the next generation of workers in 2011, posed a threat to these arrangements. In contrast to the 'failures' of the Paamese, and perhaps more typical of trends elsewhere in Vanuatu, were the 'asparagus houses' of Ambrym, triumphantly built and painted green by those who had worked on asparagus farms in Australia. Reflecting these norms, housing was a common feature of the 'success stories' widely shared in seasonal worker Facebook groups. Twenty-nine-year-old Thomas from Paama provided a typical example of a migrant 'success story'.

> Thomas travelled to Australia for the first time in 2018 through [Agent X]. He worked at [Farm Y] on a six-month long contract. At this farm he worked on mandarins and grapes. Thomas decided to travel to Australia so that he could help his mother. He wanted her to be able to sleep in a good house: 'My mother was sleeping in a *natangura* [traditional] house, and whenever it rained she would get all wet. I decided I wanted to build a good house for her, and I felt really strongly about it, so that's why I decided to join the scheme'. Last year when he came back, Thomas started building the house. The house is half finished, and he plans to complete it when he comes back from Australia this year. Once he has completed the house, he thinks he might start his own small business. (Facebook, 2020)

Craig had a similar success story to share (Fig. 9.4)

> Craig from Eratap village [Efate] is proud to show his new house built from the income he earned after attending the SWP. Inside the house he has a large TV screen with modern electrical home appliances which he also purchased from overseas before returning to Vanuatu. He now continues to be a farmer in Eratap supplying farm produce in Vila. (Facebook, 2020)

As much as anything else, development represented a visible and aesthetic transformation, primarily of consumption. In parallel, on Epi, Lamen Bay residents: solar lights and ice cream were as valuable as boosting production or creating infrastructure (Smith, 2016). Modernity was a visible phenomenon.

On Epi too, building a 'good' house was one of the main reasons that villagers participated in seasonal work. As on Paama, 'good' houses provided a symbol of success and were viewed as a means of bringing 'development' to home areas, whether villages or towns (Rockell, 2015). Being able to own both land and a house in Port Vila was the acme of

Fig. 9.4 Craig shows off his 'good' house built with SWP earnings, Vanuatu. (Source: Facebook)

development and achievement for many ni-Vanuatu. When one SWP worker heard from her husband that her remittances had finally paid off the loan to purchase land and a house there she exulted: 'I feel excited and tears running down my face, my cheeks, because I never expected .... So that's why I am come here for three years now...I've got two things that I came here for: the school fees for my children and my land' (quoted in Stead, 2019, p. 149). The durability of modern houses symbolised the provision of a secure future for families, provided a very visible status symbol and attested to the merits of the guestworker schemes. A brick house might last for several decades whereas traditional houses (Fig. 9.5) needed to be replaced every few years—usually no more than about seven years as thatch wore out and rooves began to leak—and were more susceptible to destruction during cyclones: relatively frequent in much of the Pacific. By contrast, and while the enclosed brick and cement walls of 'permanent' houses were probably less suited to tropical environments, their durable nature and the lower maintenance requirements were valued alongside the protection from natural disasters.

While a 'good' house represented the 'good life' and the future wellbeing of its occupants, some ambivalence was nonetheless attached to these new houses since money invested in building could not be shared with community members to maintain social relationships. Similarly, the need to involve a paid builder in the construction meant that the reciprocal,

Fig. 9.5 An example of traditional housing, Paama Vanuatu. (Source: Author Photo)

collective work normally involved in building a traditional materials house was slowly disappearing (Smith, 2018). Thus, in Tanna young men who had participated in the RSE expected to be paid for their role in house construction rather than viewing it as a part of a community exchange (Rockell, 2015). Similar patterns around the commodification and monetisation of labour were also evident on Paama, where participation in the schemes was very limited. Seasonal labour migration was thus speeding up—rather than creating—processes and trends already evident. Houses represented not just the 'good life' but a place where commercial relations and wage labour were gradually transcending or replacing social relations and communal work. In some respects, that too was evidence of the 'good life'.

Not only were houses constructed differently, but settlement patterns changed to accommodate this modern housing. In Lamen Bay, houses were built in a more dispersed pattern, enabling secluded, private gardens with flowers or lawns. Along with anxieties about climate change and the need to move to higher ground, this meant that houses were being constructed on land once used for food gardens, so reducing the amount of land available for food production. This expansion increased the possibility of land disputes due to the 'permanent' nature of these houses. Traditionally, constructing a 'permanent' dwelling on disputed land was discouraged as it implied the permanent ownership or use of the land. Yet, as land needs for modern housing increased, disputed land was increasingly being used. 'Good' houses—and the uneven development they implied—became a source of concern to many villagers (Smith, 2018). Remarkably similar sentiments were shared in the small eastern Caribbean states, where remittances also had similar uses, as houses could be seen as elitist and contrary to more egalitarian norms (Connell & Conway, 2000).

On the volcanic island of Ambae, building a better house was likewise a priority. There, the experience of forced evacuations following volcanic eruptions in 2018, led some seasonal workers to build houses in other locations that might provide refuge when the next eruption occurred (Bailey, 2021). New housing could offer a degree of security from natural disasters. On Emau island villagers who were able to build houses uphill, away from possible coastal erosion, did so with income from seasonal work. As at Lamen Bay, houses on higher land were more prestigious. As one villager observed:

> When they come back they build houses. It's good that the government opened this scheme. We can build well [uphill]. But when there's money, our traditional ways halt... *Waet man*'s [foreign] technology, it's OK. But I think it doesn't suit this place. (Quoted in Ruehr, 2022, p. 12)

What can be a double bind—the inflow of cash and the outflow of people—could undermine *kastom*, yet islanders felt it essential to earn money to enable both adaptation strategies in the face of climate change and broader notions of development. For Emau it was consequently argued that 'the need to accommodate a globalized capitalist system, despite documented worker exploitation and negative socio-cultural consequences, highlights the flaws in emphasizing material improvements in living conditions via economic growth over challenging unequal power relations and retaining cultural knowledge' (Ruehr, 2022, p. 12). Yet this view is too apocalyptic. Villagers sought material development—evident in the prioritising of better housing—and necessary on islands such as Emau, where climate change and coastal erosion pose existential threats to existing livelihoods.

The outcome of some villagers, but not others, being able to construct new modern houses produced an inevitable inequality, and resulted in conflicts with values of reciprocity and generosity where new houses were built with purchased materials and wage labour. At the same time, new houses were merely one particularly symbolic example of a range of social impacts that contributed to the ongoing erosion of traditional authority patterns and social structures, incipient class formation based on material wealth, the decrease of community work based on exchange, community fragmentation, the spread of individualism and new dissatisfactions. Houses and cars proved widespread examples of conspicuous consumption. Such impacts were probably most pronounced in communities with widespread and longstanding participation in the labour mobility schemes—such as Lamen Bay—and were essentially accelerated versions of what was happening more slowly elsewhere throughout the PICs.

Education

Securing children's futures via education was an aspiration of many migrant workers, and, more broadly, a goal for most Pacific islanders. When she visited Vanuatu in 2011, Kirstie was repeatedly told how important education was for a child's (and by extension the family's) future. Elliot lived on Paama with his wife and one of his three children. He explained

> The first thing that you have to invest in is school fees, because if you don't pay school fees, you ruin your children's future … Our priority is the children. Everything else is less important. After you've paid for school fees you can work on other projects. They can wait. But children only have one chance to get their education. That's it. If you miss out, you don't get a second chance, so education has become the priority. (Quoted in Petrou, 2020, p. 79)

Such sentiments were typical throughout the PICs. Indeed, the need to pay annual school fees provided motivation for workers to keep returning to New Zealand or Australia, even after house construction was completed and other investment goals had been met. An overwhelming 91 per cent of ni-Vanuatu SWP and RSE workers surveyed in 2016 used their earnings to finance education (Bailey & Wells, 2016). This was virtually unchanged from an earlier study which found that 85 per cent of Tongan and 98 per cent of ni-Vanuatu households viewed education as a 'very' or 'somewhat important' reason for participating in the RSE scheme (Gibson & McKenzie, 2014). Without guestworker earnings, many ni-Vanuatu workers felt their children's access to education would have been limited. One truck driver on Ambrym commented

> The RSE is great. I finished school too early; my family did not have enough money. But we know with the RSE scheme kids are staying in school longer and have something to aim for after school….going to work in New Zealand. RSE is great because we have no jobs to go to. Many of us, like myself drop out of school and do nothing but stay at home. (Quoted in Bailey, 2014, p. 211)

Migrant workers often paid school fees for other relatives as well as their own children. In Tonga and Samoa, young unmarried men were chosen by their families to participate in the RSE so that they could help pay school fees for their nieces and nephews (Bedford et al., 2020). In Vanuatu too, workers paid the school fees for relatives. Jenny explained

> I'll be here for six months. It's my fourth season. I got one son, but I look after so many kids—round about 13 or so. They are all my in-laws' kids, but I love kids, so yeah… and now they are in college. Back in Vanuatu the fees are very high, so with what my husband and me have, we couldn't afford it. So I have to come back to get more money so the kids can go to school. (Quoted in Dao et al., 2020)

For those ni-Vanuatu who did not participate in the seasonal worker schemes, 'fundraising' (generally by selling cooked foods to other villagers) provided a means to earn money towards school fees. Fundraising like this also provided one avenue for redistributing the new flows of cash between villagers, so that even those without a seasonal worker in their household could benefit (Bailey, 2014). Seasonal labour migration increased opportunities for education, and the benefits were not restricted to participating households.

In Lamen Bay, as elsewhere, RSE workers viewed paying school fees as a way of securing access to a 'good life' for their children who would be able to find 'good' white-collar work in town. One school principal explained that education provided insurance for families in case the RSE scheme ended.

> If this scheme in New Zealand stops, where is your investment, the money for which you worked hard? If the scheme closes tomorrow, everyone will see that you did it to invest in your children, and too, the child will be able to support himself because he has a strong foundation. But if it is just a house and solar [that parents invest in], the child will not be able to achieve that [standard of living] if the scheme stops. It is better that you support the child in their schooling. (Quoted in Smith, 2016, pp. 143–144)

The local teacher similarly told her students

> I keep telling the pupils in the classroom, "Don't you think this scheme will last forever. It will stop. And if you don't have an aim for your education, you think that when you finish school you will go to New Zealand, one day you will find it really, really hard, because one day this scheme will stop. It's not like your land, which will always be there, to make gardens on. This scheme belongs to different people" (op cit., p. 145)

At the same time, many parents in Lamen Bay feared that with increasing population putting pressure on land availability and access, their children's future as subsistence agriculturalists was not guaranteed. By investing in their education, parents hoped to secure an alternative future for their children (Smith, 2016). Yet, with high urban unemployment in most PICs, alternative employment was scarce, and a 'diploma disease' had emerged with increasingly higher qualifications required for particular jobs. More and more, primary and even secondary education was no longer enough to guarantee urban employment. Consequently, some parents had higher objectives than secondary school: Kevin Bong from

Ambrym, working in Victoria, stated in 2015 that 'I've been working hard and I'm working for my kids' education. I have never been in the university to finish my studies but I am trying my best for them to have a chance. It's the sort of experience that can change our lives, our living standards' (quoted in Lee, 2015).

Simultaneously, as in Epi, increasing levels of education were associated with a disinclination to learn traditional agricultural techniques (Smith, 2018), a process that has long been in place throughout the PICs. Attending high school in Vanuatu often involved a move off-island, or at least travelling far from home, making it even more difficult to learn such traditional techniques and respect and absorb the values attached to them. Where domestic employment is limited, it is possible that rather than a 'secure' future, even highly educated children may be left in limbo: unable or uninterested in rural semi-subsistence lifestyles, but unable to secure wage employment in town. Avoiding that future poses national development challenges.

Business Development

Establishing a small business was a common aim amongst Pacific guestworkers, and investments in small businesses that could provide ongoing economic opportunities for workers and their communities were regarded as a key aim of the schemes. In Vanuatu, the goal of establishing a small business was almost as universal as that of improved housing (Fig. 9.6). By the end of the 2010s, aspirations for some kind of business development were widespread. A 2019 study of 79 RSE and 95 SWP workers from Vanuatu found that fully 87 per cent of them had invested in their own business or one belonging to a family member. The most common businesses involved transport (31 per cent), mainly buses and taxis; agriculture (24 per cent) which included agricultural tools and machinery, and establishing cash crops such as kava and cattle; and, small corner and village stores selling either retail goods or second-hand items procured in Australia and New Zealand (18 per cent). Tools such as chainsaws could be hired out profitably. Solar panels enabled stores to install freezers and increase profitability by selling frozen goods such as chicken and ice

Fig. 9.6 A successful small business supported by seasonal worker earnings, Paama, Vanuatu. (Source: Author Photo)

cream. Given geographical limitations, a much smaller number had invested in bakeries, accommodation (rental and tourist) and guided tours for tourists (Bailey, 2019b). Some had developed valuable strategies that hinted at success; one ni-Vanuatu man was well organised:

> I will do one thing where it will help me and my family…My favourite work back home is in furniture…I have collected my tools since last year….So now I will take some more when I return and will start a furniture business. My thought is to construct furniture. Lots of people who travel to New Zealand have built houses and there is no furniture for their houses … I think furniture is the need for most people there now. (Quoted in Lepon, 2010, p. 32)

Initially, businesses were quite small: symbolised by a 'canteen' that sold a tiny range of goods and was no more than a room in the owner's house. The range of small business types that seasonal workers invested in was limited, mirrored those already in existence in villages or in the towns, but expanded business capacity and demonstrated the widespread enthusiasm for entrepreneurial development.

Despite workers' initial enthusiasm, many small businesses lasted little more than a few years. This was occasionally due to their owners continuing to participate in seasonal work to finance the businesses or repay loans (Bailey, 2019b). However, more broadly in Vanuatu, few rural businesses prospered without ongoing support from urban or other income streams; most small rural businesses had off-island partners or investors (generally kin) who helped ensure the business' economic viability (Petrou, 2020). For many seasonal workers, remoteness, distance from markets and suppliers, and inadequate knowledge of business procedures made establishing and running businesses difficult. Jealousy and the establishment of copycat businesses—which ni-Vanuatu referred to as 'monkey business'—could also hinder long-term success. Exactly the same was true in Kiribati, further reflecting the limited possibilities of business development. Stores and many transport businesses thus faced intense competition. Few small stores ran at a profit and throughout Melanesia kin were said to *kakae mani* (eat money), as social expectations enabled them to freely acquire goods from the owner. Consequently, in Melanesia, small stores habitually display signs asking that customers pay their accounts before requesting credit. In reality non-payment was a fact of life. Antoine, a 41-year-old business owner from Paama explained

> Yes, it's just family [who don't pay their accounts], so I had to make a notice like this … I feel a little bit bad … So when family makes an account, you just note it down. When they think of it, they'll come and pay. But you can't stay angry with your family, they're more important than money. (Quoted in Petrou, 2020, p. 61)

Signs such as those stating *nogat dinau* (no loans/debts) in PNG were an almost universal, and universally ignored, feature of Melanesian village stores. Under such circumstances it was difficult to run a retail business profitably, especially where small villages had more than one such store. Transport businesses fared a little better, and although vans and trucks may have earned only small incomes as businesses, they brought other benefits through, for example, providing access to markets, education and health facilities (Bailey, 2021).

Importantly, prestige rather than profit was often the aim, and small businesses were commonly used primarily as a means for facilitating traditional exchanges and maintaining social relations. Trade stores and similar small businesses that required a relatively high level of economic input to sustain them often proved incompatible with indigenous economies, whereas small businesses that could facilitate indigenous labour practices and exchanges without this threatening their financial stability were most likely to succeed. For example, sharing copra dryers could facilitate kinship-based labour exchanges and build and strengthen relationships without undermining the economic success of the business. More generally, as in PNG, business enterprises are not focused solely on making profits; the way in which they are established, managed and patronised is an expression of indigenous social and economic life. The act of doing business, whether as customer, manager or investor, reproduces social relationships within the community (Curry, 2005). Many guestworker businesses simply failed because they were intended to establish and reinforce social relationships rather than provide longer-term economic gains. Nonetheless, some businesses did survive and even prospered: businesses headed by female spouses of seasonal workers seemed to run for longer (up to eight years by 2019) than those established by the workers themselves (Bailey, 2019b) although women were less likely to establish businesses. Rather than bringing the long-term development gains envisaged by Australian and New Zealand governments, businesses were performing other social functions: conferring social prestige, reinforcing kinship ties and redistributing income. That is what was expected of them by local communities. We consider these social functions in more detail in the next section.

Social Functions and Community Objectives

While a seemingly large proportion of incomes were directed at some form of modernity—houses, education, technology, vehicles, tools and business—that challenged older forms of social and economic organisation, much income also went into both communal village activities and customary celebrations of various kinds. In Vanuatu and Tonga, community leaders expected that workers would contribute some of their earnings to community projects upon their return; ni-Vanuatu workers were expected to contribute about A$171 and Tongan workers about A$234. However, only about 6 per cent and 20 per cent of community leaders in Vanuatu and Tonga, respectively, felt they had control over how this money was spent. In both countries, donations to local churches were the most common form of community contribution (World Bank, 2017). In Samoa, where churches organised some migrant groups, they could be even bigger beneficiaries (Fatupaito et al., 2021). In Tonga, participation in the SWP scheme was sometimes motivated by a desire to make larger church donations, including the annual *misinale* contribution that supported community activities. Other church-related projects varied locally. In 2018–2019, Samoans from the village of Neiafu used RSE earnings to contribute to the construction of a new Methodist church that would cost roughly NZ$850,000 (about A$790,000) to complete (Bedford et al., 2020). In Solomon Islands, a group of nine workers made donations towards buying a laptop for their church, while workers from another village donated money towards building a new church (Dun et al., 2022). Here too, while large donors gained prestige, community projects and monetary donations contributed to a degree of equity between those who participated and those who did not.

In some villages, contributing money to community projects was a condition of participation in the schemes. RSE workers from Tanna were, at least initially, expected to contribute to community projects, and were selected on the grounds that they would do so. As a result, money flowed back into village halls and churches. While not entirely altruistic gestures, since they raised the status of contributors in the local community and the donation was expected, these contributions provided positive village outcomes (Hammond & Connell, 2009). In 2018 a group of returned workers from Epi donated some of their earnings to build a doctor's house on the island, then handing the key to the government and requesting that they send a doctor since they had been without one for decades: this is widely seen in Vanuatu as a good example of grassroots development and self-reliance—where local people understood their needs better than distant governments (Smith, 2021). Arrangements within communities however varied greatly. In Kiribati, where workers were selected via an island quota system and overall participation rates were low, workers tended to come from many different communities. Because there were often no more than a few workers from a single village, it was difficult for workers to pool incomes for home community projects. That meant minimal investment in community infrastructure. By contrast, in Samoa, village councils and the *matai* (leaders), often influenced how RSE money was invested in the community. Making contributions to the community was considered part of the *fa'a Samoa* (the Samoan way). Across the PICs, projects supported by returned RSE workers included donations to the local school and school bus (Samoa), construction of a community hall, kindergarten and community kitchen (Tonga) (Bedford et al., 2020), and health centres and schools (Vanuatu). In Samoa, 56 men from the villages of Manase and Safotu on the island of Savai'i contributed roughly US$600,000 (A$800,000) to the community and their families after 12 months in New Zealand. The money was mainly used for private purposes, including the purchase of 18 new vehicles and 17 new houses, but much was put towards village projects including a waste programme (RNZ, 8 September 2020). In some instances, workers have also contributed to national projects. In 2021, many ni-Vanuatu SWP and RSE workers made donations towards purchasing an incubator for the Port Vila Hospital, impossible without the income from guestwork.

Enforced community donations were popular across the PICs, beyond churches and community halls. In Samoa, one community organised how money would be spent before workers even departed

> The best organised village is led by a local businessman who is also a resort owner. His work is well known in Samoa as he ensures that the young men chosen to go to work in the village crew are given additional training before they go. This man is very skilled and so he can provide training that might be difficult for others but it is a very good model. The workers are taught how to manage their money, to practice their English, and there is an agreement before they go about how the money earned will be spent. This seems to be shared between the village and the family. (Union Aid Abroad-APHEDA, 2020, p. 13; see also Fatupaito et al., 2021)

This level of organisation, initiative and forward planning was rare, and would be difficult in smaller, more fragmented Melanesian villages. Indeed, not everyone agreed with such compulsory contributions, and they were sometimes contested. In Ambrym, the Lolihor Development Council, along with the local community, selected RSE workers on the basis that they would contribute a set proportion of their earnings to local development projects. During their first two seasons (2007–2009), workers contributed 10,000 vatu (approximately A$120) to the council. However, as workers and community members lost confidence in the council—'we have no faith, no trust'—and selection processes changed, contributions to the council became sporadic. Even so, workers continued to support community projects and needs through donations to churches, local schools, the health clinic and ceremonial exchanges. In addition, workers gave small amounts of money to community members who visited their house and asked for support. For some RSE workers, these donations amounted to roughly half their earnings (Bailey, 2014). Incomes from the RSE thus enabled the circulation of extra resources within the local community, but incessant demands on them accounted for workers quickly using their money.

In some instances, workers' individual projects brought benefits to the wider community. Henry, from Solomon Islands, used part of his SWP earnings to dig a borehole in his village which improved community water supply. He explained

> I'm really enjoying the programme. Really, really enjoy it because even my first year, and then go back home, and doing my house, and I'm doing borehole, paying generator, supplying up water for the whole community … It's really helpful … we have problem in the water. No water … [now] everyone … come and take water… I say [to other villagers] that "this water is Murray River. This borehole called Murray River." … The money I used to pay for this borehole is from Australia, so [we're] drinking the Murray River … Really [the SWP is] really helpful. (Quoted in Dun et al., 2022, p. 867)

Other workers from the same group helped create employment in their home villages by employing domestic helpers while they were working in Australia, or expanding their agricultural activities and employing extra labourers (Dun et al., 2022). Similarly, and as we note below, some ni-Vanuatu workers employed others to work in their gardens while they worked in Australia or New Zealand (Bedford et al., 2020).

Across the PICs, seasonal worker incomes were frequently used to meet social obligations such as life cycle events and traditional exchange: 'support[ing] the fabric of kinship obligations' (Rockell, 2015, p. 179) underpinned the use of savings and remittances in Vanuatu. However, negotiating how this new wealth fitted into community life proved difficult for some. In North Ambrym, workers' savings were generally spent or shared soon after their return; although long-term savings were negligible, money was not 'wasted'. Investing in community and kin relationships, even if the impact was largely invisible and savings could dissipate quite quickly, had long-term benefits (Bailey, 2014). Conrad, who had participated in several seasons of RSE employment, found it frustrating that while the income he had earned was based on his work as an individual: 'when you get back to Vanuatu none of the money is yours. This money is for a wedding. That money is for a nephew's school fees. This money sends your parents to the doctor'. Frustrations aside, the pattern of distribution Conrad described had enabled him to become more respected and influential amongst his kin (quoted in Cummings, 2016, p. 28). Indeed, in contrast, being seen to hoard money for oneself was believed to attract anger from spirits and result in jealousy. In North Ambrym, those who were thought to be hiding money for their own use were often teased or otherwise pressured by community members until they relented and shared their earnings in an 'appropriate' manner, both ensuring a degree of equity within the community and reinforcing kinship ties. While returned workers did keep some money aside for the proverbial rainy day, this was done discreetly: flaunting one's savings was neither encouraged nor rewarded (Bailey, 2014).

As participation became more widespread and sustained, some households became dependent on incomes from seasonal work, but not necessarily due to individual workers repeatedly participating in the scheme. Rather, dependence seemed to occur in households where intra-household transfer of RSE employment was expected; the original worker might disengage from the scheme for a year or two, but someone else within the household would replace them (Bedford et al., 2020). Such dependence was influenced by various factors, including limited opportunities for income generation locally, pressure from families or communities to ensure someone remained employed so that the opportunity for seasonal employment was not 'lost', decreased interest in subsistence agriculture as a long-term livelihood strategy, more ambitious economic goals or objectives and ongoing monetisation of village economies. Recruitment procedures that favoured experienced workers further contributed to this path dependence.

The economic benefits of guestwork were such that in Tonga and Vanuatu men reported experiencing pressure to participate in the schemes to fulfil the role of a 'good' husband who provided for his family. Whereas in the past 'good' husbands were expected to be hardworking, provide support for the family and work in the subsistence food system, there were increasing expectations that men should 'improve' their family's livelihood through guestwork. A ni-Vanuatu man explained

> One SWP male worker came back with money and other men now see him differently. Other men in the village aspire to be like him. We also want to have money like him. We aspire as well for family and to build a good house like him, buy a truck or start a business. (Quoted in Chattier, 2019, p. 56)

In some rural communities, labour migration had become not merely one livelihood option but an expected component of it. A culture of temporary labour migration was expanding and extending.

Through the monetisation of transactions and increased access to cash, seasonal work was making village life more expensive. This was an unintended consequence of the strong moral imperative to contribute to traditional exchange ceremonies. In both North Ambrym and Epi ceremonies such as weddings—already considered expensive prior to the advent of the RSE—became increasingly costly as migrant workers were able to contribute more than in the past (Bailey, 2014; Smith, 2015, 2016). For example, in Ambrym, some RSE workers used their savings from seasonal work to pay for weddings, which averaged approximately NZ$6000 (A$5580), an amount then roughly equivalent to a whole season's savings. One RSE worker, Dave, commented 'last year I made a *kastom* wedding and that was all of my New Zealand money, nothing left after that' (quoted in Bailey, 2014, p. 230). Another RSE worker, Howard, lamented:

> I'm ready to go back to work in New Zealand, all of these ceremonies are costing me too much money [...] everyone thinks that because we got to New Zealand that we have much money and expect more from us... It's a burden sometimes. (Quoted in Bailey, 2014, pp. 231–232)

In Vanuatu, wedding expenses included *braedpraes* (bride price), cattle or pigs for the bride's family, and feasts that could be attended by as many as 1200 people and last 2 weeks or more. With the financial boost from guestworker incomes, more weddings were occurring, which in turn put pressure on the wider community to contribute cash, garden produce such as yams and taro and material gifts such as fabric and mats to the celebration. For some, guestwork enabled participation in traditional ceremonies and exchanges that were previously beyond their economic means: labour migration could facilitate the continuation of traditional practices and raise the status of those involved. However, as income was increasingly required to participate effectively in traditional exchanges, and as families sought to increase their position and status in society, so inflation followed.

Increased economic expectations and pressure around weddings and other communal events and exchanges raised questions of development and equity within and between villages and households. If participation in labour migration is required to meet needs, and not everyone can participate, then inequality follows since redistribution of cash earnings is unlikely to preserve a crude socio-economic equity at village level. It was evident that, for some, even early on, the RSE had created tensions:

> themes such as jealousy, anger, sharing, borrowing, and holding back, were spoken about by workers and people in various parts of Ambrym. There are competing interests in RSE incomes. There is jealousy from those that cannot participate in seasonal programs; anger from workers that have enforced expectations of their earnings; and those who see RSE workers as 'wasting

> time and money' in New Zealand and not sharing or using their incomes appropriately. (Bailey, 2014, p. 144)

New overseas incomes disrupted village society and economy. Despite moral obligations and the potential for increased status and respect, migrants were invariably ambivalent about contributing considerable resources to the needs and demands of others. That ambivalence was typified by Conrad, who had spent most of his savings from his first season of work on consumer items for himself and his wife, but subsequently 'invested' in a culturally appropriate way in community and kinship ties, so becoming a senior, respected member of his kin group, a status that as an unemployed, urban resident young man, he would not otherwise have been able to achieve (Cummings, 2016). In Samoa, men without access to local wage employment who participated in the RSE became able to send their children to school for longer and contribute more to customary exchanges and the church than they could previously have done. Such investments and actions contributed to upward social mobility, greater respect and enhanced social status (Kerslake 2019 in Bedford et al., 2020; Bailey and Wells 2017). Donations were an investment in kinship relations that tied migrants into home communities, enabling them to retain a distinct local identity and secure customary rights to land that would ensure them a continued means of subsistence production (Smith, 2015). However, not all returned workers chose or were able to do this, especially in more hierarchical Polynesian societies where social mobility was more difficult. Gaining prestige was rarely quick or easy.

Status could be acquired even as mobility contributed to some rise of individualism, visibly manifest in the material rewards of migration. Yet new status occurred within a dwindling sense of community, accompanied by the loss of respect for traditional roles and authority. In Lamen Bay, where RSE participation was particularly high, that meant reduced cooperation within the community, the decline of the village council and its inability to police rules that governed local life, such as management of water use and deforestation. As one returned worker put it:

> Community has become a dirty word. Sometimes when you mention the word community, ears turn off because people think that "I don't want to

> do things for the community. I want to do them for my household. I want to do them for my family". (Quoted in Craven, 2013, p. 37)

As one Epi resident said: 'The people who go to New Zealand, they work for their money, they don't work for all of us. That's the problem … They never work for free. You have to pay'. Another developed the theme:

> Those that go over there, they do not recognise our struggle to keep the community in place…[the migrant] works over there for his own money, but we stay here, we stay as slaves working free for the community. Then when it comes to asking him to donate to the community he complains. (Quoted in Smith, 2015)

Such differences of opinion and practice added to the burdens of village life so that, where populations were small and stagnating as on Paama, they contributed to 'a continuous feedback chain, exemplifying the dialectic nature of mobility' (Haberkorn, 1992, p. 821). Parallel changes were however widespread within PICs and cannot be solely or even primarily attached to labour migration; labour migration merely accelerated the process. Using money in a socially appropriate way was a key issue in participating communities, but making sense of and benefiting from what was 'socially appropriate' proved challenging, and involved an ongoing process of negotiation.

Family Breakdown?

Workers' absences had impacts on family life and marital relations. Evidence (inevitably largely anecdotal) suggested that participation in seasonal labour migration could be linked to domestic violence, even before the process had begun. Some ni-Vanuatu women from Paama who challenged traditional gender roles by participating (or even just wanting to participate) in seasonal employment could be targeted by gender violence (Craig Lind, pers. comm., 2018). Similarly, it seemed that Tongan and ni-Vanuatu women who became 'too confident' as a result of their overseas experiences could experience violence (World Bank, 2018), and the RSE scheme in Vanuatu was anecdotally associated with a rise in domestic violence and relationship breakdowns (Rockell, 2015). This association between migration, domestic violence and deteriorating relationships often suggested worst-case scenarios, but it was difficult to determine the extent and veracity of rumours, gossip and whispers.

In addition to their link with gender violence, the guestworker schemes were associated with a degree of marital conflict and family breakdown. Some Samoan and ni-Vanuatu men working in the RSE scheme started new relationships or even families with New Zealanders. At least one Fijian woman became pregnant while overseas (Chand, 2021). A few i-Kiribati women had divorced their husbands due to infidelity at home while they were away (Bedford et al., 2020). In Tanna, one man related his several problems:

> The first challenge …was the issue of trust between my wife and I. The first time I went my girlfriend was unfaithful and went to live with another man. Maybe I was sending her money and other men took advance [advantage] of it making her fall into problem even after I have paid the bride price to her family. The same problem happened with three other women that I took them home on different occasions. (Quoted in Kaiapam, 2021, p. 63)

Worse was possible. An agent in Tanna observed:

> A worker got the message that his wife was having an affair with another man back in Vanuatu. This news affected him so he started drinking heavily and was missing work all the time. When he returned to Vanuatu he went to where his wife was and took her home being frustrated and angry, he instantly killed her … and was imprisoned. (Quoted in Kaiapam, 2021, p. 64)

Such extreme cases were rare but concern was considerable: as one ni-Vanuatu said: 'When ni-Vanuatu go, some of them, when they live on their own, they decide to take another wife, so another new family. This creates a problem when they return home' (quoted in Kumar, 2012, p. 68). A Tongan worker alluded to the problem of lengthy separations: 'If you go back next year to make $6000, is losing your wife worth it?

How do you rank that objectively? (quoted in Tazreiter et al., 2016, p. 112). Perceptions were otherwise more extreme than practice. 'Evidence' for affairs was usually lacking, but fears of infidelity abounded and often it was enough just to suspect an affair for break-ups to occur. The 'coconut wireless' of social media teemed with heated discussions of such issues with the blame attached to different people and places. Typical Facebook posts included:

> I want to travel, but I've seen some women post because they're concerned about their husbands… But I have a question, do you want to travel for work, or to find a partner? Because now there are lots of broken homes as a result of seasonal work. If you have a problem in your home, it would be good if the agent could screen everyone before they send them to work for the RSE or SWP… Because some people form happy couples overseas, and then when there is a complaint they blame their partner at home. If you can't travel and behave responsibly, then you shouldn't travel. That's just my opinion.
>
> It would be better if couples could travel together, then there wouldn't be any complaints or separation.
>
> This is a good idea, but what if you have children, how…? And what if they travelled together and just ended up fighting overseas instead!
>
> There were a husband and wife who travelled together to live and work in the same place. They spent the whole time fighting, and one time the man hit his wife. The woman was to blame, but they send him home.
>
> It's true, but it takes two people to make these kinds of problems, not just the person who travels. The ones that stay home are just the same. The point is that before you travel, you must decide what your goal is, so that you know what you're working for, and you must agree on it together. You can't just go and play the tourist and come back with nothing.
>
> The issue here is not the agent, we shouldn't make them responsible… It's a decision made in the home, both partners agree that one must leave to support the family. The agent asks for references from leaders like the pastor, chief, wife… I'm appealing to the homes. If you don't trust your partner, don't create problems. Everyone has the chance to work in the RSE… but married people need to be faithful.
>
> Some men and women want to go and fuck overseas, but they want to work and make money too. They don't know how to act responsibly because they have all of these strong ideas. This kind of behaviour is how you get AIDS. (Facebook, 2019)

However as Maggie Cummings (2013) observed, while women were reportedly running amok, being promiscuous and getting pregnant, while men frequently got drunk, these were exactly the sort of usually baseless stories, warning of the perils of modernity, that were familiar before either scheme was introduced. Nonetheless, in Vanuatu at least, enough children had been born from both legitimate and illegitimate love affairs in New Zealand for a new term to have been introduced: *pikinini bilong apol* (apple child), a play on a more familiar *pikinini bilong rod* (child of the road/illegitimate child) (Smith, 2016).

Concern over extramarital affairs was not restricted to Vanuatu. Responding to similar sentiments and practices, the Samoan Government described problems of infidelity and marital discord as 'manageable', and penalties such as blacklisting and fines were adopted to discourage this behaviour (Radio New Zealand, 2019). Some Samoan workers were made subject to a public 'name and shame' after marital infidelity was reported. In 2019, Samoa released a public notice of workers who had been blacklisted due to issues including extramarital affairs, theft, smoking marijuana, fighting and absconding. Prime Minister Tuilaepa explained

> The seasonal employment programme will not be and is not collateral damage due to the irresponsible actions by a handful of ungrateful Samoans [...] Time and again I have personally advised all our RSE workers before leaving to remember the importance of their seasonal employment and engagement because it's their families, children and their villages which are reaping the spin-off. Government is also not sparing any expense in securing new jobs for our people and we are the only country in the RSE Program which has a pre-departure program to prepare the lucky workers handpicked for these jobs to cope with the demands of working overseas away from their loved ones and their comfort zones. The Public Notice serves as the last straw and to re-enforce the government's zero tolerance [...] It serves as a reminder and a discouragement to seasonal workers that disobedience and any undesirable behaviour which will compromise more jobs for Samoans will be weeded out without hesitation. (PacNews, 25 June 2019)

Two years later the Samoan government was so concerned about families breaking apart after affairs that it was considering permitting only single workers to be recruited, but was rather more concerned about this giving the scheme a bad name rather than the actual family breakdowns (RNZ, 2 July 2021). Governments generally avoided such difficult social issues, although early in 2021, Vanuatu tried a different tactic, beginning a trial with PLS workers, sending married couples to Australia together. This may have gone some way to solving issues arising from the separation of partners but raised questions over the impacts of such migration on children.

In Tonga fear of marital breakdowns was one of the few real concerns of left-behind wives, dramatically expressed by one Lifuka woman:

> I thank God for bringing back my husband, and father to our two children. My children are very lucky because their father returned, unlike others, they never return. I do not know if he already met a lady in Australia on his first trip. Who knows if his second travel in the programme will be his last with us? It's better for me to eat cassava with a cup of water filled with happy than with meat full of tears… Therefore I do not want him again to travel on seasonal schemes. (Quoted in Moala-Tupou, 2016, p. 67)

Despite similar widespread concerns over infidelity, temporary absences were not always harmful. Some Tongan wives of RSE workers claimed that increased communication brought about by daily phone and video calls meant husbands and wives were working together better as a team in search of a common goal than they had before RSE participation (Bedford et al., 2020). Other women in various PICs, while experiencing new work burdens while their husbands were absent, valued the opportunity to develop their own lives and achieve a degree of empowerment (see below).

Alongside marital relationships child-parent relationships could also suffer from workers' absences. One RSE accommodation provider observed of the ni-Vanuatu workers staying at their facility:

> I can see a lot of marriage break ups as a result of coming out here and so I'm not sure how that goes with the communities back home…I have heard and seen the results of family bust ups…Having 6 months away

> from teenage children, that can't be that good. It's a long time. I mean I sit here and watch them on the phone and at times there is often a lot of shouting and tears and I think is it worth it? (Quoted in Williams, 2010, p. 86)

Ni-Vanuatu SWP workers in rural Victoria 'ached' for the children they had left behind (Stead, 2019). Children of RSE workers often cried for their absent fathers, and while they understood that they were earning money to improve their standard of living 'it doesn't ease their sadness about his absence' (Bedford et al., 2020, p. 92). Some workers were very clear about their desire not to become long-term absent parents; Timorese men working in Australia believed that SWP work was best when children were young, but did not want to remain away once their children started school (Wigglesworth, 2018). This sentiment was echoed by Tongan teachers who believed older children were more likely to suffer 'emotional trauma and learning difficulties' due to parental absence. An RSE worker commented: 'Earning money to pay school fees on its own is not enough. Children need the attention of their fathers and mothers—not just toys and screen time' (quoted in Bedford et al., 2020, p. 92). On the Tongan island of Lifuka the remittances from absent parents could enable children to benefit from school fees being paid, have better lunches and new clothes that gained them prestige in church, but children still missed the daily reassurance and supervision of a parent, experienced loneliness and were more likely to daydream in school (Moala-Tupou, 2016). In some respects, here as elsewhere, children bore as much of the burden of parental absences as lone parents.

Many workers with families—as most were—recognised that opportunities for migration represented a paradox of potential pathways to economic security linked to the challenges of absentee parenting. In Tasmania, Timorese workers described the SWP as 'a life-changing opportunity', but it was not without emotional costs. Manuela Dos Santos from Timor-Leste described the difficult balance between sacrificing time with her two children and being able to provide them with financial security: 'When we speak, they are crying but they understand me because they know I'm the one being strong for them […] Sometimes they will miss me or sometimes say, "It's all right, Mum"' (quoted in Davidson & Earl, 2020). Separation could be difficult for everyone, and while the financial rewards could be huge, they were accompanied by sacrifices.

For those at home, the impacts of separation differed according to the size, composition and location of households and the length and frequency of workers' absences. In Vanuatu, younger children had the hardest time coping emotionally with their fathers' absences, and often cried for their missing parent. Burdened with the responsibility of acting as primary carer for small children, the women in these households experienced the greatest difficulty substituting their husband's lost labour (Bailey, 2014). For older children, separation from families could mean a period of acting up. In Samoa, adolescent boys being cared for by grandparents or aunts were known to stop attending school and ignore the authority of these relatives. Some ran off to the capital, Apia, where they got into trouble (Union Aid Abroad APHEDA, 2020). Tongan women who participated in guestwork to earn school fees discovered that, in their absence, their children often dropped out of school and experimented with drug use. In turn the women decided to discontinue participation in seasonal work (Uera, 2019). More broadly, teachers in Tonga believed that the repeated absence of parents due to guestwork was causing many children to drop out of school at about age 14, before their education would equip them for the good jobs their parents hoped for (Bedford et al., 2020) and were one of their objectives in migration. In Vanuatu, children experienced discipline issues and displayed a lack of respect when their parents were away. While many women in Vanuatu developed confidence and self-esteem as they took over the role of household head, others found coping with adolescent children particularly difficult; a Tannese mother expressed such difficulties:

> They respect their father and would be at home when he was around but when he is gone to New Zealand I find it very challenging to manage them. The boys were not drinking when the father was at home but when he left, they started consuming alcohol and kava. But I only pray and ask God for help and I see that my boys are slowly coming to understand that they need to listen to me because I am alone. (Quoted in Kaiapam, 2021, p. 66)

Some *abu* (grandparents, elders) complained that it was 'taken for granted' that they would look after the grandchildren while their parents worked overseas (Rockell, 2015). Ageing grandparents could find it particularly difficult to care for grandchildren and even extended households were not always able to cope.

Negative impacts on children and parents were not unique to the SWP and RSE schemes, but were characteristic of temporary worker schemes more broadly. Mexican women, participating in Canada's Seasonal Agricultural Worker Program (SAWP), experienced grief at 'the gradual erosion of intimacy' that occurred when they were unable to interact with their children through daily activities such as making breakfast, walking them to school and the casual embraces that were taken for granted when living in the same physical space. Even periodic returns of a few months at home were not enough to make up for repeated absences; as one mother observed 'you gain something but you lose another' (Preibisch & Grez, 2013, p. 793). Gains and losses were hard to balance. The women felt empowered by their responsibilities and experiences in Canada and often believed that the sacrifice of separation was worth it, as it provided access to 'better' alternative futures for their children (Preibisch & Grez, 2010). Both men and women who participated in the SAWP missed their daily interactions with children, but women reported significantly more grief, inadequacy, guilt and doubt. Men were proud of what they had achieved, were much less likely to express self-doubt and were less concerned than women about the impacts of their absence on childrens' lives: mothers bore the moral burden in a way that fathers did not (Paciulan & Preibisch, 2013). Similarly, children tended to judge migrant fathers by their ability to provide financially and mothers by their ability to care (Dreby, 2010), and while migrant fathers could fulfil their role as breadwinners and financial providers relatively easily, women in the same position were viewed as having deserted their 'proper' caring roles (Paciulan & Preibisch, 2013). Much the same was true in the Philippines, and even though Filipina migrant mothers tended to provide more caregiving from afar than migrant fathers, children found it more difficult to have an absent mother than father (Parreñas, 2005). Guestwork had real impacts on family life, many too subtle, invisible and evolving to be documented or evaluated.

While both men and women benefitted from the Pacific guestworker schemes, the burdens of labour migration were more likely to fall on women and the rewards on men. The schemes provided opportunities for families to increase their standards of living, but these improvements were not without emotional costs. Possible longer-term impacts of labour migration on children were yet to be evaluated, but, over the short term, the repeated absence of a parent could lead to behavioural issues. For guestworkers and families at home, separation was never easy.

Labour Losses and Food Security

The absence of household members not only affected family relationships, but usually resulted in lost labour for the worker's household. The impacts varied by gender. Throughout the PICs, women usually had responsibility for domestic and unpaid care work. When Tongan and ni-Vanuatu women participated in the SWP, men often took on extra household chores but most of this domestic work was performed by female relatives. By contrast, when it was men who travelled, their wives and partners tended to take on extra responsibilities. Thus, in Tonga, wives and partners worked much longer hours around the house and garden plots while men were away, even when other family members helped out. A ni-Vanuatu villager interviewed by the World Bank (2018, p. 39) observed: 'wives who are left behind are involved in all other tasks that the husbands might have been doing before they leave'. This included both household labour, such as maintaining food gardens, as well as taking on traditional or community responsibilities normally performed by husbands. That seemed typical across Vanuatu. In the Torres Islands, a woman from Toga Island proclaimed that extra work 'is an extreme burden for me. During his [her husband's] absence, I have to look after the children, manage the garden, go fishing, or search for wild yam—all by myself' (quoted in Nef et al., 2022). In Tonga one woman started ringing the church bell, a job performed by her husband when he was at home, and something she was proud of. In Samoa, the increasing absence of men meant that women's committees had started undertaking the domestic and communal tasks normally performed by men. In some cases, this led to women's empowerment; in Samoa, for example, some women learned to drive, a task normally undertaken by men (Kerslake 2019 in Bedford et al., 2020). In Vanuatu, husbands often arranged for other men to help with planting new food gardens, however women still took on additional work including church and community commitments that 'belonged' to their husband. One ni-Vanuatu villager noted:

> Before joining the SWP, men are mostly farmers and gardeners, but when they are away, the women have more hours of work because they now have to clean the kava garden, clear the garden for copra, and take care of children. If the husband is not around then it is too much work for women, looking after kids, preparing food, cutting firewood, and so the traditional way of doing things is changing. (Quoted in World Bank, 2018, p. 40)

Women carried a heavier burden. Nonetheless, as one ni-Vanuatu woman concluded after taking on extra responsibilities including tending livestock and working in the gardens, which was hard at times, she felt it was worth the effort for the financial rewards: 'It is hard, but I'm happy [my husband] is coming back with the money' (quoted in Williams, 2010, p. 93). Another ni-Vanuatu woman explained: 'Women now take up the responsibilities of everyone—wife, husband, mother, [and] father, and it is a hard life for women when [the] husband is away. Women and kids do all the tasks that men used to do' (quoted in Chattier, 2019, p. 60). While this was associated with a gradual shift in gender roles, since absences were brief, it rarely resulted in any fundamental change in the traditional gendered relations between men and women.

In earlier times when labour migration was sometimes of limited duration, as it was to plantations within Vanuatu, men took on the extra work of planting prior to migration and returned in time for harvesting, so being away coincided with the fewest demands on the system (Jolly, 1987) but few remnants of this remain. In the early days of the schemes, recruited Samoans were said to have been encouraged to plant hundreds of taro plants before departure, to ensure that there was no effective labour shortage. By 2021 there was doubt that this had ever been effective, and the practice had disappeared (Fatupaito et al., 2021), although chiefs from Poutasi village had sought to make it a condition of participating in guestwork that workers first complete their agricultural duties. Village leaders also sought to ensure that not all men from a single family were chosen, to ensure that women were not overburdened, and also that village church choirs retained enough male voices! (*Samoa Observer*, 13 May 2009). In the 2000s, Tongan men too did 'extra' agricultural work before departing for the SWP so that work would be minimised in their absence (Blanco, 2009). Lack of labour remained a concern.

As Pacific islander women took on increased workloads in the absence of their male partners, many had little time to engage in work outside the home (Chattier, 2019; World Bank, 2018). Increases in women's workloads due to their partners' labour migration—often described as working a 'double day'—were not restricted to the Pacific. In Mexico, the wives of SAWP workers similarly took on extra 'migration work' during their husband's absence. Just as in the Pacific, that included running family businesses such as farms and small retail stores, caring for relatives and children, managing family finances including remittances and travelling to the bank and, early on, to public telephones to keep in contact with their husbands. Such additional work was particularly time-consuming for women in remote communities who had to travel long distances to access banking facilities. With little time left over, women were unable to work outside the home themselves, or had no time for craft production, so that households could become solely reliant on the husband's participation in the SAWP (Hennebry, 2014). In such circumstances, rather than diversifying household incomes and livelihoods, guestworker schemes could lead to households becoming solely and narrowly dependent on transnational income flows, a potentially precarious reliance on work that was short-term and seasonal, and that further emphasised the shift towards monetisation.

Reduction in local labour availability meant that migration could contribute to food insecurity and associated vulnerability. Early on, in both Tonga and Vanuatu, where villagers had limited access to banking facilities and thus to regular remittances, households with absentee workers experienced a decreased variety of food eaten due to a lack of cash and labour resources to purchase food from stores or tend to gardens (Rohorua et al., 2009). In Lifuka, Tonga, where men had been regular fishers, households experienced inadequate nutrition as fresh fish disappeared from diets, replaced by imports (Moala-Tupou, 2016). In both PICs people reported an increase in health complaints. As health status was self-reported, this increased incidence of health complaints may have been psychological and a reflection of emotional strain. However, lack of time and access to remittances might have meant household members could not access health care, hence ailments dragged on for longer (Rohorua et al., 2009). While such worsened nutrition was described as 'short term' and relatively minor, repeated absences and repeated periods of poor health and nutrition for several months of the year could have longer-term consequences.

In Lamen Bay, Vanuatu, where Pacific Labour Mobility (PLM) participation rates were particularly high, subsistence agriculture became increasingly difficult due to widespread and repeated absences of working-age men. One chief described the situation

> Plenty are out [working in New Zealand] and this affects community work and that is why I don't want so many from Lamen Bay and Lamen Island [participating in seasonal work] … during the fruit season the community is empty and when it is time for community work everyone is away. Plenty have no garden and they must buy food in the market … The culture of gardening has dropped badly … They are becoming more dependent on their extended families. Instead of making a garden the wife is looking after the kids, especially the young couples and the a*bu* [grandparents] has not as much strength, and next year they go back again and there is some reluctance to do the garden over and over. (Quoted in Rockell, 2015, p. 209)

Such high levels of seasonal labour migration resulted in households planting fewer crops in fewer gardens, as well as reduced quality. In addition, gardens were planted in fewer places than before. Rice imports replaced root crops. As one returned worker put it:

> Yes, my garden lies empty now that I am away in New Zealand so often. However this does not worry me. We still eat well, if not better than before. Now I am able to buy food from the store with the money I earn. (Quoted in Craven, 2013, p. 53)

The outcome was that subsistence agricultural systems had become less sustainable and secure and more vulnerable to shocks. One Lamen Bay woman commented:

> Cooking, cleaning, washing and looking after my children is a full day's work. How am I meant to go to the garden? I have no time anymore with my husband overseas. (Quoted in Craven, 2013, p. 32)

Much depended on individual villages and communities, the nature of the agricultural system, the 'norms' around seasonal work and domestic expectations as well as infrastructure and remoteness.

Labour losses were not restricted to times when workers were overseas and could extend beyond workers' physical absence; those who had worked long hours while away in Australia or New Zealand often wanted to rest and relax when they returned home and were reluctant to participate in subsistence agriculture, particularly when they had the funds to buy imported foods instead. In effect, and despite the fact that their contracts only lasted for six months, these workers were absent 'year round' (Craven, 2013). Complaints over returnees resounded:

> The workers come back and do not work for our community. They sit at home and watch DVDs, drink kava, but you will not find a [RSE] worker in the gardens. Lamen Bay is their holiday destination. All they want to do here is relax. (Quoted in Craven, 2013, p. 34)

Like so many other acerbic comments by disgruntled villagers, that was unlikely to be entirely true. However, the inability—and sometimes lack of interest—in planting and maintaining subsistence food gardens led to an increased reliance on purchased food, whether *aelan kakae* (island food) from local markets or imported foods from local stores (Rockell, 2015; Craven, 2013). This reliance on poor-quality purchased foods, and the reduction in local physical work, provided a link to non-communicable diseases (NCDs) such as diabetes and heart disease. Furthermore, it led to increased vulnerability as food deliveries to small stores were sporadic and unreliable.

With exceptionally high rates of labour participation, and long-term involvement in both schemes, Lamen Bay was atypical yet it provides an important example of what could happen in communities where participation in the guestworker schemes became so widespread that local food systems began to collapse. In Tonga, those households receiving remittances engaged in less diversification of local economic strategies, a suggestion to some that remittances were creating a disincentive effect that hampered agricultural development (Holani, 2017). Even so, issues around food insecurity and labour losses could perhaps be temporary glitches to be resolved every time workers return home. It was nonetheless paradoxical that food insecurity happened at all where community members were contributing to food systems and food security overseas.

Skill Losses and Brain Drain

From their inception guestworker schemes were meant to target and support those with limited access to other incomes, particularly people living in remote rural areas. In a context where skilled human resources are in limited supply, any loss of essential workers to low-skilled labour mobility schemes posed potential problems. While the seasonal labour schemes were something of an exception, many other migration pathways open to Pacific islanders were focused on skilled migration, with a resultant loss of skilled workers from most PICs (Voigt-Graf, 2003; Connell, 2006a, 2009). This loss of skilled workers raised questions about whether migration constituted both a 'brain drain' of unacceptable proportions, that hampered national development, and a brain-waste where some failed to gain commensurate employment in the destinations. However, it is conceivable that the extent of remittances from skilled migrants both offset training costs and enabled new investments in education (Chand, 2008). As it became evident that some skilled and semi-skilled workers had become seasonal migrants, these debates took on new meaning. In 2019, a Vanuatu parliamentary committee reported that local companies were wary of training their staff, fearing that they would later resign to participate in labour mobility programmes, hence it wanted selection to focus on the unemployed (RNZ, 12 June 2019).

Overall, data on the education, occupations and existing skills of seasonal workers are scarce. Most SWP workers had completed some secondary school, however workers from countries such as Vanuatu, where secondary school enrolment rates were relatively low, were less likely to have gone beyond primary school (World Bank, 2017). By contrast, 22 per cent of Timorese SWP workers interviewed by Ann Wigglesworth and Abel Boavida dos Santos (2016) held a university degree, while more than half of the Timorese surveyed by Annie Wu (2019) had university degrees. Overall, PIC seasonal workers were more likely to have been employed in the formal sector than the national average of their country of origin. The majority (59 per cent) of seasonal workers who were employed prior to participating in the SWP were working in the formal sector. This varied by country, and while around 80 per cent of Timorese who participated in the scheme held paid employment at home, only about 20 per cent of ni-Vanuatu were in the same position. Amongst those who did have paid employment at home, the service sector was particularly common for SWP workers across the PICs, and especially so amongst employed ni-Vanuatu (70 per cent) and Timorese (74 per cent) (World Bank, 2017). It is likely that this group of employed ni-Vanuatu workers came predominantly from urban areas. Although it is widely assumed that most seasonal workers came from an agricultural background, this now seems untrue.

Our own research with ni-Vanuatu SWP workers confirms that seasonal workers came from various backgrounds, and had often left (skilled) employment to participate in the scheme. This was true of the ni-Vanuatu workers who we met in 2012. Edna lived on Santo and, before working in Australia, had been employed in a hotel for 13 years. Alex lived in Vila and worked as a civil engineer. He had previously trained and been employed as a nurse. Joshua lived in a village on Santo where he was a farmer. Alice lived in Vila where she worked as an accountant, while her husband Keith, who was also participating in seasonal work, had a small business where he was a bus driver. Vanessa lived in Vila and was employed as a radio announcer. Charlie had worked as builder in Vila for 11 years but had recently moved back to Ambrym where he manufactured copra. Matthias lived in Vila and, although he had attended university, was employed at the wharf, offloading containers from ships. Richard lived on Tanna where he owned a small store. Most of these workers had either resigned from their employment or taken leave in order to work in

Australia. Early in 2020, Kirstie met another group of ni-Vanuatu workers employed in peri-urban Adelaide. Again some of these workers came from rural areas and had no formal employment at home, but others had held steady employment for many years including at a service station and as a supermarket manager. Some workers explained that their bosses understood the opportunities associated with guestwork and would make their jobs available for them again when they returned home. Others would be returning to unemployment, but were unconcerned; they anticipated that their skills would surely get them new jobs and they had earned much more in Australia than they would have at home. For those who planned to come back to Australia for the next season, being able to *spel* (rest and relax) at home for a few months was often more appealing than returning to paid employment.

The, perhaps unintended, recruitment of skilled workers was belated, but it had increased as the schemes expanded and their economic benefits became better understood. As early as 2009, Kirstie met a nurse, Lossie, in Tanna, who had participated in the RSE. Lossie had had a 'wonderful time' picking apples and living in a share house with other RSE workers, and was eager to return to New Zealand as soon as possible. This was at a time when medical staff were in such short supply in Vanuatu that nurses were being brought out of retirement to fill labour shortages. Lossie had found it vastly more profitable to travel to work in New Zealand horticulture. Even temporary losses of experienced employees, such as accountants, engineers and nursing staff, are potentially problematic for PICs with a substantial shortage of skilled labour, including Vanuatu (which has had to recruit nurses from Solomon Islands). A shortage of nurses has existed throughout the 2010s in most PICs (RNZ, 21 February 2020) but that has not been solely the outcome of migration.

Social costs were also related to other kinds of skill losses. In 2020 several members of the Vanuatu national football team (the main male sporting code) were employed in Australia, and thus unable to play for their clubs, or the nation, losing fitness and depleting their skills. As amateurs they experienced the need to migrate and generate income as much as anyone else. As one skilled ni-Vanuatu footballer, Junior Felix, stated:

> Currently I earn nothing from football—as [is true for] most of the footballers here in Vanuatu. Sometimes when we win an important match they pay us a prize, only few football clubs can afford to pay something reasonable to its players. Many of the players from here leave to Australia or New Zealand for apple picking. (Quoted in Oceania Football Center, 2020)

Consequently, the national team lost players who had an unfortunate and disruptive career break, and were away for too short a time and in the wrong places and season to benefit from participating in professional football tournaments while in Australia.

As in Vanuatu, concern that SWP workers were increasingly being recruited from skilled professions grew in Samoa in the 2010s and there were calls for more care to be taken in recruitment processes. The head of the local Association of Manufacturers and Exporters, Tagaloa Eddie Wilson, explained

> We have an understanding with our government that the people they select for the RSE programme are people that are unemployed themselves in Samoa [...] However we are continuing to experience, across the private sector, that we are beginning to lose the skilled workers to the RSE programmes. [By recruiting unskilled workers] the unskilled workers will then get a skill and then when they come back they can then be a source of skilled supply labour for the private sector but not the other way around. It has an impact on the private sector when we start to lose the medium management or the semi-skilled workers to the RSE programme. (Quoted in RNZ, 14 January 2020)

Wilson suggested that the scheme needed to be administered better to ensure skilled workers were not recruited. While the Labour Export and Employment Division at the Ministry of Commerce, Industry and Labour agreed that the concerns were legitimate, they also emphasised that 'finding the balance would be difficult because seasonal work offered far more attractive financial benefits than what was locally available'. In a submission to an Inquiry into strengthening Australia's relationships with PICs, Union Aid Abroad APHEDA (2020) raised similar concerns about the impacts PLM recruitment was having on skilled Samoan workers.

> One of the biggest concerns for Samoa is that some of our best educated workers like teachers and administrators leave their low paid professional job and go to work as seasonal workers leaving big gaps in our village schools and other important jobs. This is unavoidable as people need more than the very humble pay of $ST5.00 [A$2.50] per hour. Money is very tempting, but it is a concern when our teachers are leaving our schools. (Union Aid Abroad APHEDA, 2020, p. 13)

Without constraints on recruitment, and with such income discrepancies—also evident in other skilled employment categories and across the PICs—that process is likely to continue. While it is at least arguable that the needs and desires of better-paid, employed urban workers should not take precedence over those in disadvantaged rural areas, it is impossible to define who is 'most in need' of seasonal work. Nonetheless, by 2019 several PICs were experiencing shortages of skilled workers, including nurses. Fiji was importing tilers and bricklayers from south Asia; Samoa was reporting a reduced agricultural output because of labour shortages and Fiji was even importing cane cutters (RNZ, 29 October 2019, 12 June 2019, 22 November 2019): all jobs that may have been or could have been undertaken by seasonal migrants. Both a lack of skills and, unusually, an actual labour shortage existed. By then the former Vanuatu Labour Commissioner, Lionel Kaluat, was indicating:

> I think most people going to the program now are people already in employment and already acquired skills. So it's a wake-up call for the Vanuatu Government to really pull up its socks. I would like to see the government put in place legislation that would [allow] only people who are unemployed and are looking for work… that would avoid the brain drain that is currently happening. (Quoted on ABC Pacific Beat, 3 July 2019)

In late 2020, Vanuatu proposed changes to recruitment practices aiming to decentralise and shift recruitment away from urban areas and increase equity of access for more remote rural places where skilled employment was less common. In the meantime, easily accessible urban areas remained a focus for many agents.

As the previous section suggested, the loss of workers from rural areas affected food security (and posed social problems) but the extent of the losses was limited. Loss of skilled workers from (mainly) urban employment may be more critical. Even though this was temporary (and not a brain drain in the conventional sense since all returned), many did not return to their former jobs, often in the public sector, but were more likely to move into the private sector as small business owners, which potentially represented a different kind of skill loss. Yet many seasonal workers acquired new skills abroad, and it is conceivable that these made a contribution to household and national development that compensated for skill losses.

Bringing Skills Home?

Social remittances through skills transfer to home countries were widely promoted as one of the key 'wins' of PLM. Beyond 'on the job' acquisition of skills and working procedures, and incidental absorption of knowledge about Australian (and New Zealand) cultures and values, Australia initially offered Add On Skills Training (AOST) courses to first-time workers who wished to improve their skills in topics including information technology, English literacy, numeracy and first aid. For some workers, the potential to improve English language fluency and acquire other new skills was said to be a 'critical factor' in their decision to participate in the SWP. The importance of skills acquisition varied across PICs; in countries such as Vanuatu, where workers had lower levels of education, learning new skills was a factor in their desire to participate in the SWP. For the Polynesian PICs, where more workers had attended or completed high school, skills acquisition was not a factor in their decision to participate in the SWP (World Bank, 2017). For some Pacific islanders, therefore, the SWP presented opportunities to acquire skills and knowledge that they had missed out on due to limited formal schooling. In this manner, migration was hoped to enhance human capital and raise welfare (evident in the focus of remittances on school fees) within the country.

AOST had limited uptake and courses were sometimes inaccessible to workers: more than two-thirds (68 per cent) of SWP workers interviewed by the World Bank (2017) in 2015–2016 had never even been offered the opportunity to undertake AOST. Until 2018, employers were expected to organise AOST courses, but delivery was difficult in rural locations. In addition, undertaking training courses was often the last thing workers wanted to do at the end of a long working day or on their one-day-a-week off. Tired workers preferred to use any spare time to relax or undertake essential chores such as grocery shopping, laundry, attending church or communicating with families at home (Chap. 7). Others had little interest in courses which had no immediate tangible benefits; one Tongan worker declared AOST a 'waste of time', as it did not lead to pay increases or other benefits. Not surprisingly, few Tongans participated in these courses (Holani, 2017). Most workers simply saw the acquisition of income as the goal, to improve on existing livelihoods rather than envisaging a move onto something different. Nor were employers necessarily supportive of AOST, which some criticised as providing skills irrelevant to the tasks SWP workers were employed to perform. By the end of the decade, a new, more flexible training programme was being developed for implementation by DFAT's new Pacific Labour Facility (PLF) (Chap. 10).

Despite the failings of formal training courses, three-quarters (74 per cent) of SWP workers interviewed in 2015–2016 by the World Bank (2017) believed they had learned useful skills such as operating machinery, including forklifts, and acquiring skills such as picking, packing and pruning. This practical 'on the job' experience was not supported by any formal horticultural training. Workers interviewed in Australia were often optimistic about how they might employ these skills at home, but there was scant evidence that this occurred (Dun & Klocker, 2017; World Bank, 2017). Picking, packing and pruning were valuable should workers wish to later return to Australia or New Zealand, but relevant skills were often specific to crops such as stone fruit and grapes that did not grow in the tropical climates of home countries, and were not transferable to crops that did grow there. Despite migrant workers' enthusiasm for new technology and new skills, Alisi Holani (2017) found that none of the Tongan workers had been able to make use of these skills at home. When

Kirstie asked a group of ni-Vanuatu workers if they planned to transfer the tomato-growing skills they had learned in Australia to their home villages, they explained the difficulties that existed around transport and the lack of a suitable market.

Kirstie: Will you go back to Vanuatu and plant tomatoes like you do here [in Australia], or have you had enough of tomatoes now?
Eddie: In Vanuatu, we plant tomatoes, but we don't have a big market to export them to. So we just plant a small amount, just enough to make a little bit of money from them. And here we use fertiliser, but we don't use too much fertiliser [in Vanuatu]. But our tomatoes are like the ones here anyway.
Ben: Here, they make farms for tomatoes because they sell them, and Australia is a big place, so lots of supermarkets need them. But in Vanuatu, it's just a small place. So you just grow enough [for your own needs]. If you plant too many, you waste them. There's no market for them […] And if you plant tomatoes on the island and transport them to Vila by ship, the tomatoes won't be good anymore because the distance is too far.
Dora: And the ship travels overnight on the sea, it's hot.
Ben: But here it is good. We pick tomatoes, and the people in packhouse sell them to the supermarket.

Climatic conditions, limited or unsuitable transport, the lack of a market and the challenges attached to instigating large-scale commercial agricultural ventures meant that applying new skills at home was often difficult. Such enterprises could be further constrained by land ownership. Throughout the PICs, subsistence food gardens and small-scale cash crop holdings are planted on family-owned, shared or borrowed land. While it is acceptable to plant crops like kava on borrowed land and make a small profit, earning larger amounts of money can be problematic. Indeed, when land becomes too valuable through, for example, rents or high-yielding crops, land disputes can emerge (Petrou, 2020). Land tenure thus limits large-scale agricultural development. Nonetheless, some workers did return home and plant commercial crops, sometimes with new technology. Stories occasionally appeared on seasonal worker social media sites celebrating a worker's success at establishing new commercial crops. One post from Facebook in 2020 showed photos of an expansive green field (Fig. 9.7) with the accompanying text

> The seasonal worker scheme doesn't just help workers by providing an income, but it also helps them by giving them skills and enhancing their experience at commercial farming. Freddy travelled to New Zealand as an RSE worker. Now he lives on the island and applies the skills that he learned [in New Zealand].

A *Vanuatu Daily Post* (1 April 2021) article recounted a similar story of agricultural skills transfer; Atis had spent nine years working in the RSE and eventually decided to apply his knowledge to a commercial agriculture venture at home. He explained

Fig. 9.7 Successful commercial farming in Vanuatu. (Source: Facebook)

> Working abroad in huge farms has inspired me to invest the skills and knowledge I have gained in my home village. I have acquired a lot of ideas, new ways of farming and have had experience working with farming machinery such as tractors. I have managed to do some savings that have assisted me to implement an agriculture project on farming peanuts this year [...] I have allocated over one hectare of land to grow peanuts. The main objective of investing my skills and knowledge in agriculture is to show young people that farming as a business is the key to earning income at your home island. This project has assisted people in the community, my immediate families and youths through community labour and I have hired a gardener who is assisting me at the farm [...] It's time now for farmers to let go of bush knives, garden forks, spade and other farming tools and make good use of the tractor. I encourage farmers on Epi to wake up and venture into mass production by using tractor and contributing to agriculture as a productive sector. Apart from planting peanuts, I have grown pineapples too and my target for 2022 is to plant 1000 yam tubers. The purpose of mass production is to invest in agriculture as a business.

Such direct claims to success were rare, since land tenure complications often intervened, and this example had been deliberately singled out to promote the programme.

Practicalities aside, many workers had similar aspirations. Samson, a PNG worker and landowner, claimed that his experience from as many as seven seasons of participation in the SWP had enabled him to learn planting strategies that would benefit his noni plantation (Dun & Klocker, 2017). Workers from Solomon Islands similarly felt they had learned new skills in Australia and were interested in applying them at home. Some had learned about new crop varieties and wanted to try planting citrus at home, others thought they might be able to use new grafting skills on mango and hibiscus trees (Dun et al., 2022). It was impossible to know whether such plans were carried through. Agricultural 'success stories' tended to be the exception rather than the rule.

In a unique example, the establishment of Tanna Farms in 2013 was held out by the regional organisation, Pacific Trade and Investment, as an example of how skills acquired overseas could contribute to the first joint venture of its kind in the Pacific

> Tanna Farms was founded in 2013 by a New Zealander, Jon Bushell, who was the AE of Vinepower, which had recruited Tannese for RSE employment. One of those recruited was Seth Karua, who had previously been a project coordinator for World Vision for eleven years, and became the team leader for one group of workers. Bushell recognised that few development opportunities existed for returnees and, with Karua, set up a coconut mill that eventually produced and marketed niche coconut oil. Karua effectively negotiated land tenure arrangements that made this possible, while it was argued that the project would be successful because of the 'work ethic and professionalism' learned by the project's ten workers during RSE. (Gil, 2016)

However, the involvement of the New Zealand partner was crucial to marketing and to establishing a packaging factory in Vila. Entrepreneurial ambition was otherwise limited.

In addition to formal skills training, and direct on-the-job experience, it was anticipated that workers would learn 'informal' secondary skills, such as financial literacy, English language competence, work discipline and time management, and that such skills would be more useful in home communities than, for example, sorting fruit in a packhouse. Certainly, most returned workers said they had learned some new skills (Bedford, 2013; Holani, 2017). In Tonga, such 'soft' skills were useful in helping children with homework, and budgeting household incomes—and thus assisted households in being better able to cover education fees, health care costs and church donations (Moala-Tupou, 2016). Similarly, some workers from Epi, Vanuatu described how RSE participation had changed their outlook: 'I manage my time using my time wisely. I wake up in the morning and get busy all the time and I see a big outcome later' and 'In New Zealand we say time is money, so we cannot waste time. So when I came back from New Zealand, I found a big difference here' (quoted in Kumar, 2012, p. 76). Workers learned to be on time—at pain of losing valuable work—but that was driven by income needs; time as money became ingrained. As one ni-Vanuatu woman, Rose, reflected after her first season:

> The time for us, when we are at home is not really important … we just relax. In the work it is important, you have to follow the time, the timing is strictly important… I have to wake up in time in the morning to make sure I don't miss the van to go to work. I have to discipline myself… like they said [in the pre-departure training] we have to discipline ourselves, we have to behave also. (Quoted in Villafuerte, 2018, p. 156)

As she observed, that had much less meaning back home; as a representative of the Vanuatu Department of Labour put it: 'here people go around doing their own things, at their own time. It's more like a subsistence community, so you work for yourself and at any time that suits you' (quoted in Cameron, 2011, p. 98). Since many seasonal workers experienced a worker-boss relationship for the first time on a seasonal employment scheme, the notions of discipline and order were quite new. While many returnees could see virtues in working harder—the rationale behind time management—unless there were obvious objectives it was usually more congenial to return to earlier ways.

By contrast, some workers found that skills learned overseas were inappropriate and antithetical to home village contexts. Returned RSE workers in Lamen Bay (Vanuatu) believed that skills such as time management and the ideology that 'time is money' made limited sense. Rather, they considered such organisational rigour to be 'upside-down logic' and that the ability to take a break when they were tired was a superior approach to constant labouring. Speaking about the 'time management' required to labour in New Zealand, one chief on Epi, commented 'You must wake up at a certain time, work according to time, no rest […] We are still lucky here' (Smith, 2016, p. 134, 2019). Elsewhere on Epi, workers felt that island lifestyles were so different that such skills were of virtually no local value—despite guestworkers being more engaged in building houses or running small stores than before. As one returnee said: 'We work with machines in New Zealand. So everything we do we have to be careful. Back in Vanuatu there are no machines [*sighing*] we just use *lokal* hands to do things' (quoted in Kumar, 2012, p. 62). Some ni-Vanuatu workers even pointed out that the skills they learned overseas had made them docile, unfree workers; time management meant 'you have to work against the clock; it's not like "Vanuatu time" any more' with one worker claiming the most important and most difficult skill to be: 'You learn how to listen to the boss. Sometimes they give instructions that don't seem to make sense, but you learn to just do what they say' (quoted in Cummings, 2013, p. 393). Another suggested that the skill most quickly learned was English: 'Everybody learns to speak English and fast. We want to work, but we don't want to be pushed around and ripped off' (op cit., p. 394). Local values could conflict with the perceived utility of new skills; new was not necessarily better or more appropriate.

For some workers, learning new skills while in Australia was simply not a priority when they were focused on earning money, meeting weekly expenses and negotiating a foreign culture. When problems arose—as they sometimes did—skills transfer became a low priority. As one Pacific islander resident from rural Victoria told Kirstie in 2020:

> When there's not much work on, the deductions continue, so they [SWP workers] see themselves going backwards, and then you know, [to think about] what sort of practices are you taking back, what learnings are you taking [home] with you, and it's like … you're not thinking about that. You're thinking about how much debt you're going into and how much you have to pay back. Maybe if you were making a lot of money you might have the opportunity to have blue sky thinking. But if all you can see is that you're constantly in the red you ain't gonna see the blue sky. So that kinda stuff needs to be really talked about, so that when everything's going well, yeah maybe people have the opportunity to look into the future. But when things aren't going so well, how the hell are you supposed to think about educating your kids and what's possible if you continue to see every week that you're not working you're going into [debt] like A$140 just in rent?

The vague possibility of skills transfer was given little consideration when workers were focused on such immediate issues as securing work, managing finances and accommodation, and coping with personal relationships. Beneficial skill transfers from seasonal worker schemes were scarce. Crop combinations and climatic conditions—and land tenure—are so different between tropical PICs and Australia (and even more New Zealand), that realistic transfers of skills and knowledge are generally implausible. Employers could be cynical about what might have been learned: 'concerning tree culture they do learn a lot and also just the experience to have work, get out of bed in time and do something the whole day is an experience many of these workers don't have' (quoted in Brickenstein, 2017, p. 76). As David Ellerman cautiously commented: 'it is important in such schemes that the latter [North] countries not be too advanced, otherwise the experience will have little applicability back home'. Limited skills transfer was not unique to the SWP, and was mirrored by the RSE (Bailey, 2019b; Cameron, 2011; Hammond & Connell, 2009) as well as Canada's SAWP (Downes & Clarke, 2007; ILO, 2012). Formal courses were burdensome to workers who valued some free time—however valuable they might have been—and new practical skills, useful in temperate countries, were not easily or productively transferred. Nor was that the intent of an employment-oriented scheme.

Towards Modernity

At home, workers experienced benefits, but these were not always without costs. No labour mobility scheme could ever be without both. Incomes contributed to immediate housing improvements, poverty alleviation and education. While this chapter has largely focused on the impact of guestwork at the local level—where it is perhaps easier to evaluate—the PLM schemes had major national implications. One Epi returnee perceived a wider perspective

> So to my view, the RSE scheme is changing the village a lot. I think it's developing the islands. I mean the whole country is benefiting because people are bringing money back into our country and you could see when people coming from New Zealand, they're buying, they're spending, so that means the economy is growing with the money. (Quoted in Kumar, 2012, p. 77)

Income flows have been substantial, matched by the transfer of goods, but the impact of guestwork cannot simply be reduced to cash flows, numbers and quantities; development is much more than this. Social costs occurred. Many people feared guestwork was deleterious to marital relations and their children's lives. But, overwhelmingly, better economic and social wellbeing, centred on more adequate housing, was combined with the ability to pay education expenses and achieve a degree of social mobility. In Epi and elsewhere people witnessed a rise in individualism as returnees focused on their new houses and goods, rather than prioritising contributions to community activities. But this was a seemingly inevitable ongoing trend, perhaps hastened—but certainly not caused—by the larger cash flows associated with seasonal labour mobility. It is easy enough to suggest that a profusion of chainsaws and some other imported goods may not necessarily be in the best interests of development and the environment, despite their undisputed utility, but such items were proliferating well before labour migration. Likewise, youth act out, even with parental supervision, people have extramarital affairs and, ironically, increasing disdain for agricultural employment has long been evident in the PICs.

There is an inherent paradox at the heart of guestworker schemes. Clear benefits existed when people remained loyal to their traditional communities and kin groups, and supported them on their return. Yet being a labour migrant demanded the cultivation of an individualistic sensibility and work ethic, at least temporarily, and the acquisition of some new ways and goods that departed from the old order. Social remittances alone involved such uneasy liberal ideals as a 'work ethic', where 'time is money and financial literacy'. Superior material livelihoods were entirely welcome: better-equipped houses, a vehicle, new technology and a small business were rarely far from workers' minds or from eventual outcomes: the trappings of anticipated modernity, and a new aesthetics. Land purchases in urban area were a bonus. Success was relational, necessitating being able to balance relationships with kin, place and community. Youthful migrants were not easily able to do that, when they focused on domestic needs and a measure of social mobility.

New flows of wealth had evident positive impacts, but they were not unproblematic. Wealth, and the social status associated with it, changed social relationships and resulted in tensions in small communities. For families and communities, the repeated absences of young, able-bodied men caused heartache as children missed their fathers, while women faced a double burden and struggled to compensate for their husband's lost labour. Vulnerability, food insecurity and a partial transition towards food dependency occurred in some communities. And while workers gained new skills from working in Australian (and New Zealand) horticulture, the usefulness and cultural relevance of these skills were at best slight; work discipline was not easily transferable to home community contexts, and it is questionable whether returned migrants acted as agents of positive change. As more and more urban workers participated in the scheme, skills and brain drain could potentially occur, and equity and regional development be further threatened.

Overall the labour migrants themselves were usually enthusiastic about the schemes, even when they had experienced frustrations and difficulties. A survey of more than 300 SWP workers (mainly from Tonga, Vanuatu and Samoa), when asked on site whether they were satisfied with their experience in Australia recorded a level of 8.6 (on a scale of 1 to 10) with Tongans scoring 9.9 and ni-Vanuatu 6.3 (World Bank, 2017, p. 44). While such scales are blunt instruments for assessing complex experiences (and excluded outcomes), on the verge of returning home as workers contemplated new futures, optimism was palpable and almost all said they would recommend the programme to others and would themselves return (with 85 per cent saying 'for as many years as possible'). Inevitably perfunctory analysis of this kind—without rapport between analysts and subjects—usually leads to positive responses to satisfy the questioner. Thus, Lya Villafuerte (2018, p. 115) found that wives of ni-Vanuatu seasonal workers, unsure of who she was or why she was asking questions, simply responded 'all good'. Nonetheless, satisfaction was well summarised by a woman from the small Tongan island, Lifuka, where every household (in a small sample of 16) considered that seasonal labour migration had been beneficial, despite negative effects on children, and in Tonga by a previously impoverished fisherman:

> I think it is a blessing from above and it's one of the best opportunities for my husband. We haven't seen any negative impacts in my family…Look at this house; it was built with the support of my husband's pay after Cyclone Ian in early January 2014 destroyed our house…[It] really helps me to fulfil different functions that we must do such as *misinale* [annual financial contribution] for church, financial needs for children's schooling and for family daily needs… also my husband got a job instead of being a subsistence farmer since we were married. (Quoted in Moala-Tupou, 2016, p. 66)

Much the same was true in Samoa:

> I returned from my second trip and bought a second-hand car for my family to take the kids to school and take my parents to where they want to go instead of catching the bus. There were many things I dreamed of to achieve while I worked as a fisherman. My family lived in poverty and I felt sorry for my parents and the struggles we faced every day and that's the reason I wanted to go and I thank God I made the right decision. I felt so happy just by looking at what I managed to buy for my family with the money I brought with me on my return. (Quoted in Fatupaito et al., 2021, p. 57)

It is no surprise then that seasonal work was seen by many as 'simultaneously demeaning and essential' (op.cit., p. 61). Some degree of ambivalence was universal. Where development remains at least partly funded by remittances, as seems likely in many rural areas, development gains depend on continued migration since guestwork alone cannot provide the basis for alternative development trajectories. This accounts for the widespread preference for return migration. Neither in the established remittance-based MIRAB (Migration, Remittances, Aid and Bureaucracy) economies of Tuvalu, Samoa and Tonga, nor in those PICs where seasonal employment is considerable, notably Vanuatu, is remittance-led growth an assured development strategy. It is a permanently temporary fix and a band-aid solution. While it is unlikely to disappear in the near future, and at the end of the 2010s was expanding, as we discuss in the next chapter, it nonetheless points to precarity; will such schemes continue? Who will be selected, from where? As the same World Bank report that recorded the enthusiasm of the workers observed: 'there is little evidence that substantial GDP growth in sending countries has been caused by remittances. Despite the beneficial impacts that remittances have on poverty, schooling, health, and housing considerable recent research points to workers' remittances making little to no contribution to economic growth' (2017, p. 8), a conclusion valid for most PICs.

Like many others in Vanuatu, Conrad hoped that investment of his income would enable future generations of his extended family 'to choose not to go overseas in order to make ends meet' (Cummings, 2016, p. 8). As continued migration from such Polynesian states as Tonga and Samoa suggests, alongside the limited impact of migration on economic growth, that may well be unlikely. But seasonal employment has unquestionably helped Conrad and many others to benefit from a diversification of livelihoods—with costs and benefits—that would otherwise have been impossible. Becoming modern brought diverse challenges and complexities of outcomes.

Bibliography

Bailey, R. (2014). *Working the Vines: Seasonal migration, money, and development in New Zealand and Ambrym, Vanuatu*. PhD thesis, University of Otago, Dunedin.

Bailey, R. (2019a, April 26). *Pacific workers in rural Victoria*. ANU DevPolicyBlog.

Bailey, R. (2019b). *Ni-Vanuatu 'Making Business' through Seasonal Worker Programs*. ANU InBrief 2019/13.

Bailey, R. (2021). Mobilising pathways through cross-border labour schemes, DPA Seminar Series, 2 September 2021, ANU [online].

Bailey, R., & Wells, S. (2016). *Temporary Labour Schemes, the new avenue of social mobility: Case study Vanuatu*. ANU InBrief 2016/21.

Bedford, C. (2013). *Picking winners?: New Zealand's recognised seasonal employer (RSE) policy and its impacts on employers, Pacific workers and their island-based communities*. PhD thesis, The University of Adelaide.

Bedford, C., Bedford, R. & Nunns, H. (2020). *RSE Impact Study: Pacific stream report*. Bedford Consulting.

Binford, L. (2002). Social and economic contradictions of rural migrant contract labor between Tlaxcala. *Mexico and Canada. Culture & Agriculture, 24*(2), 1–19.

Binford, L. (2003). Migrant remittances and (under) development in Mexico. *Critique of Anthropology, 23*(3), 305–336.

Blanco, C. (2009). *Temporary migration, remittances and development: Early impressions of the Australian Pacific seasonal worker pilot scheme*. Honours thesis, University of Sydney, School of Geosciences.

Brickenstein, C. (2017). The development impact of seasonal worker schemes: A comparison of European and Pacific experiences. *Migration and Development, 6*(1), 60–82.

Calnitsky, N. (2016). Grape vines and orchard lines. *Left History: An Interdisciplinary Journal of Historical Inquiry and Debate, 20*(1), 11–35.

Cameron, E. (2011). *Social development outcomes of participation in the New Zealand Recognised Seasonal Employer (RSE) scheme for ni-Vanuatu seasonal migrant workers*. Master's thesis, Massey University, Palmerston North.

Chand, S. (2008). Biceps or the brain: Investing in our most valuable resource. *Islands Business, 34*(6), 40–41.

Chand, S. (2021). Luve ni Yali: Lost son of a seasonal worker. *Islands Business, 47*(5), 16.

Chattier, P. (2019). Beyond development impact: Gender and care in the Pacific Seasonal Worker Programme. *Gender and Development, 27*(1), 49–65.

Connell, J. (2006a). Migration, dependency and inequality in the Pacific: Old wine in bigger bottles? In S. Firth (Ed.), *Globalisation and governance in the Pacific Islands* (pp. 59–106). ANU E-Press.

Connell, J. (2009). *The global care chain. From the Pacific to the World.* Routledge.

Connell, J., & Brown, R. (2005). *Remittances in the Pacific: An overview.* ADB.

Connell, J., & Conway, D. (2000). Migration and remittances in island microstates: A comparative perspective on the South Pacific and the Caribbean. *International Journal of Urban and Regional Research, 24*(1), 52–78.

Craven, L. (2013). *The relationship between temporary migration and vulnerability to climate change in rural Vanuatu.* Honours thesis, University of Sydney Department of Government.

Cummings, M. (2013). Imagining transnational futures in Vanuatu. In A. Quayson & G. Daswani (Eds.), *A companion to diaspora and transnationalism* (pp. 381–396). Hoboken, New Jersey.

Cummings, M. (2016). Uncertain belongings: Relationships, money and returned migrant workers in Port Vila, Vanuatu. *Journal of Pacific Studies, 35*(3), 21–33.

Curry, G. (2005). Doing "Business" in Papua New Guinea: The social embeddedness of small business enterprises. *Journal of Small Business & Entrepreneurship, 18*(2), 231–246.

Dao, A., Green, M., & Kass, T. (2020, May 16). We feed you. *The Saturday Paper.* https://www.thesaturdaypaper.com.au/news/politics/2020/05/16/we-feed-you/15895512009836?fbclid=IwAR1X5r411aDQfb397ODGaQey8a3YtJ_hdIBAGnPLm0d4w5GXp9hE1XjDvn4.

Davidson, H., & Earl, C. (2020, October 18). The fruit pickers: Inside Australia's seasonal worker program, *The Guardian.*

Downes, A., & Clarke, K. (2007). *The Canadian seasonal agricultural workers program: The experience of Barbados.* Paper presented at the CSAWP conference, University of the West Indies, Barbados.

Dreby, J. (2010). *Divided by borders: Mexican migrants and their children.* University of California Press.

Dun, O., & Klocker, N. (2017). The migration of horticultural knowledge: Pacific Island seasonal workers in rural Australia—A missed opportunity? *Australian Geographer, 48*(1), 27–36.

Dun, O., McMichael, C., McNamara, K., & Farbotko, C. (2022). Investing in home: Development outcomes and climate change adaptation for seasonal workers living between Solomon Islands and Australia. *Migration and Development,* [In press].

Falefou, T. (2017). *TOKU TIA: Tuvalu and the impacts of climate change*. PhD thesis, University of Waikato, Hamilton.

Fatupaito, A., Utuva, L., Tauave, S., Alofipo, A., Meleisia, M., Schoeffel, P., Arthur, T., & Alexeyeff, K. (2021). Samoa's New Labour Trade. *Journal of Samoan Studies, 11*(1), 51–63.

Gibson, J., & McKenzie, D. (2014). The development impact of a best practice seasonal worker policy. *Review of Economics and Statistics, 96*(2), 229–243.

Gil, P. (2016, July 15). Tanna farms. Bringing back the coconut. *Island Life*.

Haberkorn, G. (1992). Temporary versus permanent population mobility in Melanesia: A case study from Vanuatu. *International Migration Review, 26*(3), 806–842.

Hammond, J., & Connell, J. (2009). The new blackbirds? Vanuatu guestworkers in New Zealand. *New Zealand Geographer, 65*(3), 201–210.

Hennebry, J. (2014). Falling through the cracks: Migrant workers and the Global Social Protection Floor Initiative. *Global Social Policy, 14*(3), 369–388.

Holani, A. (2017). *Enhancing the development impact of labour mobility in Pacific Island countries: The role of the seasonal worker program and the PACER plus*. PhD thesis, University of Adelaide.

International Labour Organisation. (2012). *Pacific seasonal migration to New Zealand and Australia: Literature review*. ILO.

Jolly, M. (1987). The forgotten women: A history of migrant labour and gender relations in Vanuatu. *Oceania, 58*(2), 119–139.

Kaiapam, S. (2021). Impacts of labour migration on migrants and their families: A case study of Vanuatu. MA thesis, University of the South Pacific.

Kumar, R. (2012). *Migration and development: The development impacts of short-term labour migration of Ni-Vanuatu workers to New Zealand*. MA thesis, University of the South Pacific.

Lee, T. (2015, May 2). Pacific Seasonal Worker Program: Vital employment for Vanuatuans, assured labour for farmers, *ABC Landline*.

Lepon, D. (2010). *'It does not go as well as it could'. The views of Melanesian migrant farm workers of the cultural, economic and social benefits and costs of seasonal work in New Zealand.* Master of Social Sciences thesis, University of Waikato.

Moala-Tupou, L. (2016). The social impacts of seasonal migration on left-behind children: An exploratory study from Lifuka, Tonga. Master of Social Science thesis, University of Waikato, Hamilton.

Nef, D., Kotra, K., Stauffacher, M., Six, J., & Kruetli, P. (2022). Vulnerability to food insecurity in a telecoupled world: Insights from Vanuatu. *Frontiers in Sustainable Food Systems* 6:818586.

Oceania Football Center. (2020). Seasonal working: A tough reality for football players from Vanuatu. https://oceaniafootballcenter.org/2020/06/21/.

Paciulan, M., & Preibisch, K. (2013). Navigating the productive/reproductive split: Latin American transnational mothers and fathers in Canada's temporary migration programs. *Transnational Social Review, 3*(2), 173–192.

Parreñas, R. (2005). *Children of global migration. Transnational families and gendered woes*. Stanford University Press.

Petrou, K. (2020). *If everyone returned, the Island Would Sink: Urbanisation and migration in Vanuatu*. Berghahn.

Preibisch, K., & Grez, E. (2010). The other side of el otro lado: Mexican migrant women and labor flexibility in Canadian agriculture. *Signs: Journal of Women in Culture and Society, 35*(2), 289–316.

Preibisch, K., & Grez, E. (2013). Between hearts and pockets: Locating the outcomes of transnational homemaking practices among Mexican women in Canada's temporary migration programmes'. *Citizenship Studies, 17*(6–7), 785–802.

Radio New Zealand. (2019, April 8). Extra-marital affairs issue for Samoa seasonal workers 'manageable'. *RadioNZ*. https://www.radionz.co.nz/international/pacific-news/386617/extra-marital-affairs-issue-for-samoa-seasonal-workers-manageable

Rockell, D. G. (2015). *Pacific Island labour programmes in New Zealand: An aid to Pacific Island development?: A critical lens on the recognised seasonal employer policy.* PhD thesis, Massey University, Palmerston North.

Rohorua, H. T., Gibson, J., McKenzie, D., & Martinez, P. G. (2009). How do Pacific island households and communities cope with seasonally absent members? *Pacific Economic Bulletin, 24*(3), 19–38.

Ruehr, S. (2022). Beyond the vulnerability/resilience dichotomy: Perceptions of and responses to the climate crisis on Emau, Vanuatu. *Island Studies Journal, 17*(1), 157–176.

Smith, R. (2015, November 12). The 'hidden abodes' of temporary migration programs. FocaalBlog. www.focaalblog.com/2015/11/12/rachel-smith-the-hidden'abodes-of-temporary-migration-programs.

Smith, R. (2016). *The goal of the good house: Seasonal work and seeking a good life in Lamen and Lamen Bay, Epi, Vanuatu*. Unpublished PhD Thesis, University of Manchester, Manchester.

Smith, R. (2018). Changing standards of living: The paradoxes of building a good life in rural Vanuatu. In C. Gregory & J. Altman (Eds.), *The quest for the good life in precarious times: Ethnographic perspectives on the domestic moral economy* (pp. 33–55). ANU Press.

Smith, R. (2019). Be our guest/worker: Reciprocal dependency and expressions of hospitality in ni-Vanuatu overseas labour migration. *Journal of the Royal Anthropological Institute, 25*(1), 1–19.

Smith, R. (2021). Declarations of self-reliance: Alternative visions of dependency. *Citizenship and Development in Vanuatu. Oceania, 91*(2), 236–256.

Stead, V. (2019). Money trees, development dreams, and colonial legacies in contemporary Pasifika Horticultural Labour. In V. Stead & J. Altman (Eds.), *Labour lines and colonial power: Indigenous and Pacific Islander Labour Mobility in Australia* (pp. 133–158). ANU Press.

Tazreiter, C., Weber, L., Pickering, S., Segrave, M., & McKernan, H. (2016). *Fluid Security in the Asia Pacific* (pp. 103–127). Macmillan.

Uera, M. (2019). *A Pacific/ Tongan perspective. Pacific Update 2019*. USP.

Union Aid Abroad APHEDA. (2020). Inquiry into strengthening Australia's relationships with countries in the Pacific region: Submission 20. Submission to Joint Standing Committee on Foreign Affairs, Defence and Trade—Foreign Affairs and Aid Sub-Committee: Canberra.

Villafuerte, L. (2018). *Seasons of change: Ni-Vanuatu and the Recognised Seasonal Employers (RSE) scheme*. PhD thesis, University of Auckland.

Voigt-Graf, C. (2003). Fijian teachers on the move: Causes, implications and policies. *Asia Pacific Viewpoint, 44*, 163–174.

Wigglesworth, A. (2018, June 12). Why Timorese workers love the Seasonal Worker Program. ANU DevPolicy Blog.

Wigglesworth, A., & Boavida dos Santos, A. (2016). *Timorese workers in Australia: A study of seasonal work experiences, skills and future options*. Labour Mobility Assistance Program.

Williams, A. (2010). *The experiences of the recognised seasonal employer scheme and Ni-Vanuatu temporary workers in Te Puke, New Zealand*. Master of Planning thesis, University of Otago, .

World Bank. (2017). *Maximizing the development impacts from temporary migration*. World Bank.
World Bank. (2018a). *The social impacts of seasonal migration: Lessons from Australia's Seasonal Worker Programme for Pacific Islanders*. World Bank.
Wu, A. (2019). *Timorese participants in the Australian seasonal workers programme: Migration, remittances and development*. PhD thesis, University of Sydney.

10

A New Phase: Stepping up a Gear?

As the Seasonal Worker Programme (SWP) scheme developed and was widely regarded as successful, interest developed in both Australia and the Pacific Island Countries (PICs) towards expanding the scheme into other sectors. That was prompted by increased demand in Australia for skilled and semi-skilled workers, and occurred alongside the establishment of the Australian Pacific Training Coalition (APTC)—formerly the Australian Pacific Technical College—a programme to support and deliver Technical and Vocational Education and Training (TVET) in the Pacific region. APTC courses met Australian standards, and it was therefore anticipated that APTC would train semi-skilled workers for Australian labour markets. Target skills and industries under APTC included such activities as hospitality, painting, vehicle maintenance and care giving (Chap. 3). Rather conveniently, the SWP offered a potential model for resolving some of Australia's non-seasonal labour shortages through temporary migration. At the same time, the PICs continued to seek superior migration opportunities in metropolitan states. This chapter examines the earliest phase of the expansion of guestworker opportunities beyond agriculture through what was to become the Pacific Labour Scheme (PLS), but where both this and the SWP scheme would

K. Petrou, J. Connell, *Pacific Islands Guestworkers in Australia*,
https://doi.org/10.1007/978-981-19-5387-3_10

eventually become subsumed under a more integrated and consolidated Pacific Australia Labour Mobility (PALM) scheme. Due to the recency of the PLS scheme, yet to be evaluated, and the unexpected intervention of COVID-19, this chapter is primarily indicative of directions of change, pointing to new challenges, some familiar problems and Australia's wider, more long-term engagement with Pacific labour migration.

New Beginnings

By the start of the 2010s, it was readily evident that several Australian employment sectors were at least intermittently short of labour, both skilled and unskilled. For most of this century, skilled migrants were able to enter Australia on 457 visas, where they were nominated by specific businesses to fill existing vacancies. These migrants could bring their families and were allowed to stay in Australia for up to four years, after which it was common for many to apply for (and be granted) permanent residence. In 2019, the 457 visas were replaced by Temporary Skill Shortage (TSS) 482 visas. TSS visas removed the need for employer sponsorship, as long as potential migrants were skilled in an occupation deemed to be in shortage. This too offered a partial model for what might be extended to semi-skilled Pacific workers (and stimulated lengthy debate on what might be meant by 'skilled' and how shortages might be defined).

Without specific migration pathways for semi-skilled workers, both New Zealand and Australia faced similar labour problems to those that had brought the RSE and SWP schemes into being, but now in industries far beyond horticulture and fruit picking. In New Zealand, the dairy and meat processing industries requested that the then Department of Labour extend the provisions of the RSE to their operations. Hitherto, steady supply and demand throughout the year meant that both these sectors had fewer problems in estimating labour requirements than the horticultural industry. In Australia, the same industries experienced similar long-term shortages. Some such labour needs were filled by new migrants and refugees who had been effectively directed to work in regional areas through the terms of their settlement. In several places, like

Young in New South Wales, Biloela in Queensland and the Limestone Coast in South Australia, decentralised meat processing and abattoir industries were particularly short of labour and characteristically absorbed a significant refugee workforce (e.g. Stilwell, 2003; Feist et al., 2014; Hugo, 2014; Boese, 2015). Simultaneously, recent Pacific islander migrants commonly worked in various agricultural activities in the Riverina and elsewhere (Nishitani & Lee, 2019). This had other wider benefits for regional development, since—like seasonal workers—new migrants contributed to regional areas, primarily through their work but also from their local expenditure and more indirectly their impact on social capital and the dynamism of small town economies.

Outside agriculture too, even before the Pacific Seasonal Worker Pilot Scheme (PSWPS) had concluded, positive evaluations of the programme suggested that it might be extended into other areas, such as forestry and fishing, construction (including plumbing), and hospitality and food services (TNS, 2011). Some of these sectors, such as hospitality, were already experiencing a degree of dependence on migrant workers, including working holiday makers (WHMs), and even employed a number of temporary skilled 457 visa holders. In both New Zealand and Australia, there had been earlier discussion of a more general extension of temporary migration into the hospitality industry, and care services, where demand for workers was considerable, but required some relevant skills (Callister et al., 2009a, 2009b). The World Bank's sponsored review of SWP labour mobility and of the labour requirements of both the Australian and New Zealand aged care sectors gave strong support and impetus to recruitment of aged care workers from PICs (Curtain et al., 2016; World Bank, 2018). Rather like the food service industries, a high proportion of all caring jobs in Australia were held by new migrants from a range of 'non-traditional' source countries especially in south Asia (Negin et al., 2016). Many Pacific islanders had already taken up employment in these caring industries at home or abroad (Connell, 2009) and had relevant experience and skills. Diverse labour shortages, and migrant interests and aptitudes, raised wider questions over whether seasonal temporary employment opportunities, as in the SWP, could be scaled up to incorporate more sectors and more people. Extending the SWP into other, non-seasonal areas, including activities such as meat processing and tourism, meant the potential for longer periods of employment and higher skill requirements. Nonetheless, and with far less media attention and public debate than occurred during the lead up to the SWP, extending opportunities into other, non-seasonal employment sectors, which experienced ongoing labour shortages, became the distinctive component of the new phase of guestwork. Significantly, the possible role of APTC in all of this was scarcely discussed in policy circles.

Late in 2011, the first 'breakthrough' into non-farm employment occurred when Australia announced that workers from Timor-Leste were to be invited to work in the tourism industry in rural Western Australia, focusing on Broome, for 'seasonal' employment under the terms of the SWP, but with longer contracts. This more obviously expanded the geography of the scheme far beyond eastern Australia, broadened it to include Timor-Leste (and also indicated how Timor-Leste was frequently perceived as akin to a PIC) and widened its sectoral orientation. The first few Timorese workers, all men, arrived in 2011 to work with a pearling company in Broome; they came for an anticipated two years to work as deckhands and divers, and to gain technical skills to fill labour gaps in Timor-Leste. The Australian programme coordinator argued that it was important for these 'cadets' to learn new skills that could help industries like pearling develop at home: '[Timor-Leste] is very dependent on developing the petroleum industry and offshore industry so we are very keen to find ways of developing new industries... Aquaculture is a very important and strategic industry to help us generate employment' (ABC News, 6 January 2011). Eight more 'cadets', including three women, were expected to follow. Timorese tourist industry workers were soon more numerous however, mainly in Broome itself but also in other parts of northern Western Australia, but usually on six-month contracts to cover staff shortages during peak tourism seasons. It took several more years for what began in Broome to extend beyond Western Australia, to northern Australia where job shortages were most evident, and rather later to Kangaroo Island (South Australia). Once again expansion was centred on Australian needs.

The Northern Australia Worker Pilot Program (NAWPP)

The second phase of expanding contemporary Pacific labour mobility to Australia, both geographically and by sector, came in 2015 as Australia introduced a two-year 'microstate visa' for northern Australia, specifically for Kiribati, Tuvalu and Nauru—the 'orphan microstates'—to provide employment opportunities for up to 250 people or around 50 per year over five years. None of these three PICs had significant numbers participating in the SWP scheme. At that time, northern Australia claimed a labour shortage in the hospitality industry of 38,000 workers, which was expected to increase to 120,000 by the end of the decade. That shortage was partly due to the decline in working holiday visas, but also related to a mismatch between WHMs who wanted short-term work, and employers who wanted longer-term commitment. Pacific guestworkers provided a potential solution.

The Pacific microstates' Northern Australia Worker Pilot Program (NAWPP), as the scheme was called, sought to pursue more flexible foreign worker arrangements in high demand areas, especially hospitality, through a Northern Territory Designated Area Migration Agreement (DAMA). The Northern Territory, with its significant tourist industry, had the lowest unemployment rate in Australia, hence sourcing local labour was unusually difficult.

The DAMA allowed Northern Territory employers to take on skilled and semi-skilled overseas workers for positions that could not be filled locally; these included several care positions. The NAWPP, modelled on the SWP, used the same approved employer framework, but permitted employment in any low skilled industry. A second priority of NAWPP was that of finding employment for graduates of APTC, and so justify one of the main goals of establishing APTC, expanded opportunities for labour mobility. However, the migration of APTC graduates was quite limited (Chap. 3). Despite the apparent potential linkage, therefore, the NAWPP was consequently largely unrelated to the APTC.

Early in 2016, Kiribati became the first country to sign a memorandum of understanding with Australia. Mulpha, a tourism and hospitality employer based at Hamilton and Hayman Islands, off the Queensland coast, was the first to sign up to the new scheme, and its representatives travelled to Kiribati and interviewed between 80 and 100 people. After shortlisting about 30, 18 i-Kiribati workers eventually arrived on Hayman Island in October 2016 and were employed in housekeeping, gardening and maintenance, stewarding and food and beverage provision. A further group, mainly women, went to Hamilton Island in 2019. Despite pressure from Kiribati's outer island MPs to include participants from their islands, most of the jobs were allocated by the Ministry of Labour to women who spoke English and had previous experience working on cruise liners. Three years later, on the Pacific Trade Invest website, the Commissioner, Caleb Jarvis, elaborated,

> Although Allan and the team on Hayman had never heard of Kiribati, I knew Kiribati was a great fit and the workers could excel on Hayman. They are used to tropical temperatures and remote island living conditions and possess good English skills and work ethic. Over the last three years the i-Kiribati recruits have flourished; many have achieved their Certificate III in Hospitality and have used the opportunity to invest in their families' future. Like Retati Tekaai, who was initially recruited as a housekeeper. The oldest of six children, she has been able to pay for her younger siblings' schooling.

Some workers, including 13 from Kiribati, were able to earn a Certificate Three in Hospitality during their employment, returning home not only with earnings and new skills but also with internationally recognised qualifications. Nonetheless, overall participation rates remained low, reaching only 80 workers by the time the scheme was discontinued and replaced by the PLS. Nauru only signed a Memorandum of Understanding in August 2017, and sent one worker in 2018, while Tuvalu sent its first nine workers in 2018. Despite the very small numbers, the NAWPP was regarded as a success and one that could usefully be expanded.

The Pacific Labour Scheme

In July 2018, NAWPP was discontinued and replaced by the intendedly nation-wide PLS. A purpose designed Pacific Labour Facility (PLF), accountable to the Australian Department of Foreign Affairs and Trade (DFAT), facilitated this expansion but added a bureaucratic complication, as the SWP continued to be administered by the Department of Education, Skills and Employment (DESE). A certain geographical complexity was also introduced, as the PLF was based in Brisbane, rather than Canberra.

The PLF partnered with APTC to train Pacific workers with the specific but hitherto unsuccessful goal of meeting Australian labour market gaps. The PLS was much wider in range than previous Pacific guestworker visas in terms of both employment types and contract duration. It targeted the same countries as the SWP and allowed approved employers in low and semi-skilled industries to recruit workers on contracts for one to three years. Workers had to be aged between 21 and 45 (whereas there was no official upper limit on the age of SWP workers), and upon their return home, PLS workers were expected to use their new skills, along with their savings, to pursue opportunities for self-employment. All sectors in rural and regional Australia were eligible to participate in the scheme, but in practice this was limited to tourism, hospitality, aquaculture, forestry, agriculture (labour hire), aged care and meat processing. The inclusion of female dominated industries—such as hospitality and aged care—was expected to increase Pacific women's participation in labour mobility, and it was initially expected that the care sector would be the largest growth sector within the PLS.

Despite workers being able to stay for up to three years, the scheme did not allow them to bring their partners and families. This contrasted with similar visas, notably the skilled 457 visa, where accompanying families were permitted. The intent was to avoid any leakage of PLS workers into permanence—a possibility that was not considered an issue for 457 visa holders (who were unlikely to be Pacific islanders). However, it was also evident that many PLS workers were likely to be employed in remote areas where accommodation was inadequate, access to services (such as education and childcare) was poor and partners might be unable to work, so making family life expensive and difficult (possibly discouraging remittances) unless support was available through employers or government. Logistics aside, SWP workers could experience social problems arising from family separation over periods of several months (Chap. 10), and it seemed likely that the much longer periods of separation in the PLS could potentially lead to more significant social problems, as has occurred for families of i-Kiribati seamen (Borovnik, 2007).

The PLS began slowly, with employers preferring backpackers where that was feasible (and to avoid the bureaucracy associated with hiring through the PLS), and in the first year only 203 workers participated. About 50 of these were i-Kiribati workers who had arrived in Australia under the NAWPP and were subsequently absorbed into the PLS. Furthermore, in the first year only a handful of employers participated across six industries: hospitality, caregiving, agriculture (labour hire), forestry, meat processing and fishing/aquaculture (Fig. 10.2). However, some 50 employers had gained approved employer status by July 2019, so interest in the scheme within Australia was considerable. In the early days, employment was heavily concentrated in Queensland, where 85 per cent of the workers were employed. The presence of workers from Kiribati, thanks to the 'head start' provided by the NAWPP, and, to a lesser extent, Tuvalu, and the absence of workers from Tonga (Fig. 10.2), distinguished the PLS from the SWP. Vanuatu remained an important labour source, but did not dominate numbers to the same extent as in the SWP. Hospitality and agriculture (through labour hire companies) accounted for the most jobs. Early in 2019 11 APTC graduates from Kiribati had gained employment as aged care workers in central western Queensland. Some 40 per cent of PLS workers were then women (much higher than in the SWP) with clear gender patterns evident in employment by industry; all aged care employees were women while all workers in meat processing, forestry, fishing and aquaculture were men (Fig. 10.1)

While hospitality and aged care showed promise in the early days, the structure of employment changed quite quickly, as meat processing numbers grew rapidly. By March 2020 there were 650 PLS meat workers in Australia, absolutely dominating the scheme, to the chagrin of the PLS who had anticipated a more skilled and less masculine workforce. Agriculture was a distant second with 144 workers (mainly employed by

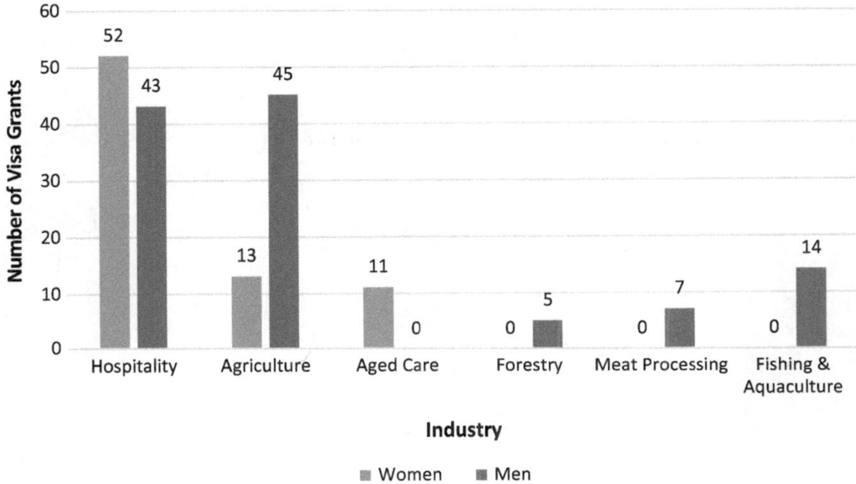

Fig. 10.1 PLS workers by gender and industry of employment in June 2019. (Source: PLF)

labour hire companies) and hospitality a close third employing 137 workers (Fig. 10.3). The shift to meat processing had also brought in other countries, notably Fiji and Tonga (Table 10.1). The few workers in aged care were mostly from Kiribati, reflecting what a group of i-Kiribati women had told John in 2019: 'what would we do if we have to go to Australia; we cannot farm so we would not get jobs but we could work in old people's homes'. None of the women with whom John spoke were at all enthusiastic about leaving Kiribati however. By contrast, social media posts about early PLS information sessions in Vanuatu showed images of packed rooms and long lines (Figs. 10.4(a) and (b)): the PLS was an attractive prospect for many.

By early 2020, women had begun to gain employment in meat processing, and a handful of men had taken up employment in aged care, but clear gender divisions remained. Moreover, female participation had fallen to about 20 per cent; women were back to comprising a similar proportion of participants as in the SWP scheme, despite the somewhat token Australian preference for greater gender equity. By the start of 2021 women's participation had dropped further to 17 per cent (Carnegie & Jeffries, 2021), and without increased recruitment in industries such as

Table 10.1 Total PLS visas granted by gender and PIC, March 2020

	Gender	Fiji	Kiribati	Nauru	Papua New Guinea	Samoa	Solomon Islands	Timor-Leste	Tonga	Tuvalu	Vanuatu	Total
Visa grants	Female	31	87	8	2	13	12	18	12	4	40	227
	Male	205	50	14	47	171	127	65	56	46	140	921
Total visas granted		236	137	22	49	184	139	83	68	50	180	1148

Source: PLF and APTC

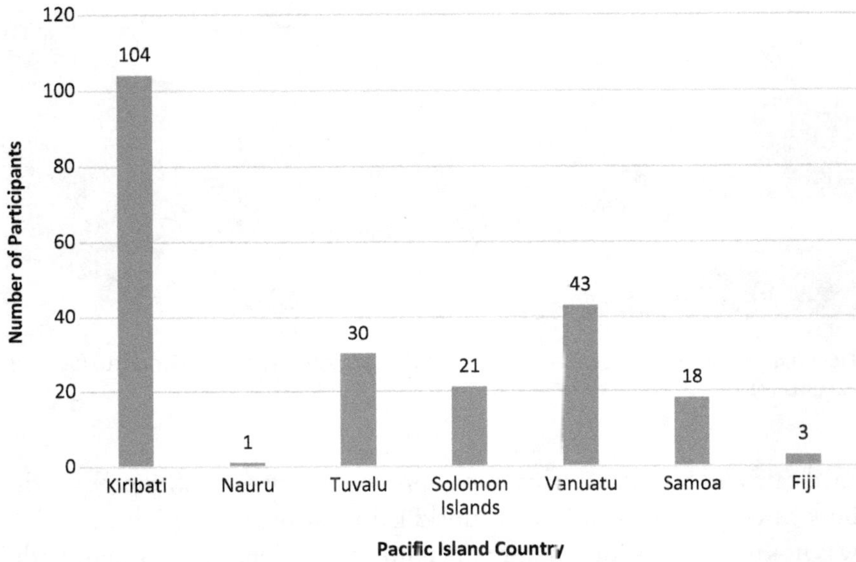

Fig. 10.2 Participation in the PLS by country of origin, June 2019. (Source: PLF)

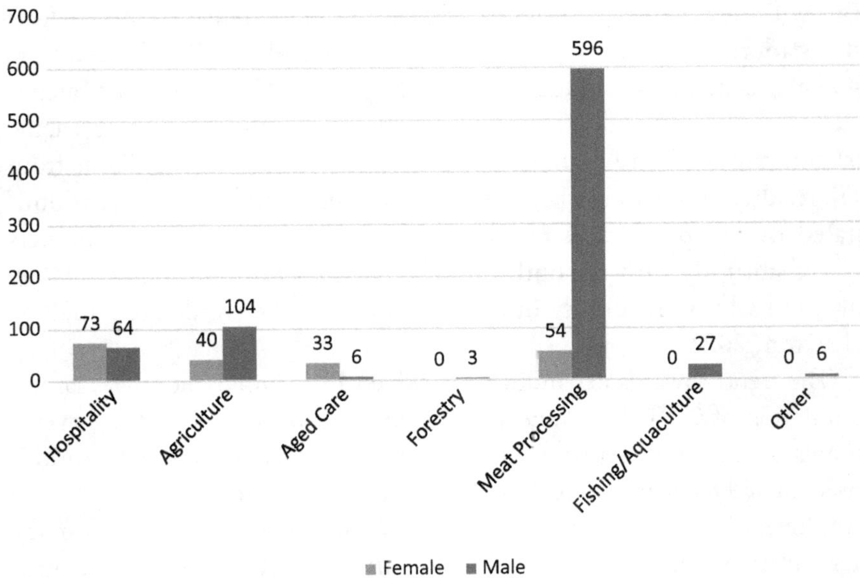

Fig. 10.3 PLS workers in Australia by industry of employment and gender, March 2020. (Source: PLF)

Fig. 10.4 **(a)** and **(b)** Crowds at early PLS information sessions in Vanuatu. (Source: Facebook)

aged care and hospitality, this trend appeared likely to continue. With the bulk of employment in low- or unskilled meat processing jobs, the PLS was not meeting its intention to open up other skilled employment pathways, including those where training was provided through APTC.

After a brief pause, the PLS continued to expand during the COVID-19 pandemic so that by January 2021 some 1633 people were employed through the scheme (with 98 others waiting in quarantine) and by June this had grown to 2891 workers. As in the standard SWP horticulture scheme, demand was much greater during COVID-19 (as other categories of workers dwindled) and charter flights continued to bring Pacific islanders to work in Australia. The relative ease of organising flights from Fiji resulted in Fijian worker numbers booming, and by 2021 they dominated overall participant numbers—although this growth in numbers halted abruptly with the outbreak of COVID-19 in Fiji in 2021. Many of those who were already in Australia prior to the pandemic remained (Lawton, 2021).

The trend towards expanding low-skilled PLS employment was largely a function of COVID-19 and the power given to Australian state governments to dictate the entry of migrant workers. State governments prioritised their Departments of Primary Industry to coordinate the approved employer (AE) recruitment process in response to strong pressures from the agriculture and horticulture lobby groups, including labour hire companies, centred on narratives of 'fruit rotting on trees and on the ground' and rising food prices (Chap. 7). The focus on meat processing and agriculture increased to such an extent that by early 2021, 69 per cent of all workers were in meat processing and 26 per cent were in agriculture, with barely 5 per cent employed in other activities. Agriculture and horticulture had grown to become the second most important sector, in response to employers seeking greater stability of their agricultural workforce, assisted by the new flexibility of employment structures under the PLS. In effect agricultural employment under the PLS was little different from the SWP, despite some movement into sectors where demand for labour was year-round. The very brief account of Allen's work experience, from the PLS webpage (see below), could equally describe SWP employment.

From late 2020, recruitment of PLS workers into aged care effectively stalled, as unlike the agriculture and horticulture industries, the aged care sector was without a strong voice at state level (aged care being the responsibility of the national government). In addition, aged care employers had less capacity to cover worker quarantine costs. Faced with restrictions on the numbers of people who could be approved to come to Australia, they were doubly disadvantaged. Closed international borders and restricted domestic movement, along with localised lockdowns, left tourism and hospitality in crisis, and several PLS tourism workers returned to their home countries. Others were redeployed to aged care or meat processing (with one i-Kiribati woman moving to Longreach, in far western Queensland, where her main job became skinning kangaroos, a skill with minimal transferability to Kiribati).

Despite expectations that aged care would be a significant source of PLS employment (Doyle & Sharma, 2017) and the fact that publicity for the scheme often had pictures of smiling Pacific islanders caring for contented older Australians, a slow start and the arrival of COVID-19 meant that by 2021, aged care numbers were few, even though two-thirds of nursing homes were understaffed in 2016 (which resulted in a lack of adequate care, a situation that became evident when COVID-19 arrived). Likewise, despite the much heralded success of Timorese in Broome and i-Kiribati workers in Queensland, and clear labour shortages, tourism employment was yet to make a breakthrough

Familiar Themes of Success and Failure

Formal evaluations or academic analysis of the PLS are lacking. However, stories about PLS workers featured on the Pacific Labour Mobility website and Facebook page covered many familiar themes of community connections through church and sport. For example, on Palm Sunday 2021, a group of PLS workers attended a special service at St Albans Church in Innisfail (Queensland), along with employers, community members and the Vanuatu High Commissioner, Samson Vilvil Fare. The morning provided an opportunity to connect with the community through sharing foods and cultural performances. In Tamworth (NSW), Fijian workers had formed a rugby team, 'Bula Brothers', and practised twice a week after work. Jeremaia Marawa, a local community member, explained, 'It's a chance for them to work on their physical health and to connect with other workers in the area' (Pacific Labour Mobility Facebook page, April 2021). Another group of Fijian PLS workers 'saved' the local rugby club in Singleton (NSW), that had been struggling for four years to recruit players. One of the club's directors even suggested that the new recruits would become 'almost cult heroes' (Pacific Beat, 22 February 2021).

On the PLS website, worker success stories were eulogised and marketed, as the example of Angelica Fisheries demonstrates. In 2021, three Australian tuna operators were involved in the PLS with workers from Tuvalu placed in Cairns, Mooloolaba and Ulladulla. One of these owners and operators, Steve Basile, sang the scheme's praises:

> Angelica Fisheries on the New South Wales south coast joined the Pacific Labour Scheme in 2018. The business had struggled to find local labour and now employs 3 workers from Tuvalu. One of the owners, Steve Basile, says being able to access workers from Pacific island countries saved the business. In 2018, Angelica Fisheries joined the Pacific Labour Scheme (PLS) and travelled to Tuvalu to interview workers for their tuna fishing fleet. Steve says "If you're having trouble getting local employees for your business, I'd recommend seeing what the PLS can offer. There are some really nice, reliable, hard-working people out there who really want to work, so you've got nothing to lose. When I come down to the boat at 2 or 3 in the morning to go fishing, you know they're there. You don't have to

> ring them up, chase them up and wait for them. It's just peace of mind. One of Steve's workers, Talimanatu Uilese, is now earning 10 times more than his usual wage in Tuvalu. When Uilese finally saved enough money, he was given 5 weeks off so he could return home and build the house he had dreamed of

As with the SWP, PLS workers were valued for their reliability and the highly flexible nature of their labour; they were always there to be called upon.

Other than their extended stay in Australia, the aspirations and potential outcomes of workers from the PLS were much the same as for SWP workers. The same Talimanatu Uilese mentioned above had travelled from Tuvalu to be employed as a deckhand on a commercial fishing boat in coastal New South Wales. With a wife and four children at home, his objective was to build a house when he eventually returned: 'The main reason why I am here is for my children's future. It is very hard for me [to be away] but the other side, it is better for me to earn money than stay back home without doing anything'. Australia's then Assistant Minister for the Pacific, Anne Ruston, claimed, 'We see this as very much a win-win-win. Obviously we are getting a great source of fantastic labour … but, equally, we can be training these workers when they come to Australia so that when they return home to their countries they have got the kind of skills that are going to assist in capacity building in their own country' (Pacnews, 11 November 2018). Similar positive stories were placed on the Pacific Labour Mobility Facebook page:

> Allen Tumbu from Papua New Guinea (PNG) works in Perth, Australia through the Pacific Labour Scheme (PLS). His job picking and harvesting broccoli has helped him to support his family—even allowing him to pay for his brother's medical school fees. "I would like to say thank you, PLS program. We PNG boys have learnt a lot of things here in Perth. We are proud. Being in Australia is such a blessing to my family and me," he said. (April 2021)

Another story explained,

> Retati Tekaai from Kiribati works in the Australian tourism sector through the Pacific Labour Scheme. The oldest of 6 children, she uses her income to support her family and wider community back home. Last year, Retati paid for a public toilet block to be built in her village, and is now saving to help build a kindergarten. "Being on the Pacific Labour Scheme has helped me achieve what I want to do in life and go forward with my goals without relying on anyone," she said. "I want to encourage all women who think they cannot achieve their goals." (March 2021)

These few stories, intended to promote the PLS, obviously highlighted its positive impacts, while underlining the commonality of structural goals, experiences and outcomes across the SWP and PLS.

One difference between the two schemes was their arrangements for pastoral care. Unlike the SWP where pastoral care was devolved to employers, for PLS workers, pastoral carers and support staff were provided via the PLF itself. A greater effort was made to ensure these staff were Pacific islanders who could understand workers' cultural context and speak their languages. Support services included regular welfare checks and access to a 24-hour support hotline. Australian authorities were thus taking a much more structured and involved part in the programme.

Despite more formalised pastoral care, some PLS workers faced similar challenges to their SWP counterparts around accommodation and wages. In June 2020, for example, a group of Samoan meat workers in Inverell were paying a total A$1350 per week to rent a four-bedroom, one-bathroom house. Similar properties were being advertised online for A$340 rent per week. Since accommodation was potentially longer-term, higher standards were expected, but as for the SWP, housing was scarce in places. DFAT had to fund PLS Accommodation Brokers in two regions—the Cassowary Coast and Sunraysia—with the task of identifying adequate accommodation solutions. However, the unavailability of appropriate housing stock remains a regional issue.

Other PLS workers experienced confusion over payslips and deductions. Ellen from Vanuatu was employed in meat processing and lamented,

> The problem is I don't really know how much I owe the company and how long it will take for all these deductions [to come out]. With this amount of rent that I'm paying, I don't think it's fair and I do regret coming here. I just want to go home (Quoted in Hermant, 2020)

This too was a tale familiar from the SWP. What was also becoming more quickly apparent was that the PLS, offering the option of staying longer, was potentially contributing to a skill drain, dramatically evident in an IT worker in Samoa becoming a worker in an abattoir: 'I quit my job because I know abattoir is better money than what I'm doing in Samoa' (quoted on ABC Pacific Beat, 3 July 2019). Similar losses of skilled workers were evident in several small and medium-sized enterprises in Samoa, to the consternation of their owners (Fatupaito et al., 2021). Late in 2021, with 25 i-Kiribati nurses employed in Australia through the PLS scheme, their loss was particularly evident in the national hospital, with the remaining staff working very long hours but unable to meet local needs (Marazita, 2021). Retired nurses, some retired for more than 10 years, have been recalled to work.

Despite the lack of assessment, small numbers, the advent of COVID-19, the difficulty in expanding beyond meat processing and no possibility of knowing what the long-term impact of the scheme might be, the PLS was still seen as a success. Consequently, the case for more pathways to permanent migration was pursued further. Thus, with reference to Timorese, Michael Rose (2020) argued,

> What we will have in PLS workers who make it through their first three years is a cohort of people who are ready to settle in Australia. They will have established roots and routines. They will have acquired much-needed skills. They will have demonstrated the fortitude to work in a way few others are willing to. As permanent residents and highly experienced workers they will command higher wages than they do under the PLS, but for any regional employer this will be offset by the ease of managing them as regular employees rather than as part of a tightly regulated government program. They are not, having settled and acquired skills that are suited to regional Australia, likely to 'flee' to the city. They will, however, bring new life to regional communities that are crying out for it.

Others have been equally positive (e.g. Curtain & Howes, 2021), but, as long as it was possible to secure cheap—and skilled and semi-skilled—workers from elsewhere, such a change faced political opposition and remained unlikely, whatever its merits. It went against the long-established Australian reluctance to open up pathways to permanency for Pacific islander migrants.

Towards Permanence?

Once again a temporary employment scheme had emerged, this time with minimal debate, to address Australia's growing labour shortages. Globally too, temporary negotiated transnational schemes were expanding and extending into non-agricultural areas with similar labour deficits such as tourism and aged care. Around the world, demand in these sectors is likely to continue to grow and will not be met by domestic workers, just as in Australia. At the start of 2019 an estimated 42,000 vacancies existed outside Australian state capitals, across a range of occupations and skill levels, but with health care prominent (Houghton, 2019). As with horticultural employment (Chap. 7), these shortages were emphasised further during COVID-19 (e.g. Harari, 2021). A long history accompanied the possibility of increased numbers of PIC (and other) migrants being involved in care-giving at a time when the unsatisfied demand for caregivers was steadily increasing and a high proportion of all caring jobs were held by new migrants. However, the dominance of meat processing as a source of PLS employment indicated the strength of certain sectoral lobbies in pressing immediate economic issues over longer-term social welfare concerns.

Addressing the seasonality of labour demands continued to pose problems where new PLS migrants were employed in agriculture and tourism, as had also occurred in similar programmes in the United States (Terry, 2016), while other familiar issues such as inadequate accommodation surfaced yet again. Nonetheless, by 2020, the PLS seemed to be working, although barely 500 workers had signed on at the end of 2019 before the COVID pandemic temporarily slowed the take-up rate. By 2021 numbers were growing again as Australia's labour shortages worsened and controlled guestworker migration was once again allowed. At a small scale, workers provided solutions to regional labour shortages. While the scheme was dominated by employment in meat processing and agriculture, some workers including hospitality workers on Hayman Island, and aged care workers in western Queensland, were employing and learning diversified and useful skills. Despite small numbers, the loss of some workers from the PICs, notably nurses from Kiribati (migrating to a less skilled but better rewarded care industry), left a deficit at home. While the Ministry of Health had 'little concern about brain drain' as the nurses were expected to return at the end of their contracts (Marazita, 2021), the temporary loss posed problems for national health care and would surely recur with another cohort.

With significant labour shortages in various economic sectors, Australia and New Zealand continue to represent an important—and in many ways untapped—market for Pacific labour migration. As the PLS demonstrated, an unmet market for agricultural labour remains. The Australian fishing industry and caregiving became interested in temporary work schemes; although several relevant companies have become registered employers, few workers were taken on. The proliferation of labour hire and agricultural companies among the PLS-approved employers indicated that the scheme had not strayed far from the agricultural sector. At the same time, few workers from PICs had gained TSS 482 visas where job vacancies existed for more skilled migrants. In terms of present economic and public health circumstances, and continued perceptions that significant unskilled migrant employment would be a threat to the wages and working conditions of low-income Australians (as it was early on for the establishment of the PSWPS), significant diversification seemed unlikely to be imminent. Nonetheless, the Australian government was moving towards a more elaborate and wide-ranging PALM scheme that would be uncapped and demand driven, and intended to be the primary means of filling unskilled and semi-skilled workforce shortages in rural and regional Australia. Once again Australian employment took precedence in an international migration scheme.

Slowly, and without any new legislation, the PLS represented a small shift towards greater permanence, a potential first step towards opening pathways to more permanent migration for semi-skilled and unskilled workers. Employing Pacific islanders on longer contracts is possible, but the outcomes remain uncertain. As the migration of i-Kiribati nurses has shown, even temporary migration from sectors other than agriculture can pose problems. Allowing workers more rights as migrants, and more freedom to move around, leave unscrupulous employers and even change occupations would mean that workers were better off and would bring SWP and PLS visa conditions more in line with other migrant visa categories—and with International Labour Organisation (ILO) recommendations about best practice for guestworker schemes. Likewise, allowing families to join workers (in reasonable circumstances) would reduce social costs and enable Australia to not merely rhetorically claim that it is part of a 'Pacific family', but that the real families of unskilled workers are also welcome. So far that has proved a step too far. We can now review the wider consequences of this evolution in structures and the relationships between Australian and other schemes.

Bibliography

Boese, M. (2015). The roles of employers in the regional settlement of recently arrived migrants and refugees. *Journal of Sociology, 51*(2), 401–416.

Borovnik, M. (2007). Labor circulation and changes among seafarers' families and communities in Kiribati. *Asian and Pacific Migration Journal, 16*(2), 225–249.

Callister, P., Badkar, J., & Williams, J. (2009a). Paid caregivers & domestic workers: Some policy issues in meeting future demand in New Zealand. *Policy Quarterly, 5*(3), 38–43.

Callister, P., Badkar, J., & Williams, J. (2009b). The future supply of caregivers for the elderly: Should Australia and New Zealand look to the Pacific? *Pacific Economic Bulletin, 24*(3), 162–172.

Carnegie, M., & Jeffries, L. (2021). Pacific Labour Scheme: Preliminary examination into the implications and impacts of family accompaniment presented at Pacific Migration Research Workshop, 17th June 2021, ANU [online].

Connell, J. (2009). *The global care chain. From the Pacific to the World.* Routledge.

Curtain, R., Dornan, M., Doyle, J., & Howes, S. (2016). *Pacific possible. Labour mobility: The ten million dollar prize.* World Bank and ANU.

Curtain, R., & Howes, S. (2021). *Helping APTC trades graduates to migrate to Australia under the TSS.* ANU Development Policy Centre Policy Brief No. 20.

Doyle, J., & Sharma, M. (2017). *Expanding employment pathways for Pacific Islanders in the Australian and New Zealand aged care sectors*. World Bank.
Fatupaito, A., Utuva, L., Tauave, S., Alofipo, A., Meleisia, M., Schoeffel, P., Arthur, T., & Alexeyeff, K. (2021). Samoa's New Labour Trade. *Journal of Samoan Studies, 11*(1), 51–63.
Feist, H., Tan, G., McDougall, K., & Hugo, G. (2014). *Enabling rural migrant settlement: A case study of the Limestone Coast*. Australian Population and Migration Research Centre, University of Adelaide.
Harari, F. (2021, June 5). Help wanted. *Weekend Australian Magazine*, 12–16.
Hermant, N. (2020, November 6). Workers from Pacific Islands paying hundreds to live in cramped quarters in regional Australia. *ABC News*. Retrieved April 10, 2021, from https://www.abc.net.au/news/2020-06-11/pacific-island-scheme-wages-deducted-high-rent-inverell/12336278
Houghton, K. (2019). *The future of regional jobs*. Regional Australia Institute.
Hugo, G. (2014). Immigrant settlement in regional Australia: Patterns and processes. In R. Dufty-Jones & J. Connell (Eds.), *Rural change in Australia. Population, economy, environment* (pp. 57–82). Ashgate.
Lawton, H. (2021, May 27). Pacific Labour Scheme: Expanding while borders are closed. ANU DevPolicyBlog.
Marazita, J. (2021, December 7). Pacific nurses in the desert: Kiribati brain drain is outback Australia's gain. *The Guardian*.
Negin, J., Coffman, J., Connell, J., & Short, S. (2016). Foreign-born aged care workers in Australia: A growing trend. *Australasian Journal on Ageing, 35*(4), E13–E17.
Nishitani, M., & Lee, H. (2019). Becoming 'Overstayers': The coloniality of citizenship and the resilience of Pacific Farm Workers. In V. Stead & J. Altman (Eds.), *Labour lines and colonial power: Indigenous and Pacific Islander Labour Mobility in Australia* (pp. 159–184). ANU Press.
Rose, M. (2020, January 28). *The Pacific Labour Scheme needs a pathway to permanency*. ANU DevPolicy Blog.
Stilwell, F. (2003). Refugees in a region: Afghans in young, NSW. *Urban Policy and Research, 21*(3), 235–248.
Terry, W. (2016). Solving seasonality in tourism? Labour shortages and guest worker programmes in the USA. *Area, 48*(1), 111–118.
TNS. (2011). *Final evaluation of the Pacific Seasonal Worker Pilot Scheme*. TNS.
World Bank. (2018). *Expanding employment pathways for Pacific Islanders in the Australian and New Zealand aged care sectors*. World Bank.

11

The New Blackbirds?

This chapter places the temporary seasonal employment of Pacific islanders in Australia within a series of themes relating to guestwork as it has existed—and continues to exist—across time and space. We trace similarities and differences between the Seasonal Worker Programme (SWP), its nineteenth-century predecessor and other guestworker schemes elsewhere. Underlying this is the straightforward question of whether temporary labour migration programmes have succeeded in contributing to development in source communities. That then becomes a more complex questions of whether development primarily relates to economic growth, poverty alleviation, freedom and/or the Sustainable Development Goals (SDGs), and who and where the beneficiaries are (and might there also be losers?). Can guestwork generate more extended development, or does it encourage dependence on an uncertain migration? Has the Australian scheme learned from similar guestworker schemes elsewhere and is it an improvement on these other schemes? In short, has the SWP achieved a triple win as promised, and how similar or different is the current scheme in its successes (and failures) to the blackbirding days of the nineteenth century?

We suggest that most framings of a triple win—elaborated on below—are too simplistic, although there are unquestionably substantial gains, in

K. Petrou, J. Connell, *Pacific Islands Guestworkers in Australia*,
https://doi.org/10.1007/978-981-19-5387-3_11

contexts that vary enormously through time and space. Such framings tend to ignore some of the social costs of migratory employment in what can be precarious contexts. The last decade alone is too short for definitive conclusions about the SWP, but everything points to parallels with other schemes, where the costs and benefits are uneven and inegalitarian; some workers, some farmers, some communities and some households do relatively well, while others miss out. Development itself can be too subtle and transient to balance its multiple elements and reach definitive conclusions. This chapter nonetheless presents a brief overview of the benefits and costs of the SWP at different scales and in different places, and considers what contribution this has made to development in the Pacific Island Countries (PICs). It thus examines the past and present of this and other schemes, before, in the final chapter, we turn to the future of both the SWP and Pacific Labour Scheme (PLS) schemes, which have become, quite by chance, part of a new era of Pacific change and development.

The New Blackbirds? Plus ça Change …?

An historical focus offers a distinctive perspective on the SWP (and, as we show later, on the Seasonal Agricultural Worker Program [SAWP] and other schemes) as part of broader ideological, economic and political shifts in the global economy that have gained ground in this century. Across continents and eras it can easily seem that farmers, growers, plantation owners and intermediaries take advantage of their privileged and powerful positions and are complicit in some degree of exploitation, however inadvertent, of temporary workers who have migrated to earn a better livelihood and support themselves and their families back home. It is not unreasonable to ask if there is any legacy from, or similarities with, the blackbirding era of a century ago (Chap. 3). Clearly there are obvious differences—no workers now migrate against their will and contact with home is usually easy enough—but there are intriguing parallels in the 'classic' problems of temporary labour migration schemes. Workers have come from many of the same places, but now travel to work in a wider range of destinations and activities. Contemporary schemes still deny workers the right to remain in Australia or bring family members with them, just as did past guestworker schemes. After a hundred years, Australia, Vanuatu and Solomon Islands have in some respects turned back the clock towards a 'new blackbirding'.

In the nineteenth century relationships were so unequal and migrant workers so exploited that more than a century later, it had not been forgotten: in July 2021, Jack Dempsey, the mayor of Bundaberg, offered the first official apology to Australia's South Sea Islander community. By contrast, at much the same time, Vanuatu, along with other PICs, sought out and enthusiastically entered into a new, more short-term labour migration scheme with Australia (and New Zealand) where there were more obvious reciprocal benefits. Yet, regardless of islander aspirations, migration was wholly linked to host country labour demand: the schemes began and partly ended (in the first phase) in response to metropolitan needs. Such labour requirements, rather than incidental considerations of aid delivery, have almost wholly dominated the operation of the SWP.

During both eras, the earliest workers moved out of some degree of poverty. Alien recruiters and local agents preferred agricultural workers, and sought to exclude educated, or even skilled, workers (and the urban unemployed), a situation true also of American recruitment of Caribbean cane cutters and recruitment of Javanese for Sumatran plantations (Hahamovitch, 2008; Stoler, 1985). In the nineteenth century, plantations sought to avoid workers who might be 'tainted' by radical ideas or previous experiences. Such preferences and prejudices partly explain the gradual northward movement of the labour reserve frontier in Melanesia into more remote areas where potential workers were entirely unfamiliar with Australia. Only belatedly therefore did blackbirders recruit from more impoverished inland areas, a contrast to contemporary schemes where there was initially a deliberate attempt to do exactly that, at least in Vanuatu. But by the time the SWP began, 'impoverished' areas were quite different from the past, and in the twenty-first century, rural workers knew something of Australia and of agricultural employment.

In both phases, labour migrants were actively recruited in their home islands, promises were made of 'full' employment and substantial earnings (not always realised), and choice of work was impossible. During the first phase of labour migration, elite Melanesians, such as Kwaisulia in

Malaita, rapidly took on lucrative roles as middlemen. Where, infrequently, choice was available to them villagers sought a continued moral order and economy by selecting 'worthy' recipients, or local leaders, or occasionally exiling troublemakers. The new phenomenon of migration was incorporated into local cultural practices. Islanders of higher status were more likely to become involved once the rewards became more evident, just as in contemporary Vanuatu urban residents increasingly took part in the SWP as the economic benefits of participation became widely known. During the nineteenth century, circumstances in destinations were often difficult: 'by European standards the work was long and hard and the pay meagre, [yet] most islanders found their new life exciting and stimulating, exposed as they were to a totally novel cultural environment' (Howe, 1974, p. 383). Though their labour power was controlled, and exploitation rife, all workers had moved from places of limited economic development that were no 'paradise lost' (Brass & Bernstein, 1992). Wages were miserable, but they were novel and welcome, and could be exchanged for useful goods, that enabled status at home.

Despite increasing flows of information in the twenty-first century, in both eras few early migrants had realistic expectations of what lay ahead of them, in terms of incomes, employment or social conditions, yet labour migration was usually a rational and conscious choice (cf. Baak, 1999). Many workers expressed a wish to return for a second contract as it became evident that only limited development could be achieved through a single employment stint. Incomes, and their utility, were less than expected, but they enabled improved material wellbeing. New goods were welcomed in both periods and were oriented towards basic needs (although, equally, guns in early days and chainsaws in the current era could easily be criticised). Skills were gained but were rarely transferable to quite different Pacific agricultural systems even though skills transfers have been (at times) emphasised as one aim and virtue of contemporary schemes. Certainly, during New Zealand's guestworker scheme of the 1980s, the Fijian government emphasised the need for skill acquisition and training over the potential for remittances (Levick & Bedford, 1988). Yet skills transfers have always been limited (Macpherson, 1981; Dun & Klocker, 2017), and perhaps increasingly counteracted by skill losses. However, the experience probably benefited many workers and their extended families in subtle ways, and new incomes were invested in education. Labour migration enabled some people to escape, or at least seek to escape, from agriculture, but newly learned commercial agricultural skills could not be easily transferred to more remunerative business development or urban employment.

The reproduction of labour remained in Pacific island villages and towns and employers played no part in education, health or aged care. During the first era, migration from labour reserves was no threat to the authority of colonial or national elites (despite constantly concerned missionaries). Discursive representations of Melanesian workers portrayed and patronised them as outsiders who were docile but an inferior commodity. It was considered a metropolitan duty to 'help the Pacific', though in both phases labour migrants were largely left to formulate their own development strategies. Labour was treated as a commodity only in the earliest years of nineteenth-century labour migration. The blackbirds were rather more free—and recruitment less fraudulent or violent—than some initial literature suggested. Yet, bonded labour was still bonded, even with more enlightened conditions. Working conditions were poor and inadequately regulated with long hours being normal. Blackbirds transported to contemporary times might well have recognised familiar labour lines, although none so demanding as the cane plantations. Intermediaries could be unscrupulous and self-interested, from Kwaisulia to the local recruiters and labour hire firms of modern times. Governments and administrations played little part in regulation and often distanced themselves from it. What might have been seen as teething problems never really went away. It was in few people's interests to ensure that they did, and workers (who were easily replaceable) were never included in that few. Institutional regulation, interest and empathy were often remote.

By the time of the SWP, migrants were no longer as alien and socially marginal as they were in the nineteenth century. Languages could be shared. Women were somewhat more likely to be recruited into the SWP as the 'subtlety' of agricultural employment changed and a 'female touch' of nimble fingers was required for jobs such as those in the packing sheds. Churches and other civil society organisations played a continued part in socialisation, so reducing the purely labour aspects of migration, while enabling some unscrupulous employers to ignore the duty of care. Likewise, labour regulations and occupational safety were more effective and accommodation superior to the blackbirding era. Wages were much higher in the twenty-first century, but as consumer tastes altered, so too did the nature of agricultural work with consequences for longer working hours. The duration of stay became restricted to fewer than seven months, and migrants had limited ability to control conditions since their stay was briefer than a century earlier. By the nineteenth century, workers were already resisting exploitation creatively and energetically (Graves, 1993), directions that took both practical and electronic form in the twenty-first century.

SWP workers' experiences in Australia were informed by racialised histories of the types of labour for which Pacific islanders were considered 'suitable' (Stead, 2019, 2021a). From the workers' perspective, understandings of the SWP were influenced by past experiences of blackbirding; 'blackbird' and allusions to histories of slavery were common tropes on social media whenever dishonest workplace practices or unfavourable employment conditions came to light. In 2021, in response to a news article about SWP worker exploitation, comments on the ni-Vanuatu Seasonal Worker Facebook group included:

> Pacific Islanders laid the foundations for a successful cotton and sugar cane industry 150 years ago in Queensland. Some of their graves are there in the sugar cane fields in Bundaberg. The actions by the farmer or contractor in Bundaberg is totally uncalled for. Vanuatu Government through the Vanuatu High Commission must vigorously pursue this mistreatment dished out to the Ni Vanuatus. Further the Vanuatu Government must push for review into this modern day slavery program before anymore harassment and mistreatment of Ni Vanuatu workers.
>
> The SWP and PLS are just a cover up. The reality is that both schemes are stage two of blackbirding.

Such accusations were far from unanimous, however, and sometimes led to considered exchanges such as the following post from 2017:

> I'm a new member who has never tried the Recognised Seasonal Employer (RSE) scheme, but do you guys think the scheme is a type of slavery or just a way of getting money where you make the sacrifice and work in another place. No offence admin, I just want to know what people think.

> Slavery is not a good word to label the scheme with ... If you reflect back to the time the scheme started, you'll find lots of interesting stories. If you go and visit all of the islands, you'll see big changes and the standard of how people live has improved. Kids can go to school up until their senior years or even go to university. It's true, you could argue that we can make money at home too, but in reality we're sick of this kind of work ... If you follow the rules of the scheme properly, you'll never come across any of the problems that are affecting lots of workers today. But yes, in my opinion the scheme is good.

As Victoria Stead (2021a) notes, many SWP workers enthusiastically sought out employment through the SWP and wanted to continue participating in the future. Yet they were also perfectly aware of the fundamental asymmetries inherent in the scheme and the historical relationships that foreshadowed it—historical experiences that, for some, were still part of living memory.

Temporary labour migration has had limited long-term development impact in most of the source regions, hence the universal acceptance of a second phase. At a time when Pacific poverty and hardship are increasing, migration is part of a global trend where remittances substitute for aid and where national investment in rural and regional areas is reduced. But, as in Indonesia, the 'very success' of circular migration may be 'counterproductive in the longer term because it diverts attention from the only strategy that will ultimately assist the poor in rural areas'; guestworker schemes are merely a 'stop-gap measure' that diverts resources and attention from investment in (long-term) rural and regional development (Hugo, 1982, p. 76). Yet, where a poverty of opportunity prevails and choice of livelihoods is limited, labour may be further reduced to a commodity, its autonomy weakened and its price static. The contemporary rise of even more temporary agricultural work—that is precarious and casual—represents a shift towards again treating labour as a pure commodity—with limited ability to control its time, organise collectively or ensure a guaranteed income. What is abundantly clear from both eras of Pacific guestworker schemes is that the limitations of labour migration are evident in the minimal subsequent structural and domestic changes, and the impossibility of converting household gains into sustainable development or into more effective structures of national development. Uneven outcomes are consequently inevitable. After more than a century much has remained the same: economic gains were most evident in the destinations as surplus value remained in the metropoles.

An Uneven World?

Not only do parallels exist between the two phases of Australian guestworker schemes, but they exist elsewhere, logically and most obviously in Canada, but also in many European destinations. These international schemes offer similar possibilities and problems as the Australian schemes, and have been both widely welcomed and widely criticised. The most basic underlying problem of all guestworker schemes derives from the narrow 'economistic' perception that views migrant workers as mere units of labour that can be moved at will from place to place. Yet migrants are, of course, people and any analysis of the impact of labour migration requires a broader approach: a much wider conceptualisation of costs, benefits and contingencies that goes beyond mere economics.

Some of the earliest studies of seasonal workers in agricultural systems in the countries of the north recorded difficult local circumstances for workers. In 1950 Puerto Rican guestworkers in Michigan were complaining of overcrowded living quarters, no beds or bathrooms, inadequate food, a lack of drinking water and no pay (or receiving piece rates rather than the hourly rates paid to Americans), a situation where they required help from the American Red Cross. Their complaints were directed at the Puerto Rican government that had organised contracts and failed to support them (Meléndez, 2017). Circumstances seemed little better three decades later for Caribbean seasonal migrants in Florida and Ontario—all of whom were men, but where workers were a 'docile labor force unable to press for any improvements', and a legal 'sub-class', since those who failed to meet productivity quotas, filed a complaint or displeased employers for some reason were dismissed, sent home and permanently blacklisted. Systematic undercounting of hours worked was not unusual (Wood & McCoy, 1985, p. 255). In Canada circumstances seem to have generally been better than in the United States despite limited unionisation and local people's disregard for migrant workers (Cecil & Ebanks, 1991, 1992). In both countries, employers sought to employ returnees who were less injury- and sickness-prone and more skilled and productive. Workers' remittances and savings were oriented to housing (building, modification and repair), children's education, consumer goods (such as clothes and appliances) with little aimed at investment. Neither study found any Caribbean entrepreneurial activity, and both concluded that few migrants had obtained relevant skills for such activities: more basic needs had first to be satisfied. With only limited changes these descriptions mirror accounts of this century's SWP and the many other seasonal worker schemes.

These early studies of Caribbean seasonal workers were not designed to be critical, but they nonetheless pointed to the flaws in the schemes and to teething problems that might be remedied. Yet time and again subsequent studies reported how workers' health suffered due to hazardous and stressful working conditions, sub-standard housing, lack of social support, limited access to legal, health and social welfare services, and, not infrequently, direct exploitation. Rather later, seasonal migration from Jamaica was found to have brought benefits to home communities through greater capitalisation of farms, the transfer of new seeds and skills, and the ability to hire labour, enabling agricultural expansion. In considerable contrast, on the much smaller island of St Vincent, where seasonal workers' remittances took precisely the same pattern as in the Pacific (Connell & Conway, 2000), purchase of land was of greater significance enabling people to establish better housing, focus on less labour-intensive crops or simply abandon agriculture (Thomas-Hope, 2017). The SAWP differed from the SWP in the very small number of women in the scheme (seemingly never more than 3 per cent) and the extremely high rate of return. In a sample of mainly Jamaican and some other Caribbean island workers in 2005, 65 per cent had worked for at least six years on the scheme, 18 per cent had participated more than 16 times, and just 2 per cent were on their first contract (Beckford, 2016). Even by the mid-1980s Barbadian participants had averaged seven years on the scheme and a third of them had also worked as cane cutters in Florida (Manning, 1986). Overtime (beyond the daily ten hours) was paid only at regular rates while voluntarily working on holidays or Sundays increased workers' chances of being invited back, effectively an 'undercurrent of coercion' (Beckford, 2016, p. 166) where problems of health and safety, accommodation, social isolation, transport and a lack of rights were never far away. In a group of Jamaican workers in southern Ontario, where many migrant workers experienced 'abusive work environments, exploitative scheduling, and inadequate housing', and lived under the constant threat of repatriation, they framed their experiences in the language of slavery. Workers routinely referred to abusive employers as 'slave masters' or 'slave handlers', and the SAWP as a slavery system. Ironically, that allowed them to invoke their strength as the descendants of slaves, with the pride, ability and resilience to survive (Mayell, 2016, p. 85). For Caribbean migrants, the SAWP offered limited fulfilment and minimal household or national development. Remarkably few Caribbean follow-up studies have been undertaken, especially on the smaller eastern Caribbean states, whether on selectivity, socio-economic impacts or social remittances.

Despite the SAWP often being represented as the model for the SWP, it has drawn considerable criticism because of the limited legal rights and protection afforded to workers, relating to labour relations, health and safety standards. Even where such rights existed, workers lacked the capacity to gain access to them. The practice of repatriating injured workers and preferring returnees for future participation in the programme increased employer control, the dependency that migrant workers had on the programme and the likelihood that workers would not seek treatment for medical conditions or injuries. The scheme was criticised for rendering workers exploitable, isolated and precarious, where this involved geographical, cultural and linguistic isolation, and limited access to health care and transportation (Hennebry et al., 2016) alongside racial and gender-based discrimination and segregation (Perry, 2012; Preibisch, 2010; Hennebry, 2014), inadequate human rights, and poor quality accommodation within a structure of power and surveillance over migrant workers' lives marked by 'draconian' social control (Cohen & Hjalmarson, 2020). Regulation, control and power were emphasised by the frequent remoteness and isolation of much work and accommodation and the immobility of workers without independent access to transport (Reid-Musson, 2017). Human nature, quite different cultural contexts and the desire to return, meant that workers were reluctant to complain about their employers, injuries or safety standards for fear of repatriation. Workers were rendered uniquely vulnerable because of the closed relationship with employers, leading to the precariousness of employment (Hennebry & Preibisch, 2012; Hennebry & Williams, 2015; Binford, 2019). Thus, partly because of the primacy of the SAWP, a range of critical studies pointed to distinctive flaws in the model.

It may be in the nature of critical academic inquiry to find flaws and detect and even typify 'bad apples', but as studies of the SAWP mounted so did the litany of problems described above (Binford, 2019; Bridi, 2015, 2022; Cohen & Hjalmarson, 2020; Preibisch, 2007, 2010; Preibisch & Hennebry, 2013; Silverman & Hari, 2016; Strauss & McGrath, 2017). In a social context, the 'guests' were not made particularly welcome, despite the necessity of their being workers. However, if Canada's scheme—a supposed model of best practice—had flaws, these were usually just as evident elsewhere, suggesting that many were inherent to seasonal migration schemes. Whether some degree of exploitation is unavoidable in temporary labour migration schemes remains questionable. Certainly, similar issues have recurred across all labour mobility schemes over different time periods. However, analysts usually took similar perspectives and unsurprisingly found similar problems. The same 'standard' problems occurred in a multiplicity of seasonal agricultural work schemes whether in the United Kingdom (Rogaly, 2008; Potter & Hamilton, 2014; Scott, 2015; Consterdine & Samuk, 2018), Austria (Horvath, 2014), France (Mésini, 2014), Greece (Kasimis et al., 2010; Papadopoulos et al., 2018), Italy (Garrapa, 2016), Poland (Górny & Kaczmarczyk, 2018; Polkowska & Filipek, 2020), Norway (Rye, 2018; Rye & Andrzejewska, 2010), Portugal (Sampaio & Carvalho, 2016), Sweden (Eriksson & Tollefsen, 2013; Hedberg & Haandrikman, 2014; Axelsson & Hedberg, 2018; Carmo & Hedberg, 2019; Stenbacka, 2019), Spain (Hoggart & Mendoza, 1999; Hartman, 2008; Lindner & Kathmann, 2014; Villadóniga, 2020), Canada and the United States (Hahamovitch, 2011; Beckford, 2016; Binford, 2013; Weiler et al., 2020) and more broadly in Europe (Rye & Scott, 2018). Similar circumstances existed in many places distant from the global north, as in Malaysia, Laos and Myanmar (e.g. Anderson & Franck, 2019; Kusakabe

& Myae, 2019; Anderson, 2021), India (Rai, 2018; Shah & Lerche, 2020; Choithani et al., 2021), South Korea (Seo & Skelton, 2017) and Kenya (Gibbon & Riisgaard, 2014). Within countries too seasonal schemes existed, as in Turkey, the 'original' source of guestworkers, where workers now moved from more impoverished eastern provinces for agricultural employment (Turhan et al., 2015). Analogous circumstances were present in various related service sectors, such as meat packing in Ireland (Maher & Cawley, 2016) and agricultural services in the United Kingdom (Agar & Manolchev, 2020), where factories were as remote as the fields, and employment was rather less likely to be seasonal, but was otherwise indistinguishable from work in agricultural packing houses (e.g. De Castro et al., 2020; Rye & O'Reilly, 2021). Many such schemes proved just as flawed as the Canadian scheme, yet employers continued to prefer overseas migrant workers who would be 'reliable, flexible and compliant' and grateful (Rogaly, 2008, p. 500) with limited proficiency in host country languages (and hence less easily able to complain). Like Thai berry pickers in Sweden, ethnic stereotyping meant many seasonal migrants were regarded as 'industrious, reliable, ignorant, and good workers' (Carmo & Hedberg, 2019, p. 107). Racism, or at least paternalism, was never far from the surface, taking many forms, sometimes overt, but often more subtle and, to a degree, structural (Preibisch & Binford, 2007). Where information exists on who the workers were and how they were chosen, selectivity was apparent not just in terms of latent skills, experience or youth, but also according to personal networks (that included agents and brokers), socio-economic status and political connections (e.g. Beckford, 2016; Carmo & Hedberg, 2019). The limited duty of care on the part of employers and the lack of a sense of a moral economy were omnipresent.

One other similarity was prominent. An abiding theme, especially for workers from the Caribbean, was that however difficult and demoralising, sometimes dangerous and exploitative guestwork might have been, it 'seemed worth the risk, if not for themselves then for their children' (Hahamovitch, 2011, p. 7). The impoverishment of the source countries and the challenges to national development were reflected in the aspirations of rural Jamaicans: 'Even men who were badly injured or who alleged wage fraud were often eager to try again' (ibid). Enthusiasm for employment that could provide a decent life for family members sometimes bordered on denial and desperation.

However, as in the analyses of the SAWP, few studies recorded material rewards from labour migration or positive (or other) changes and development in home societies: a missing perspective where benefits to the participants were most likely to be at home. Where the many studies made more than passing reference to possible or anticipated development benefits, they were based more on the aspirations of the workers and their dreams and goals rather than what actually transpired and had been achieved. Quite simply, as Cindy Hahamovitch (2011, p. 11) phrased it, 'Far more ink has been spilled over workers who struck or protested than over those who worked quietly and contentedly'. There were probably plenty of the latter; hence, one more similarity was that, like the Caribbean islanders, most workers sought to return and many returned over and over again (e.g. Carmo & Hedberg, 2019). Gruelling and exploitative these alien livelihoods may have been, they nonetheless offered incomes and possibilities rarely available at home.

Exceptions to the usual recruitment priorities do exist. Spain established an agricultural worker scheme with Colombia where preference was given to workers from areas experiencing environmental hazard. The beneficiaries included displaced communities, female household heads and people from high-risk zones of natural disasters, that is areas prone to volcanic activity (de Moor, 2011). While the scheme apparently benefited these volcanic areas and could be adapted to recruit people from areas of particular environmental risk, numbers were small and such a scheme could not respond immediately to particular hazard events, especially substantial ones. Indeed, a rise in applications to participate in the SWP immediately followed Cyclone Pam in Vanuatu in 2015, including from workers with no previous interest in seasonal employment, as people sought to earn money to rebuild houses (Bailey & Ng Shiu, 2016), but no obvious mechanism or means to enable participation at short notice existed, in a scheme based on metropolitan labour shortages.

Despite variations in time and space, with few exceptions, structural similarities in guestworker schemes were widespread as were institutional responses (or their absence) which in some circumstances brought further direct comparisons between Australia and Canada (Newman, 2013; Walsh, 2014; Brickenstein, 2017; Robillard et al., 2018) and elsewhere (Corrado et al., 2016; Polkowska, 2019). Such parallels are unsurprising and have coincided with a context where somewhat similar forms of work, that are temporary, informal and associated with a gig economy, have resulted in widespread precarity as a growing phenomenon associated with recent, often temporary migrants (e.g. Ye & Yeoh, 2018; Chacko & Price, 2021). Beyond seasonal agricultural employment, multiple circumstances exist where precarious livelihoods are no less common, complex and diverse.

Precarious Work, Precarious Lives

Precarity is associated with marginality, uncertainty and restrictive access to the benefits of employment. For guestworkers, precarity is partly a consequence of inter-related visa and employment contexts. Precarity is not a fixed status, but in the contemporary era, it has become widespread. Early on, precarity was described as 'instability, lack of protection, insecurity and social or economic vulnerability [but] the boundaries around the concept are to some extent arbitrary' (Rodgers & Rodgers, 1989, p. 5). Unlike vulnerability, precarity goes beyond mere insecure employment to encompass weak social relations (Anderson, 2010) within structures of governance and management. Discussions of precarity thus extend to a focus on understanding how 'late capitalism', structured by global economic and labour market forces, affects already poor or vulnerable migrant groups (Gibbon & Riisgaard, 2014). These groups have been described as forming a precariat consisting of 'people who have minimal trust relationships with capital or the state … [precarity] has none of the social contract relationships of the proletariat, whereby labour securities were provided in exchange for subordination and contingent loyalty, the unwritten deal underpinning welfare states' (Standing, 2011, p. 2), or, indeed, underpinning a moral economy. Within guestworker schemes, precarity was part of a flexibility of labour arrangements and employment practices that favoured employers, and would not have been possible with a domestic workforce (Preibisch, 2010; Bridi, 2022).

Migrant farm workers experience precarity through the intersection of their employment and immigration status, involving restrictive visas and curtailed citizenship rights, so minimising any identification with destination country institutions and social organisations. Guestworkers are dependent on their employers for temporary residency and have limited access to social services and rights. Employment is, by definition, temporary, and job security is non-existent. Workers may experience income insecurity and have little control over employment conditions, or the location and temporality of employment. Guestworkers may be prevented from joining unions or punished for doing so. If dissent is voiced, employers may threaten to prevent workers from returning for a subsequent year, and even to change the source country, a practice described as 'country surfing' (Preibisch & Binford, 2007). That practice has drawn both acquiescence and concern in participating PICs.

Precarious labour conditions are characterised by 'flexible' contracts that may be exploited both inside and outside the workplace, and which preclude anything outside wages (such as holiday pay, pensions and medical coverage). Precarity is closely linked to exploitation; however, what exactly constitutes 'exploitation' is open to debate; for some, exploitation is defined by exclusive benefits to the exploiter, while, for others, individuals may consent to their exploitation (Lenard & Straehle, 2010). This raises questions such as are conditions exploitative or simply unfair and unequal? Might exploitative conditions be mitigated or justified by overall benefits? Why should any trade-off between restrictions and constraints on migrant workers, and the redistributive benefits that migration enables, be necessary only for unskilled workers? More obvious exploitation of guestworkers through underpayment and illegal wage deductions, inadequate pastoral care, unsafe working conditions and inappropriate accommodation have appeared regularly in the media and academic literature (e.g. Bauder, 2005; Enoka, 2019; Doherty, 2017; Wang & Connell, 2021). Canadian media traced similar narratives, where negative components of the system were justified because they were assumed to be better than the conditions workers face at home, so legitimising the 'exploitative and coercive labour practices experienced by migrant workers on Ontario's farms' (Bauder, 2005, p. 52). Of course, not every worker experiences problems, and when they do they are probably nowhere near serious enough to engage the local and national media, so that only the worst cases merit media attention. But problems with guestworker schemes can be ever present since, as Joanna Howe et al. (2019) argued for the SWP, its administration was not done in a publicly accountable way, nor was it transparent.

The same disregard for migrant rights, working conditions and abuse inherent in temporary migration schemes recurs in other forms of short-term migrant employment (Piper, 2010; Surak, 2015). For temporary migrant workers—including those employed through guestworker schemes—precarity is enhanced by reluctance to complain about conditions. Thus, temporary Italian migrant workers in Australia were reluctant to complain, either individually or collectively, whether employed in food service industries or on farms, fearing employer reprisals, so downplaying the problem by preferring to see the current job as only a temporary phase in the life course (Campbell et al., 2019). Here and elsewhere visa conditions 'breed short-termism, resignation and compliance rather than voice, mobilisation and protest' which limits the ability of unions to involve temporary workers (Underhill et al., 2015, p. 296), especially where small groups of workers are dispersed through rural areas.

Precarity is particularly acute for seasonal agricultural workers and is exacerbated by their geographical isolation. Inadequate accommodation and access to transportation, with limited access to services, circumvention of regulations and welfare provision 'beyond regulatory reach' (Axelsson & Hedberg, 2018; Reid-Musson, 2017; Lozanski & Baumgartner, 2022), provide contexts that have enhanced worker concerns about their rights, but made it more difficult to achieve them. Consequently, and not only in Australia, the same loosely exploitative conditions repeatedly recur, despite media reports (or, of course, academic fieldwork and analysis).

In Australia the rise of precarity, both within the SWP and across Australian horticulture more generally (and with parallels beyond that), is a function of the shift in the political economy of horticulture towards more neoliberal or 'pro-market' governance arrangements evident in the intensification of supply chain pressures (such as demands from supermarkets for 'quality'), the emergence of labour market intermediaries (who can enable employers to escape legal responsibilities) and the reduced presence of trade unions (Wright et al., 2017; van den Broek et al., 2021). As Joanna Howe has pointed out, 'When stories of exploitation come out you actually have supermarkets and retailers crack down harder on growers, putting greater pressure on them to do the right thing, but without allowing any give on the price of fresh fruit and vegetables' (quoted in McGlone, 2021). That pressure is then transferred to the workers.

During the COVID-19 pandemic circumstances were particularly difficult, especially for workers who had stayed on, with state border closures contributing to uncertainties surrounding finding work. Despite being legally forced to stay employed through the SWP, more than 1100 workers left their assigned workplace in search of adequate and more secure work, and away from wage theft and bad working conditions, so deemed to be 'absconding' from their official employers, most of which were labour hire companies. As Victoria Stead (2021b) observed, absconding was a word more usually associated with escaping from custody. That brought strong opposition from the Australian Department of Foreign Affairs and Trade (DFAT) which printed posters in six different Pacific languages stating that the visas of workers who absconded would be cancelled, they would not be allowed to work in Australia again ('this may include your family and community members') and their actions would bring 'shame to your family's reputation'. SWP workers have no freedom to move between employers without consent, and can be deported for doing so. Australia's militaristic Border Force sought out the miscreants and targeted a Christian missionary in Bundaberg believed to have helped the workers. As the usually conservative national newspaper, *The Australian*, observed, if charged and convicted on the grounds of 'referring unlawful non-citizens for work', he would be 'the only casualty to date in Australia's abysmal treatment of its guest workers' (Rice, 2021). With a lone exception, where a labour hire contractor paid a fine of A$490, the Fair Work Ombudsman's office, legally tasked with doing so, has not enforced legal standards. While the posters were eventually withdrawn, the notion that migrant workers should never step out of line, despite mistreatment or lack of work, was omnipresent.

Seasonal worker programmes thus illustrate the fundamentally 'unfree' character of temporary employment. In the United States H2A temporary work permits provide minimal legal and wage protection, so placing migrants in situations of complete dependence on labour contractors and employers. By contrast, the SAWP recruits migrants through government agencies in the migrants' home countries, but consular officials are discouraged from taking action to protect their country's migrants when employers simply switch labour sources from one country to another in a 'race to the bottom' for the most compliant workers. Above all, migrant rights are harmed when the legal authorisation to work depends on their satisfying the demands of a single employer and lapses after the completion of the job. Changing employers was not permissible, but dismissal and deportation were feasible. In Canada, the 'race to the bottom' was typified in the shift to a predominantly Mexican workforce for many growers reflecting demands for the 'most hardworking, reliable, and flexible work force' (Preibisch & Binford, 2007, p. 5). Mexicans were easier to exploit than Caribbean islanders who spoke English. In Ireland too an association between poor knowledge of English and wages below the minimum made language an element in the 'dual-market economy' (Maher & Cawley, 2016). Inequitable wage rates may be accepted because of comparisons with area of origin rather than destination, low bargaining power and lack of knowledge of rates and labour legislation. In various circumstances this has extended as far as wage theft.

Wage Theft

Wage theft is widespread in Australian workplaces, well established amongst international students and backpackers, across multiple nationalities, and in various areas of employment, but mainly in services, the hospitality industry and the gig economy, partly because of the outcome of their growing use of temporary migrant workers and the lack of proper regulation. Working holiday makers (WHMs) and workers in the food industry are particularly likely to be exploited in this way (Tan & Lester, 2012; Clibborn & Wright, 2018). WHMs in agriculture, often working for a three-month period, experience similar situations and problems as SWP workers, including extreme and consistent underpayment, overcrowded and isolated accommodation, and abuse from both labour hire contractors and farmers (e.g. Cavanough & Wherrett, 2020; Wang &

Connell, 2021). Repeated incidents of wage theft have been documented in recent years, especially in restaurants, including several run by 'celebrity chefs', and small chain franchise stores (such as 7-Eleven and Chemist Warehouse) where workers are often temporary, casual and not permanent residents.

Exploitative behaviour was much easier to carry out in remote out-of-sight rural areas where temporary migrants were unfamiliar with circumstances and potential remedies. However, its widespread extent suggests that remoteness was unnecessary, and particular problem areas that were identified, including Bundaberg and Coffs Harbour, were not nearly as remote as many SWP sites. Much wage theft also occurs in inner city areas, yet even in these locations, most exploited workers were aware they were underpaid, but were unable or unwilling to challenge this because of unfamiliarity with the process, it was too difficult and slow or they were fearful of losing their visas—effectively a large, silent underclass of underpaid migrant workers (Berg & Farbenblum, 2017, 2020; Farbenblum & Berg, 2018; Fels & Cousins, 2019). As an enquiry noted in 2020,

> The hospitality, franchising, retail, manufacturing, construction and professional services sectors are riddled with employers cheating staff, some by accident, some by design. Most of the abuse is at the expense of the lowest paid individuals… that do not have time or energy to personally fight for their rights (Ferguson, 2020, p. 2)

Over a period of several years as many as 88 per cent of advertised positions on foreign language sites in New South Wales offered rates of pay below the minimum wage, especially in the construction and hair and beauty sectors (Unions NSW, 2020). Government inquiries into the agricultural sector brought similar results with misuses of piece rates, illegal labour hire arrangements, intimidation and widespread underpayment of workers (Australian Government, 2019; Fair Work Ombudsman, 2019). In the horticultural sector almost all advertised positions were offered at piecework rates, which would not have allowed workers to earn the minimum wage and in some cases they would have earned less than A$1 an hour (Unions NSW, 2021). That brought the potential for eventual legal reform late in 2021 when the Fair Work Commission ordered that every farm worker in the country be entitled to the minimum casual pay rate of A$25.41 an hour. Exploitation extended into working conditions, notably those of food delivery drivers, several of whom were killed on city roads in 2020, having had inadequate information and training on risks and safety.

Wage theft and oppressive conditions also occur in other contexts where Pacific islanders work overseas, notably in the Asia-Pacific fishing industry and in some other marine employment areas. Thus, the Forum Fisheries Agency has criticised foreign fishing companies that 'often require crew to stay for up to a year with poor pay and conditions and harsh penalties for dissent' that have contributed to several deaths at sea (quoted in Komaisavai & Magick, 2019, p. 10) in circumstances that have changed little in this century and been graphically described as akin to slavery (Borovnik, 2011; Able Seaman 2010). In multiple migratory contexts, islanders—and others—have been exploited in different ways, from unscrupulous selection procedures to fraudulent wages, emphasising the widespread insecurity and frustrations of SWP (and other) migratory employment, the links between precarious work and precarious lives, despondency over achieving fairer and more equitable outcomes, the challenges of resistance and agency while temporary and the growing need for more effective compliance, management and regulation. The ubiquity of fraud both makes it not at all improbable or unusual in seasonal labour schemes and particularly difficult to challenge.

Weapons of the Weak

Guestworkers have often been portrayed as 'selfless' fathers and mothers, ideal citizens who make willing and necessary sacrifices to improve the lives of their families at home. This view is not only too simple, but such perceptions may legitimise or condone exploitative labour practices (Bauder, 2008). If nothing else, workers resented unfair conditions and, where they could, employed whatever tactic was feasible to change their circumstances and those of others who might follow. Resistance to the restrictions put upon workers—the 'weapons of the weak' (Scott, 1985)—has taken similar forms in Australia, Canada and elsewhere, and been boosted by contemporary telephony and social media. Nowhere have there been the kinds of strikes or protests, let alone riots, that might characterise protests by urban, national workers; opposition and frustration have taken more subtle, everyday forms of defiance and challenge. Resistance was not necessarily consciously political nor part of a broader social movement—challenging the nature of the guestwork system—but simply directed towards obtaining a more just and dignified existence through modified working and living conditions.

The diverse forms of resistance prevalent in the SAWP—lying, falsifying hours, working slowly (if not on piece rates) or giving away produce—do not seem to have occurred in Australia. It is certainly culturally rather less likely that Pacific island workers would challenge hierarchical work structures, wages and accommodation, however bad they might be, coming from countries, other than Fiji, where unions are rare, challenges to 'industrial' authority few and dissent against leaders even rarer. It may also be because Pacific islanders are not moving away from the extremes of poverty or disintegrating agricultural systems that have displaced Mexican workers (Cohen, 2019; Bridi, 2022) and that they are more willing to tolerate unpleasantness and indignity for a few months rather than risk losing their place in the programme. Much like academic critics, no workers sought abolition of the schemes. Reaction and resistance demonstrate that migrant workers, rarely with considerable education or overseas experience, can at least subtly challenge states and systems and rectify their more egregious facets. But resistance sits uneasily with self-preservation, and few wanted to be seen as victims, while resistance and speaking truth to power are barely possible where an unregulated 'system' acts against their interests and lacks effective sanctions on aberrant farmers and agents.

As we showed earlier (Chap. 8) near universal access to social media has greatly benefited SWP workers' ability to challenge and resist exploitation and its widespread use for that purpose emphasised that many forms of mistreatment continue to exist. Social media critiques were often subtle and low key, but raised persistent and widespread concerns, that linked tens of thousands of people between widely scattered workplaces and across countries. Social media provided the rare ability to speak out in general terms without concern. As elsewhere workers strived for dignity and humanity (Cohen & Hjalmarson, 2020) but in less organised, less varied and more electronic form, than where workers were more permanent. Social media activism did not create structural changes to the scheme, although comments and discussions sometimes centred on issues around potential pathways to citizenship and similar matters. Rather, through exchanging information, social media allowed workers to navigate existing systems and avoid being caught out by unscrupulous practices such as the docking of holiday pay or merely understanding administrative processes such as submitting tax returns and claiming superannuation. But, in addition, social media provided a sense of a wider 'seasonal worker' community, various forms of employment in Australia and an avenue for celebrating shared successes with those at home.

Winning and Losing?

In light of the above, it is reasonable to ask whether it is possible (or even sensible) to attempt an adequate evaluation of the SWP and its ability to deliver a triple win (even were there to be adequate data available). Defining what constitute wins and losses is hard enough, while balancing positives and negatives is quite impossible even within single households, as circumstances change over time. It is naively simplistic to fall back on standard quantitative measures of improved national and household incomes which fail to account for welfare, let alone freedom and even 'happiness' issues. Importantly, measures of success are influenced by how, where and whom 'development' is defined and measured.

Standard triple-win assertions are premised on migrants (and their households) gaining in incomes, skills and welfare; sending economies benefiting from possible reductions in under-employment and increases in capital gain as it 'trickles up'; and recipient farmers and host economies benefitting by satisfying labour market needs relatively efficiently and cheaply. There may be more than a triple win, as so many ancillary 'intermediate' activities seemingly benefit—accommodation providers, money transfer operators, pastoral care workers, thrift shops, travel agents and other urban service sector businesses. Builders, contractors and hardware stores all benefit from the housing implications of migration. Fuel suppliers welcome car and taxi purchases. It is hard to differentiate three clear wins as all are inter-related. Yet there are also costs to winning evident in household stresses, the pains and strains of precarity, environmental change, reduced food security and unequal power relations leading to exploitation and distortions to national development strategies. Importantly, migration is as much a social as an economic process, where the costs of social reproduction and the 'invisible economies of care' that produce the workers (Shah & Lerche, 2020) are undertaken in the PICs.

Clearly no simple cost-benefit analysis can be applied to guestwork. In an age of multiple SDGs, not all of which can be satisfied simultaneously, how can the undeniable benefits of the SWP be balanced against social costs (and benefits), and how do these relate to—and accentuate or reduce—the uneven structural conditions that underpin the programme? Indeed various otherwise detailed studies of the SWP and the RSE never specify how they may or may not contribute to sustainable development. However development is measured, there are always both positives and negatives, in such a diverse region involving many countries. Any scheme is hard to evaluate where immeasurable, gendered changes of status occur. In practice too little detailed research exists to make a fine-grained examination of trends and outcomes. A few partial quantitative surveys exist, but qualitative assessments must be derived from multiple personal observations, secondary literature and an uneven scatter of a few good, analytical studies. That is not enough despite some obvious directions of change.

Agricultural work is tedious and often demanding—if not necessarily exploitative. There are however exploiters: farmers, agents and accommodation providers (appropriately 'bad apples') or simply people with inadequate cultural understanding, or unrealistic views of harvest size and timing. There are also bad workers. Contracts can easily be violated and working conditions unfair, and we have cited many examples where this has occurred. No such multicultural, complex and evolving programme as the SWP can be flawless. That however does not suggest that all contracts are violated and that this occurs routinely, which would be very far from the truth. Employers who are appreciative and supportive of workers and may lend them vehicles or incorporate them into family-like relationships, alongside a degree of respect, do not make the news.

Seasonal labour migration involves devoting a short period of a lifetime (at least initially) to an intense focus on making money—sometimes to the exclusion of comfort and alongside loneliness and homesickness, the loss of intimate interpersonal relationships amidst the challenges of adjustment to a foreign environment. For some, this may mean the loss of rights and the need to put up with fraud and social marginalisation. Yet many workers return repeatedly for another spell, some perhaps gritting their teeth as they do so. For others, time away brings camaraderie, excitement and new experiences—none tangible or measurable. For kin at home a similar set of problems and possibilities exists as work (and leisure) is restructured, and absences coped with or even enjoyed, and where parallel sacrifices may be made. Relationships between farmers, workers, agents, unions, churches, two states (one being Australia) and also local governments—as intermediaries—are all important and all vary considerably. Crucially, the SWP is ultimately a demand and market-driven scheme, albeit also at the behest of the PICs, while glossed as part of Australia's aid programme. Its genesis was political and it is far from value free: Australian agricultural interests take precedence. We can now turn to a tentative summary of how all this might be assessed.

Equity

After a slow start, in recent years the SWP has provided employment opportunities for around 10,000 workers per year, but these opportunities have been unevenly distributed between and within PICs. Two-thirds of eligible PICs—Fiji, Kiribati, Nauru, Papua New Guinea (PNG), Solomon Islands and Tuvalu—have barely participated (although Fiji and Kiribati are key participants in the newer PLS). Within the PICs, attempts to target opportunities towards the relatively poor in less developed parts of the countries increasingly became rhetoric rather than reality. Indeed inequity in access to the scheme is evident within countries, islands and villages, as some households or communities have benefited from the scheme and others missed out. In some countries, including

Vanuatu, there has been a shift towards recruiting workers from urban areas. Everywhere, men have overwhelmingly dominated the scheme, and there seems to be an increasing trend towards employing more skilled workers (Chap. 6). The gravitation of skilled workers into seasonal employment has occurred, exactly as in the SAWP where skilled tradesmen with businesses in Jamaica became a large proportion of participants (Hjalmarson, 2016). Seasonal employment has thus come a little closer to resembling parallel flows of longer-term skilled migrants. A further consequence is that labour migration increasingly tends to derive from and add to ongoing processes of urbanisation and the gradual concentration of the population on the main island of each PIC. As temporary migration—under the newer PLS scheme (Chap. 10)—is increasingly for longer time periods and involves workers engaged in tasks outside agriculture, these tendencies have become reinforced, with repeat return migration offering the semblance of a transnational community.

Economics

While it is impossible to differentiate the remittances of seasonal workers from other remittance streams, or account for the totals (since much is transferred in back pockets), the substantial income gains for PICs are evident. Australia is the source of about a quarter of remittances to the Pacific, but that includes flows from resident islanders. Annual remittances from SWP workers however have accounted for about A$36.5 million for Tongans and A$8 million for ni-Vanuatu (Chap. 7). In Tonga remittances make up as much as 38 per cent of GDP (although most of that is from overseas residents) compared with 16 per cent in Samoa and 1.3 per cent in Vanuatu (World Bank, 2021). The contribution of remittances to the economic development of PICs, of households and of nations, has been the key reason why their governments support labour export. Every phase has been described or perceived as a 'labour trade'.

In Tonga, Samoa and Kiribati in the late 2010s remittances were more valuable than exports (although in Tonga and Samoa the majority of remittances probably came from more permanent overseas residents, and in Kiribati from seafarers). In Vanuatu, where remittances represented about two-thirds of the value of exports, remittances most probably did come from SWP and RSE workers: the only PIC where this was true. Tuvalu had a similar ratio of remittances to exports—but from diverse sources. In almost every PIC overseas development assistance was considerably more than remittances (indicating why the main remittance receiving PICs have been referred to as Migration, Remittances, Aid and Bureaucracy [MIRAB] states).

While aid delivery may be more valuable in monetary terms for most PICs and is focused on issues of national development (such as infrastructure provision), its expenditure usually exhibits a substantial urban bias. The remittances and savings of seasonal workers (and other migrants) go directly to households (and to a lesser extent to local institutions such as churches and schools, although institutions tend to be the beneficiaries of more established migrants). Returning workers and the immediate recipients of guestwork earnings may well have a better sense of what would improve their lives, and benefit their own development goals, than do distant aid donors and governments. The volume and use of remittances, at least initially targeted mainly at welfare goals, strongly suggest a major contribution to poverty alleviation, especially in earlier years when remittances had a community component and relatively poor people were more likely to be deliberately selected (Chap. 6). As the remittances of more long-term migrants have done, demonstrably so in Fiji and Tonga (Brown et al., 2014), seasonal worker remittances reduced poverty and acted as social protection where state support was limited. However, as workers became more likely to come from urban areas and left formal employment, that role and contribution became rather less.

Whether the immediate benefits from remittances translate into sustainable long-term economic and social gains remains in question. Most studies of remittances focus on those of long-term established migrants who are not necessarily about to return; however much they perceive remittances as social and economic insurance against returning, they have no immediate targets of their own (hence such remittances sometimes being described as 'altruistic'). Evidence from Fiji, Samoa and Tonga indicates that remittances sent over a long period of time, firstly, do enable sustained development (and also dependence), and, secondly, over time, as most households at home have access to remittances, equity becomes more evident than the uneven development that follows selective initial migration (Brown et al., 2014; Brown & Connell, 2015). However, such circumstances do not exist in Vanuatu—where there are few established overseas ni-Vanuatu—or in most other participating countries. The remittances and savings of short-term migrants are likely to have a different impact. Beyond small businesses minimal evidence exists for economic development as the direct outcome of guestworker migration, and sustainable development may even be threatened (certainly where food security is reduced, environmental degradation occurs or land acquisition marginalises others). However, seasonal migration is not the sole culprit for such ills.

In Australia (and New Zealand), the economic benefits have been most obvious through the presence of relatively cheap (usually on minimum wages or piece rates), pliant and generally hard-working, reliable workers where local labour is otherwise unavailable. That situation was made dramatically evident during the COVID-19 epidemic, when the value of the lost SWP labour was crudely calculated in various contexts and could be contrasted with the earnings of the seasonal workers, revealing that the main economic gains were in Australia (Chap. 8). Here especially migration of guestworkers could be seen as 'a form of development aid given by poor to rich countries' (King, 2018). While the impact on the overall national economy is small, in some agricultural regions, such as the Riverina and around Bundaberg, its impact is significant. Workers necessarily spend money in Australia; however much it might go to second-hand charity stores (and so benefit Australian NGOs—that may support projects in the PICs!) this expenditure contributes to the goods and services sectors (including accommodation and transport providers) in various small Australian towns. In a broad sense seasonal employment contributes to capital accumulation in Australia, and since local workers are not displaced, beyond rare criminality amongst migrants, there are no obvious costs to Australia.

Society

If economic benefits and costs are difficult to assess, the social and cultural outcomes of the SWP are still more intangible and immeasurable.

These too have changed over time as more people became involved. While *kastom* may be maintained through directing some remittances into maintaining social obligations (such as contributing to funeral and wedding expenses, sometimes producing inflation), the social changes that have followed labour migration are almost always directed towards some form of modernity. Many workers—and their families—have gained status through the proceeds of labour migration, as they acquire a better house and more goods, enjoy the 'conspicuous consumption' of consumer durables, distribute some newfound wealth and perhaps relish the confidence, maturity and vision that may come from overseas experience and employment. But such benefits are not at all straightforward and are often accompanied by a degree of ambivalence.

Households experience social costs where (usually) women take on the 'double burden' or 'double day' of extra work and care, as it is usually men who migrate. This can impact relationships, but evidence is mixed as to how such absences might empower women (or not). Children may find themselves without a parent. However, the outcomes of this are poorly documented, and costs are likely to be less acute than the well-documented care deficit problems in Asian-sending countries where migrants are away for years on end (e.g. Parreñas, 2005a, 2005b). Early evidence from Tonga and Samoa suggests that female-headed households in PICs challenge stereotypes of poor, disempowered and isolated female-headed households elsewhere in the world, by being relatively well-off, largely because they benefit from remittances from male migrants (Gailey, 1992; Stewart-Withers, 2011; Moala-Tupou, 2016). That has probably become more true of SWP households where return is certain.

Incentives to work have shifted where young adults reject work opportunities at home, and disengage from communal work, since they are perceived to be less lucrative than guestwork. Agricultural production on the Tongan island of Vava'u benefited when COVID-19 prevented young men taking up employment in SWP (Vainikolo, 2021), but that was by default. Otherwise the outstanding irony of the SWP is that it discourages domestic agricultural employment. Acquisition of useful skills and new approaches to life and livelihoods are few, probably trivial and unlikely to be gained in a single period overseas (Chap. 9). Workshops on re-integration are nascent at best, and whatever skills are obtained make little if any real contribution to capacity building.

The potentially transformative impacts of seasonal migration and their association with the 'migration-development' mantra are thus linked to such liberal ideals as 'work ethic', financial literacy, thrift, discipline and democracy, a continued reflection of 1960s modernisation narratives, linked to economic themes of productivity and entrepreneurship (R. Smith, 2015, Villafuerte, 2018). Workers characteristically aspired to the acquisition of a superior house as both a material and a moral achievement. Yet, beyond housing, it is here that the SWP has been least influential as temporary and tentative work and life disciplines usually returned to a local rhythm in accordance with social organisation and with nature, as the social capital and social order that attended such activities as running trade stores and taxi services faded fast. 'Money trees', development dreams and colonial legacies converged and sometimes collided in regional Australia (Stead, 2019). Evaluating their cultural context at home is notoriously difficult.

Whose Development?

Development is a relational concept. Uneven costs and benefits cast doubts on the notion of mutually beneficial and equitable wins. Certainly there are widespread beneficiaries at both ends of the system: farmers, agents, workers, households, nations, and multiple goods and service providers. Writing of the Turkish *gastarbeiters* in Germany, Philip Martin (1991, p. 18) concluded 'there should be no illusions about equally shared benefits from migration. However unequally shared benefits do not deny that there are benefits'. That allows no easy conclusion. Perhaps we must simply note that all those involved seek to ensure that the SWP continues—however much worker continuity is based on frustration with local development in the PICs, the associated need to return over and over again, and Australian farmers' anxiety over labour supplies. On balance the SWP benefits most people most of the time, but the extent to which a demand-driven scheme can equitably and effectively meet PIC needs is quite debatable.

Necessarily, after the SWP's first long decade, this book offers a short-term perspective and can say little about the long-term development implications in the PICs (or indeed in Australian horticulture). As Graeme Hugo (1982, p. 76) described for Indonesian circular migration some decades ago,

> On the one hand, there is no doubt that from a short-term perspective, certain economic benefits usually accrue to the individual movers, their families, and to some extent their villages of origin. On the other hand, non-permanent migration most likely also preserves and perhaps exacerbates existing inequalities, and the widespread occurrence of this movement may in fact divert attention from the long-term need for a more equitable investment of total resources in rural areas and peasant agriculture.

The same temporal problems evident in the SWP broadly reflect, rather than contrast with, what has occurred in the past and in parallel schemes elsewhere. These suggest that lessons can be learned, but not easily implemented while the basic structure of guestworker schemes is inflexible and unequal, and the advantages accrue to the more powerful. The remarkable similarities in structures between schemes are less surprising where all such schemes have been devised and implemented, however advertised, to meet the labour needs of farmers (and related employers). In every case, these 'spatially and temporally confined workers risk becoming part of a new model of indentured servitude' (De Genova, 2009; Castles & Ozkul, 2014, p. 41). Employment is time and place bound and consequently insecure and inflexible, whereas workers are expected to be flexible or at least on hand. The PICs and other sending countries, especially smaller states, have become more fully incorporated into the international circuit of capital as labour became organised as a commodity and, in turn, as they remitted capital (Rosewarne, 2010). A perpetual reserve army appears constantly in waiting.

Any 'triple win' must therefore be seen in a much broader perspective both globally and historically, as a widespread structural conjuncture in time and place. In the economic circumstances of this century, circular migration has grown. The potential economic benefits are clearly substantial, but can also be uncertain; local development opportunities remain scarce, workers may be exploited (as elsewhere in the labour force) and families are divided. Recognising workers as actors with agency does not entail denial of their structural exploitation, so beautifully expressed by a ni-Vanuatu migrant: 'Sometimes we feel like slaves, but we believe that there is a Judge God that will turn these hardships into a blessing in many of our children's lives, if this scheme continues' (quoted in R. Smith, 2015): the costs and sacrifices of seasonal migration should benefit the next generation.

The developmental rhetoric of 'win–win–win' for recipient and sending countries, and the migrants themselves, contrasts with these structural realities and the inherent precarity and vulnerability of employment. Such precarity is increasingly evident in national economies with the rise of the 'gig economy' and the casualisation of other sectors of the workforce, with growing numbers of migrants in low-paid, insecure jobs with irregular hours and uncertain pay in sectors including cleaning, care, construction and hospitality. The expanded PLS scheme seeks involvement in such sectors, with potential for contrast with 'decent work' that combines rights with dignity: a better complaints mechanism, access to unions, and the possibility of eventually gaining permanence and citizenship (Lenard & Straehle, 2010, 2011; see Chap. 12). Present circumstances ensure that migrant workers live in an all-too-flexible state of 'permanent temporariness' especially as return migration becomes more frequent. A possible direction for action was summarised by Graeme Hugo:

> Destination countries should design their migration policies and programs to be development-friendly, and origin countries should ensure the capital and expertise that migrants bring back are put to good use. Most importantly, destination and origin countries must cooperate to build coherent systems that work for the benefit of all three parties—destinations, origins and migrants. (2013, p. 1)

While manifestly true, this is overly simplistic—a better, more cosmetic and acceptable organisation—but which takes no account of who might initiate such changes and whether these are the same people and institutions who might benefit from restructuring: a wicked puzzle. Reconfiguring labour laws and protections to recognise non-continuous service of workers offers one possible key (Newman, 2013), but such legislative changes are improbable in an age that is shifting away from government intervention and towards austerity and frugality in welfare provision.

Geographically and structurally beyond effective civil, cultural, political, social and economic rights—which may produce a consciousness of relative deprivation, anxiety, anomie and alienation—seasonal incomes come at a cost. Yet any attempt to abolish schemes like the SAWP and SWP would meet strong opposition from the participating workers who are most involved. It cannot therefore be described as forced migration or even a response to the lack of decent local work, but simply one reliable means of achieving reasonable if limited goals, and a modicum of new status. Demand for seasonal work in the PICs (and elsewhere) exceeds supply, and workers seek to return. Remittances have brought beneficial outcomes. Can what appears to be a triple win be sustained over time or will what appears a growing path dependency on labour migration eventually be shattered? A growing dependence on remittances, where those from the SWP are just one part, has reflected the PICs' broader dependence on remittances, aid and rental incomes (notably from fisheries), which arguably has contributed both to their becoming 'mired in MIRAB', with domestic policies being driven by external influences, and to the PICs moving towards a 'remittance trap' where their economies are 'stuck on a lower growth, higher emigration treadmill', where remittance income ironically contributes to training skilled labour for export (Abbott & Pollard, 2020). That has contributed to a loss of labour for domestic development, especially in agriculture, as in Samoa, Tonga and Vanuatu, but also from the public service and business (Chap. 6), reduced the possibility of the PICs wresting control over their national economies, and discouraged private sector development.

However, in small island states choices are truncated, a poverty of opportunity prevails and migration may seem the most plausible solution to development dilemmas for both countries and people. As Cindy Hahamovitch (2011, p. 234) observed of the United States, 'For guestworkers here and around the world, slavery and freedom are not at different ends of a spectrum: they are simultaneous conditions'. Guestworkers may be underpaid and exploited, but return repeatedly because it is the best chance they have, and one of few choices to get ahead at home. In what may well have been a grand gesture, a returned migrant worker in Vanuatu (where some had been to New Zealand as many as five times by 2012) declared:

> Migration and culture do not pull in different directions; migration is becoming part of our culture. When our children finish school, their first desire is to go to New Zealand [on the RSE]. Before you can build a house you are proud of, you must go to New Zealand. Everyone here has been to New Zealand, or desperately wants to go—it is where we find our opportunity (Quoted in Craven, 2013, p. 50)

It is why so many others, especially from the Polynesian states, have migrated permanently. It is not always much of a choice, but the prevalent culture of migration associated with longer-term Polynesian migrants (and Irish, Mexicans and Jamaicans …) has spilled over into a temporary migration gradually evolving towards greater permanence.

Debates on the relationship between migration and development were overwhelmingly optimistic about its positive impact on development, especially in the 1960s as the volume of remittances kept rising (e.g. Galbraith, 1979), but that was questioned in the following two decades, as migration did not necessarily improve but rather undermined people's income or security by being inadequate to support new livelihoods but enough to perpetuate dependency (hence the increasing recourse to return migration). Nonetheless, it experienced a resurgence in policy discourse in this century (Faist, 2008; King, 2018), especially as aid fatigue set in; governments in sending countries were found wanting and remittances could bypass such problems. Such debates were largely centred on more permanent and more skilled migration, but much of seasonal migration still occurs where migrants are moving from relative poverty, do not normally constitute a significant skill drain and soon return with considerable capital to regional areas. Consequently, the positive relationship between migration and development appears more evident if unlikely to be transformative. If development is viewed as freedom (Sen, 2001), one-off or repetitive seasonal migration can scarcely lead to development in many contexts, despite migrants' considerable, albeit short-lived, financial success. Development is always both endpoint and process. As Victoria Stead observed, 'development yields its own kind of cruel optimism, promising futures while working simultaneously to foreclose them, to render them (through the logic of temporary visa regimes) spatially elsewhere' (2021a, p. 15). Put another way, development is always elusive spatially and temporarily.

It would, almost certainly, be too much to ask that the SWP, beyond other similar schemes, contributes significantly to long-term development or to triumph the poor as agents of their own development. But certainly it is a safety valve, with immediate gains, that should not conflict with longer-term development. Even the long-term migration context in such states as Ireland and Pakistan, Mexico and the Philippines, with decades of permanent or temporary migration, suggests that such a dependency is most unlikely to be influenced by changing circumstances at home. Still growing populations in small PICs, especially in Melanesia, render this even more unlikely. Seasonal work remains work that resident Australians are largely unwilling to undertake, hence the recourse to temporary PIC workers, a conjuncture that has intensified and broadened. But that raises questions of whether such employment will continue, whether PIC residents will always seek it out and who might compete for employment in a future international context. Whether ultimately positive or negative, its future hangs on a wide range of factors elaborated on in the next chapter.

Bibliography

Abbott, D., & Pollard, S. (2020, March 2 and 3). Mired in MIRAB: Migration and remittances. ANU DevPolicyBlog.

Able, S. (2010). Letter. *SPC Fisheries Newsletter, 133*, 27–28.

Agar, C., & Manolchev, C. (2020). Migrant labour as space: Rhythmanalysing the agri-food industry. *Organization, 27*(2), 252–271.

Anderson, B. (2010). Migration, immigration controls and the fashioning of precarious workers. *Work, Employment and Society, 4*(2), 300–317.

Anderson, J. (2021). Managing labour migration in Malaysia: Foreign workers and the challenges of 'control' beyond liberal democracies. *Third World Quarterly, 42*(1), 86–104.

Anderson, J., & Franck, A. (2019). The public and the private in Guestworker Schemes: Examples from Malaysia and the US. *Journal of Ethnic and Migration Studies, 45*(7), 1207–1223.

Australian Government. (2019). *Report of the Migrant Workers' Taskforce.* Commonwealth of Australia.

Axelsson, L., & Hedberg, C. (2018). Emerging topologies of transnational employment: 'Posting' Thai workers in Sweden's wild berry industry beyond regulatory reach. *Geoforum, 89*, 1–10.

Baak, P. (1999). About enslaved ex-slaves, uncaptured contract coolies and Unfreed Freedmen. *Modern Asian Studies, 33*(1), 121–157.

Bailey, R., & Ng Shiu, R. (2016). *New responses to natural disasters through Seasonal Labour Mobility Programs.* ANU InBrief 2016/12.

Bauder, H. (2005). Landscape and scale in media representations: The construction of Offshore Farm Labour in Ontario, Canada. *Cultural Geographies, 12*(1), 41–58.

Bauder, H. (2008). Foreign farm workers in Ontario (Canada): Exclusionary discourse in the newsprint media. *Journal of Peasant Studies, 35*, 100–118.

Beckford, C. (2016). The experiences of Caribbean Migrant Farmworkers in Ontario, Canada. *Social and Economic Studies, 65*(1), 153–188.

Berg, L., & Farbenblum, B. (2017). *Wage theft in Australia: Findings of the national temporary migrant work survey.* Migrant Worker Justice Initiative.

Berg, L., & Farbenblum, B. (2020). *International students and wage theft in Australia.* Migrant Worker Justice Initiative.

Binford, A. (2019). Assessing temporary foreign worker programs through the prism of Canada's Seasonal Agricultural Worker Program: Can they be reformed or should they be eliminated? *Dialectical Anthropology, 43*(4), 347–366.

Binford, L. (2013). *Tomorrow we're all going to the harvest: Temporary foreign worker programs and neoliberal political economy.* University of Texas Press.

Borovnik, M. (2011). Occupational health and safety of merchant seafarers from Kiribati and Tuvalu. *Asia Pacific Viewpoint, 52*(3), 333–346.

Brass, T., & Bernstein, H. (1992). Introduction: Proletarianisation and deproletarianisation on the colonial frontier. *Journal of Peasant Studies, 19*(3), 1–40.

Brickenstein, C. (2017). The development impact of seasonal worker schemes: A comparison of European and Pacific experiences. *Migration and Development, 6*(1), 60–82.

Bridi, R. (2015). Migrant workers in Ontario's tobacco belt: An examination of workplace dynamics. *Human Geography, 8*, 54–67.

Bridi, R. (2022). Spaces of alienation in the tobacco fields: The case of migrant workers in Ontario, Canada. *GeoJournal, 87,* 131–146.
Brown, R., & Connell, J. (2015). Migration and remittances: A multidisciplinary synthesis. In J. Connell & R. Brown (Eds.), *Migration and remittances* (pp. xiiv–lxviii). Edward Elgar.
Brown, R., Connell, J., & Jimenez-Soto, E. (2014). Migrants' remittances, poverty and social protection in the South Pacific: Fiji and Tonga. *Population, Space and Place, 20*(5), 434–454.
Campbell, I., Tranfaglia, M., Tham, J., & Boese, M. (2019). Precarious work and the reluctance to complain: Italian temporary migrant workers in Australia. *Labour and Industry, 29*(1), 98–117.
Carmo, R., & Hedberg, C. (2019). Translocal mobility systems: Social inequalities and flows in the wild berry industry. *Geoforum, 99,* 102–110.
Castles, S., & Ozkul, D. (2014). Circular migration: Triple win, or a new label for temporary migration? In G. Battistella (Ed.), *Global and Asian perspectives on international migration* (pp. 27–49). Springer.
Cavanough, E., & Wherrett, C. (2020). *Blue Harvest. Wage theft & other labour infringements in the NSW Mid-North coasts's 2019/20 berry harvest.* The McKell Institute.
Cecil, R., & Ebanks, G. (1991). The human condition of West Indian Migrant Farm labour in Southwestern Ontario. *International Migration, 29*(3), 389–405.
Cecil, R., & Ebanks, G. (1992). The Caribbean Migrant Farm Worker Programme in Ontario: Seasonal expansion of West Indian Economic Spaces. *International Migration, 30*(1), 19–37.
Chacko, E., & Price, M. (2021). (Un)settled sojourners in cities: The scalar and temporal dimensions of migrant precarity. *Journal of Ethnic and Migration Studies, 47*(20), 4597–4614.
Choithani, C., van Duijne, R., & Nijman, N. (2021). Changing livelihoods at India's rural–urban transition. *World Development, 146,* 105617.
Clibborn, S., & Wright, C. (2018). Employer theft of temporary migrant workers' wages in Australia: Why has the State Failed to Act. *Economic and Labour Relations Review, 29*(2), 207–227.
Cohen, A. (2019). 'Slavery hasn't ended, it has just become modernized': Border imperialism and the lived realities of migrant farmworkers in British Columbia, Canada. *ACME: An International Journal for Critical Geographies, 18*(1), 130–148.
Cohen, A., & Hjalmarson, E. (2020). Quiet struggles: Migrant farmworkers, informal labor, and everyday resistance in Canada. *International Journal of Comparative Sociology, 61*(2–3), 141–158.

Connell, J., & Conway, D. (2000). Migration and remittances in island microstates: A comparative perspective on the South Pacific and the Caribbean. *International Journal of Urban and Regional Research, 24*(1), 52–78.

Consterdine, E., & Samuk, S. (2018). Temporary migration programmes: The cause or antidote of migrant worker exploitation in UK agriculture. *Journal of International Migration and Integration, 19*, 1005–1020.

Corrado, A., de Castro, C., & Perrotta, D. (Eds.). (2016). *Migration and agriculture. Mobility and change in the Mediterranean area*. Routledge.

Craven, L. (2013). *The relationship between temporary migration and vulnerability to climate change in rural Vanuatu.* Honours thesis, University of Sydney Department of Government.

De Castro, C., Reigada, A., & Gadea, E. (2020). The devaluation of female labour in fruit and vegetable packaging plants in Spanish Mediterranean agriculture. *Organization, 27*(2), 232–250.

De Genova, N. (2009). Conflicts of mobility, and the mobility of conflict: Rightlessness, presence, subjectivity, freedom. *Subjectivity, 29*, 445–466.

de Moor, N. (2011). Temporary labour migration for victims of natural disasters: The Colombia-Spain Model. In M. Leighton, X. Shen, & K. Warner (Eds.), *Climate change and migration: Rethinking polices for adaptation and disaster risk reduction* (pp. 92–103). UNU-EHS.

Doherty, B. (2017, August 3). Modern day slavery in focus: Hungry, poor, exploited: Alarm over Australia's import of farm workers. *The Guardian*. https://www.theguardian.com/global-development/2017/aug/03/hungry-poor-exploited-alarm-over-australias-import-offarm-workers.

Dun, O., & Klocker, N. (2017). The migration of horticultural knowledge: Pacific Island seasonal workers in rural Australia—A missed opportunity? *Australian Geographer, 48*(1), 27–36.

Enoka, A. (2019). Under the Gaze: A study of the portrayal by the New Zealand print media of Pacific Island workers in the Recognised Seasonal Employer (RSE) scheme, 2007–2012, PhD thesis, Massey University, Palmerston North.

Eriksson, M., & Tollefsen, A. (2013). Disciplining temporary workers. Thai seasonal labourers and the Swedish berry industry. In M. Geiger & A. Pécoud (Eds.), *Disciplining the transnational mobility of people* (pp. 185–206). Palgrave Macmillan.

Fair Work Ombudsman. (2019, April 23). Agri-Labour to pay migrant workers $50,000. Australian Government Fair Work Ombudsman. https://www.fairwork.gov.au/about-us/news-and-media-releases/2019-media-releases/april-2019/20190423-agri-labour-eu-media-release#twitter.

Faist, T. (2008). Migrants as transnational development agents: An inquiry into the newest round of the migration-development nexus. *Population, Space and Place, 41*, 21–42.
Farbenblum, B., & Berg, L. (2018). *Wage theft in silence*. Migrant Worker Justice Initiative.
Fels, A., & Cousins, D. (2019). The migrant workers' taskforce and the Australian government's response to migrant worker wage exploitation. *Journal of Australian Political Economy, 84*, 13–45.
Ferguson, A. (2020, February 22). Scourge of wage fraud demands action, *Sydney Morning Herald*, 2.
Gailey, C. (1992). A good man is hard to find. *Critique of Anthropology, 12*, 47–74.
Galbraith, J. (1979). *The nature of mass poverty*. Harvard University Press.
Garrapa, A. M. (2016). The citrus fruit crisis. Value chains and "just in time" migrants in Rosarno (Italy) and Valencia (Spain). In A. Corrado, C. Castro, & D. Perotta (Eds.), *Migration and agriculture: Mobility and change in the Mediterranean area* (pp. 111–127). Routledge.
Gibbon, P., & Riisgaard, L. (2014). A new system of labour management in African largescale agriculture? *Journal of Agrarian Change, 14*(1), 94–128.
Górny, A., & Kaczmarczyk, P. (2018). A known but uncertain path: The role of foreign labour in Polish agriculture. *Journal of Rural Studies, 64*, 177–188.
Graves, A. (1993). *Cane and labour. The political economy of the Queensland Sugar Industry 1862–1906*. Edinburgh University Press.
Hahamovitch, C. (2008). 'The Worst Job in the World': Reform, revolution and the secret rebellion in Florida's cane fields. *Journal of Peasant Studies, 35*(4), 770–800.
Hahamovitch, C. (2011). *No man's land. Jamaican Guestworkers in America and the Global History of Deportable Labor*. Princeton University Press.
Hartman, T. (2008). States, markets, and other unexceptional communities: Informal Romanian labour in a Spanish agricultural zone. *Journal of the Royal Anthropological Institute, 14*, 496–514.
Hedberg, C., & Haandrikman, K. (2014). Repopulation of the Swedish countryside: Globalisation by international migration. *Journal of Rural Studies, 34*, 128–138.
Hennebry, J. (2014). Falling through the cracks: Migrant workers and the Global Social Protection Floor Initiative. *Global Social Policy, 14*(3), 369–388.
Hennebry, J., & Preibisch, K. (2012). A model for managed migration? Re-examining best practices in Canada's Seasonal Agricultural Worker Program. *International Migration, 50*(21), e19–e40.

Hennebry, J., & Williams, G. (2015). Making vulnerability visible: Medical repatriation and Canada's migrant agricultural workers. *Canadian Medical Association Journal, 187*(6), 391–392.

Hennebry, J., McLaughlin, J., & Preibisch, K. (2016). Out of the loop: (In) access to health care for migrant workers in Canada. *Journal of International Migration and Integration, 17*, 521–538.

Hjalmarson, K. (2016). Race, labour and the postmodern plantation: Jamaican Migrant Farmworkers in Canada's Seasonal Agricultural Worker Program. MA thesis, University of British Columbia, Okanagan.

Hoggart, K., & Mendoza, C. (1999). African immigrant workers in Spanish agriculture. *Sociologia Ruralis, 39*(4), 538–562.

Horvath, K. (2014). Securitisation, economisation and the political constitution of temporary migration: The making of the Austrian seasonal workers scheme. *Migration Letters, 11*(2), 154–170.

Howe, J., Clibborn, S., Reilly, A., van den Broek, D., & Wright, C. (2019). *Towards a durable future: Tackling labour challenges in the Australian Horticulture Industry*. University of Adelaide.

Howe, K. (1974). Review of Peter Corris. *Passage, Port and Plantation, Journal of the Polynesian Society, 84*(3), 382–384.

Hugo, G. (1982). Circular migration in Indonesia. *Population and Development Review, 8*(1), 59–83.

Hugo, G. (2013). *What we know about circular migration and enhanced mobility*. Migration Policy Institute Policy Brief No. 7.

Kasimis, C., Papadopoulos, A., & Pappas, C. (2010). Gaining from rural migrants: Migrant employment strategies and socioeconomic implications for rural labour markets. *Sociologia Ruralis, 50*(3), 258–276.

King, R. (2018). Is migration a form of development aid given by poor to rich countries? *Journal of Intercultural Studies, 39*(2), 114–128.

Komaisavai, P., & Magick, S. (2019, July). Keeping our seafarers safe. *Islands Business, 45*(7), 10–13.

Kusakabe, K., & Myae, A. (2019). Precarity and vulnerability: Rubber plantations in Northern Laos and Northern Shan State, Myanmar. *Journal of Contemporary Asia, 49*(4), 586–601.

Lenard, P., & Straehle, C. (2010). Temporary labour migration: Exploitation, tool of development, or both? *Policy and Society, 29*, 283–294.

Lenard, P., & Straehle, C. (2011). Temporary labour migration, global redistribution, and democratic justice. *Politics, Philosophy and Economics, 11*(2), 206–230.

Levick, W., & Bedford, R. (1988). Fiji labour migration to New Zealand in the 1980s. *New Zealand Geographer, 44*(2), 14–21.
Lindner, K., & Kathmann, T. (2014). Mobility partnerships and circular migration. In J. Gertel & S. Sippel (Eds.), *Seasonal workers in Mediterranean agriculture: The social costs of eating fresh* (pp. 121–129). Routledge.
Lozanski, K., & Baumgartner, K. (2022). Local gastronomy, transnational labour: Farm-to-table tourism and migrant agricultural workers in Niagara-on-the Lake, Canada. *Tourism Geographies*, (in press).
Macpherson, C. (1981). Guest-worker movements and their consequences for donor and recipient countries: A case study. In G. Jones & H. Richter (Eds.), *Population mobility and development: Southeast Asia and the Pacific* (pp. 257–277). ANU Development Studies Centre Monograph No. 27.
Maher, G., & Cawley, M. (2016). Short-term labour migration: Brazilian migrants in Ireland. *Population, Space and Place, 22*, 23–35.
Manning, F. (1986, January 10). West Indians complete another season on Ontario farms. *Caribbean Chronicle,* 1589.
Martin, P. (1991). *The unfinished story: Turkish labour migration to Western Europe*. ILO.
Mayell, S. (2016). *Up-rooted lives, deep-rooted memories: Stress and resilience among Jamaican agricultural workers in Southern Ontario*. MA thesis, McMaster University, Hamilton.
McGlone, T. (2021, April 11). Seasonal work on Australia's farms: 'No one wants to do this sort of work'. *Guardian*.
Meléndez, E. (2017). *Sponsored migration. The state and Puerto Rican Migration to the United States*. Ohio University Press.
Mésini, B. (2014). The transnational recruitment of temporary Latino workers in European agriculture. In J. Gertel & S. Sippel (Eds.), *Seasonal workers in Mediterranean agriculture: The social costs of eating fresh* (pp. 71–82). Routledge.
Moala-Tupou, L. (2016). The social impacts of seasonal migration on left-behind children: An exploratory study from Lifuka, Tonga. Master of Social Science thesis, University of Waikato, Hamilton.
Newman, A. (2013). The legal in/Security of temporary migrant agricultural work: Case studies from Canada and Australia. *Deakin Law Review, 18*(2), 361–389.
Papadopoulos, A., Fratsea, L., & Mavrommatis, G. (2018). Governing migrant labour in an intensive agricultural area in Greece: Precarity, political mobilization and migrant agency in the fields of Manolada. *Journal of Rural Studies, 64*, 200–209.

Parreñas, R. (2005a). *Children of global migration. Transnational families and gendered woes*. Stanford University Press.
Parreñas, R. (2005b). Long distance intimacy: Class, gender and intergenerational relations between mothers and children in Filipino transnational families. *Global Networks, 5*(4), 317–336.
Perry, A. (2012). Barely legal: Racism and migrant farm labour in the context of Canadian multiculturalism. *Citizenship Studies, 16*(2), 189–201.
Piper, N. (2010). All quiet on the Eastern front? Temporary contract migration in Asia revisited from a development perspective. *Policy and Society, 29*(4), 399–411.
Polkowska, D. (2019). New dimensions of precarious work among migrants: A qualitative meta-analysis of current research. *Asian and Pacific Migration Journal, 28*(4), 440–456.
Polkowska, D., & Filipek, K. (2020). Grateful precarious worker? Ukrainian migrants in Poland. *Review of Radical Political Economics, 52*(3), 564–581.
Potter, M., & Hamilton, J. (2014). Picking on vulnerable migrants: Precarity and the mushroom industry in Northern Ireland. *Work, Employment and Society, 28*(3), 390–406.
Preibisch, K. (2007). Local produce, foreign labor: Labor mobility programs and global trade competitiveness in Canada. *Rural Sociology, 72*, 418–449.
Preibisch, K. (2010). Pick-your-own labor: Migrant workers and flexibility in Canadian agriculture. *International Migration Review, 44*(2), 404–441.
Preibisch, K., & Binford, L. (2007). Interrogating racialized global labour supply: An exploration of the racial/national replacement of foreign agricultural workers in Canada. *Canadian Review of Sociology and Anthropology, 44*(1), 5–36.
Preibisch, K., & Hennebry, J. (2013). Buy local, hire global: Temporary migration in Canadian agriculture. In C. Straehle & P. Lenard (Eds.), *Legislating inequality: Canada's temporary migrant worker program* (pp. 49–72). McGill-Queen's University Press.
Rai, P. (2018). The labor of social change: Seasonal labor migration and social change in rural western India. *Geoforum, 92*, 171–180.
Reid-Musson, E. (2017). Grown close to home™: Migrant farmworker (Im) mobilities and unfreedom on Canadian family farms. *Annals of the American Association of Geographers, 107*(3), 716–730.
Rice, S. (2021, November 10). Australia's 'slave trade' disgrace. *The Australian*, 4.
Robillard, C., McLaughlin, J., Cole, D., Vasilevska, B., & Gendron, R. (2018). 'Caught in the Same Webs'—Service providers' insights on gender-based and structural violence among female temporary foreign workers in Canada. *Journal of International Migration and Integration, 19*, 583–606.

Rodgers, G., & Rodgers, J. (1989). *Precarious Jobs in Labour Marker Regulation: The growth of atypical employment in Western Europe*. ILO.
Rogaly, B. (2008). Intensification of workplace regimes in British horticulture: The role of migrant workers. *Population, Space and Place, 14*, 497–510.
Rosewarne, S. (2010). Globalisation and the commodification of labour: Temporary labour migration. *Economic and Labour Relations Review, 20*(2), 99–110.
Rye, J. (2018). Labour migrants and rural change: The 'mobility transformation' of Hitra/Frøya, Norway, 2005–2015. *Journal of Rural Studies, 64*, 189–199.
Rye, J., & Andrzejewska, J. (2010). The structural disempowerment of Eastern European migrant farm workers in Norwegian agriculture. *Journal of Rural Studies, 26*, 41–51.
Rye, J., & O'Reilly, K. (Eds.). (2021). *International labour migration to Europe's rural regions*. Routledge.
Rye, J., & Scott, S. (2018). International labour migration and food production in Rural Europe: A review of the evidence. *Sociologia Ruralis, 58*(4), 928–952.
Sampaio, D., & Carvalho, R. (2016). From the Al-Maghrib to the Al-Gharb: An anatomy of the recruitment and labour incorporation of Moroccan agricultural workers. In A. Corrado, C. Castro, & D. Perotta (Eds.), *Migration and agriculture: Mobility and change in the Mediterranean area* (pp. 231–245). Routledge.
Scott, J. (1985). *Weapons of the weak: Everyday forms of peasant resistance*. Yale University Press.
Scott, S. (2015). Making the case for Temporary Migrant Worker Programmes: Evidence from the UK's rural guestworker ('SAWS') scheme. *Journal of Rural Studies, 40*, 1–15.
Sen, A. (2001). *Development as Freedom*. Oxford University Press.
Seo, S., & Skelton, T. (2017). Regulatory migration regimes and the production of space: The case of Nepalese workers in South Korea. *Geoforum, 78*, 159–168.
Shah, A., & Lerche, J. (2020). Migration and the invisible economies of care: Production, social reproduction and seasonal migrant labour in India. *Transactions of the Institute of British Geographers, 45*, 719–734.
Silverman, S., & Hari, A. (2016). Troubling the fields: Choice, consent, and coercion of Canada's seasonal agricultural workers. *International Migration, 54*(5), 91–104.
Smith, R. (2015, November 12). The 'hidden abodes' of temporary migration programs. FocaalBlog. www.focaalblog.com/2015/11/12/rachel-smith-the-hidden'abodes-of-temporary-migration-programs.

Standing, G. (2011). *The Precariat, The New Dangerous Class*. London: Bloomsbury Academic.
Stead, V. (2019). Money trees, development dreams, and colonial legacies in contemporary Pasifika Horticultural Labour. In V. Stead & J. Altman (Eds.), *Labour lines and colonial power: Indigenous and Pacific Islander Labour Mobility in Australia* (pp. 133–158). ANU Press.
Stead, V. (2021a). Precarity's reach: Intersections of history, life, and labour in the Australian horticultural industry. *Journal of the Royal Anthropological Institute, 27*(2), 303–320.
Stead, V. (2021b, November 10). Australia needs better working conditions, not shaming, for Pacific Islander farm workers. *The Conversation*.
Stenbacka, S. (2019). Responsibilities, caring practices and agriculture: Farmers' perspectives on recruitment and employer–employee relationships. *Sociologia Ruralis, 59*(2), 255–274.
Stewart-Withers, R. (2011). Contesting a Third World development category: Female-headed households in Samoa. *Women's Studies International Forum, 34*, 171–184.
Stoler, A. (1985). *Capitalism and confrontation in Sumatra's Plantation Belt, 1870–1979*. Yale University Press.
Strauss, K., & McGrath, S. (2017). Temporary migration, precarious employment and unfree labour relations: Exploring the 'continuum of exploitation' in Canada's temporary foreign worker program. *Geoforum, 78*, 199–208.
Surak, K. (2015). Guestworker regimes globally: A historical comparison. In L. Talani & S. McMahon (Eds.), *Handbook of the International Political Economy of Migration* (pp. 167–184). Edward Elgar.
Tan, Y., & Lester, L. (2012). Labour market and economic impacts of international working holiday temporary migrants to Australia. *Population, Space and Place, 18*(3), 359–383.
Thomas-Hope, E. (2017). Migration, small farming and food security in the Caribbean: Jamaica and St. Vincent and the Grenadines. *International Migration, 55*(4), 35–47.
Turhan, E., Zografos, C., & Kallis, G. (2015). Adaptation as biopolitics: Why state policies in Turkey do not reduce the vulnerability of seasonal agricultural workers to climate change. *Global Environmental Change, 31*, 296–306.
Underhill, E., Groutsis, D., van den Broek, D., & Rimmer, M. (2015). Organising across borders: Mobilising temporary migrant labour in Australian food production. *Journal of Industrial Relations, 62*(2), 278–303.
Unions NSW. (2020). *Wage theft. The shadow market*. Unions NSW.

Unions NSW. (2021). *Wage theft. The shadow market. Part Two: The horticultural industry*. Unions NSW.
Vainikolo, L. (2021). Small and isolated Vava'u, Tonga: From weakness to strength during COVID-19? In Y. Campbell & J. Connell (Eds.), *COVID in the Islands: The Pacific and the Caribbean* (pp. 207–218). Palgrave Macmillan.
van den Broek, D., Wright, C., Howe, J., & Reilly, A. (2021). Pro-market governance, migration status and worker vulnerability: The case of Australian horticulture. *Economic and Industrial Democracy, 42*(4), 1305–1325.
Villadóniga, J. (2020). Unstable places. Modern-day slavery of the 21st century in the Strawberry sector in Huelva in Spain. *Migraciones Internacionales, 11*(4), 1.
Villafuerte, L. (2018). *Seasons of change: Ni-Vanuatu and the Recognised Seasonal Employers (RSE) scheme*. PhD thesis, University of Auckland.
Walsh, J. (2014). From nations of immigrants to states of transience: Temporary migration in Canada and Australia. *International Sociology, 29*(6), 584–606.
Wang, Z., & Connell, J. (2021). Taiwanese working holiday makers in rural and regional Australia: Temporary transnational identities and employment challenges. *Australian Geographer, 52*(2), 191–208.
Weiler, A., Sexsmith, K., & Minkoff-Zern, L.-A. (2020). Parallel precarity: A comparison of U.S. and Canadian Agricultural Guestworker Programs. *International Journal of the Sociology of Agriculture and Food, 26*(2), 143–163.
Wood, C., & McCoy, T. (1985). Migration, remittances and development: A study of Caribbean cane cutters in Florida. *International Migration Review, 19*(2), 251–277.
World Bank. (2021). *Pacific labor mobility, migration and remittances in Times of COVID-19*. World Bank.
Wright, C., Groutsis, D., & van den Broek, D. (2017). Employer sponsored temporary labour migration schemes in Australia, Canada and Sweden: Enhancing efficiency, compromising fairness? *Journal of Ethnic and Migration Studies, 43*(11), 1854–1872.
Ye, J., & Yeoh, B. (2018). Disciplining deserving subjects through social assistance: Migration and the diversification of precarity in Singapore. *Annals of the American Association of Geographers, 108*(2), 478–485.

12

Hosts and Guests

A modest scheme that brings a few thousand workers from some Pacific Island countries to some agricultural districts in Australia may be insignificant in many ways. Indeed, until COVID-19 and associated worker shortages thrust it briefly into the media spotlight, the Seasonal Worker Programme (SWP) was totally unknown in most parts of Australia (however much fruit and vegetable prices and availability influence consumption and nutrition). By contrast, the SWP is highly significant in some Pacific Island Countries (PICs), notably Vanuatu, where it is regarded as an invaluable component of development. The scheme is underpinned by economic issues—and specifically the needs of Australian farmers—but bolstered and modified by political and strategic trends. It is a small part of contemporary geopolitics, with Australia the definitive broker and defender of the scheme. Guestworker migration is political, even as the state may seem peripheral or even invisible, but where the reality of the market requires disposable, inexpensive foreign labour. This final chapter examines the multiple factors and many possibilities that shape temporary labour migration schemes in the region, to consider their possible future.

As previous chapters have shown, for all its faults and benefits, labour migration has become of increased value to Australia in this century,

K. Petrou, J. Connell, *Pacific Islands Guestworkers in Australia*,
https://doi.org/10.1007/978-981-19-5387-3_12

incorporating temporary skilled workers, grey nomads, working holiday makers (WHMs) and belatedly including seasonal workers from the Pacific Island Countries (PICs). More recently temporary labour migration has become less seasonal and extended beyond agriculture and horticulture into various processing and service activities: a similar pattern to that in other relatively developed countries (Chap. 11). That evolution has been welcomed in the PICs where economic growth has been slow and population growth more rapid. Universally seasonal worker schemes appear to have achieved some 'triple wins'—for workers, and for countries of origin and destination—but neither on any significant scale or without costs. Increasing the effectiveness of the SWP will be difficult (and even contradictory) not merely because of vagaries in agricultural systems, difficulties in predicting medium-term demand for labour, or even unpredictable 'black swans' such as COVID-19, but because the scheme—like its parallels elsewhere—is part of a neoliberal trajectory where cheap, flexible labour is invaluable. But the SWP scheme is also at risk of political shifts: the PICs are potentially vulnerable to any scheme dictated by demand, and not underpinned by any legal agreements. That emphasises the power imbalance between the PICs and Australia (as also with New Zealand). In turn that provides a basic challenge to long-term sustainability where the future is inherently uncertain.

For Australia, the SWP has been a source of some regional soft power and a form of creative aid that responds to PIC pressures (and which may also have reduced the resort to illegal workers). Moreover, while the SWP (and the Pacific Labour Scheme [PLS]) is dependent on Australian demand, the PICs pushed for its establishment and most, but especially Vanuatu, have been eager participants. As in similar schemes elsewhere, PICs play an active role and 'adopt brokerage strategies to package and commodify surplus labour' (A. Smith, 2015, p. 286) in a competitive environment. From the earliest days of the Recognised Seasonal Employer (RSE) scheme the PICs sought to implement marketing strategies appropriate to New Zealand and Australia, to 'promote each country's strengths and comparative advantage to gain market share and entry into new markets' (Ramasamy et al., 2008, p. 173). That extended into PICs developing and marketing their supposedly unique 'brand' with even the World Bank initially referring to 'labour export' rather than a more politically correct labour mobility (Hao'uli, E., 2013), just as the Work Ready Vanuatu website was stating 'Vanuatu seasonal workers are highly praised among employers for their excellent work attitudes, willingness to work hard and their flexibility' (quoted in Villafuerte, 2018, p. 183). In that vein workers were constantly invoked to be positive representatives of their countries. Marketing was underpinned by a moral imperative that has continued throughout the schemes. When the first meat workers went from Solomon Islands, they were told by a Solomon Islands High Commission representative: 'This is a big opportunity—future groups depend on everyone doing an excellent job. There are lots of jobs in Australian abattoirs and we want Solomon Islanders to have the best reputation as meat workers. Future growth depends on you' (PacNews, 17 July 2019). The first group of Fijians were similarly urged by the Permanent Secretary for Employment and Industrial Relations, Osea Cawaru, to be 'ambassadors' since the 'key to opening more doors for fellow Fijians' was in their hands so that they would 'make Fiji proud by being productive workers, dedicated and diligent with the right attitude values and work ethics, as this will create a positive image of Fiji as we continuously search for more employment opportunities for our fellow Fijians' (PacNews, 1 August 2019). Such constant invocations were marked by, and responded to, both the small size of the market and the occasional threats by employers to recruit from different source countries (Chap. 7). The market focus emphasised that any triple, or multiple, win primarily represented a short-term economic perspective, focused on the material benefits of remittances and the value of agricultural output in Australia (Chap. 11).

Guestwork has become part of a much wider quest for modernity that has everywhere brought changes to material life. It is embedded in hybrid forms of development, where 'global' capitalism is entangled in local social relationships. Since the earliest days of migration, remittances have enabled the purchase of multiple consumer durables, from tools to solar panels, from fast food to smartphones, that have made lives and livelihoods more attuned to the rhythms of urban and international worlds. That material process, partly symbolised in Fig. 9.1, has been accelerated by governments, external agencies and the media. Where, three decades ago, on Ambae, much like elsewhere in the Pacific: 'islanders have one foot somewhat tentatively in the capitalist market-place and the other squarely in a lifestyle based on values and relationships that are very different from those underlying a capitalist way of life' (Rodman, 1993, p. 183), the balance has inexorably shifted to the former, simply marked by greater individualism and what older islanders refer to as a lack of respect. Meanwhile selection of participants in the scheme has become a little less focused on equity and outcomes have come closer to a brain drain. That this transition has been urban-led has been particularly evident in the quest, in Vanuatu, for urban land and housing ownership in Port Vila. It is further manifest in the ubiquitous depopulation of outer islands, where modern development and services are relatively absent, and in the declining interest in domestic agricultural work.

While the SWP brings benefits for both sending and receiving countries, it also draws upon histories of colonialism and constitutes a widening of international inequalities (Phillips, 2009; Rogaly, 2021). Victoria Stead (2019a, p. 153) astutely observes that SWP participation can 'be experienced as reflective, and indeed constitutive, of deeply entrenched inequalities that run through and characterise Australian coloniality in relation to the Pacific region'. In 2020, Australia's Deputy Prime Minister even asserted that islanders were essentially in Australia to serve, denying notions of reciprocity. While inequalities are inevitable and define other relationships, including trade, that does not deny advantages to PICs, especially where development opportunities are scarce. Indeed, with few alternatives, seasonal migration represents one means of diversifying livelihoods and generating income (even if the macro-economic impacts are small) and the benefits too few to enable sustainable development. However, as we indicated (Chap. 11), the economic benefits from seasonal work are concentrated in Australia, with remittances a fraction of total earnings, so that the SWP makes a valuable aid contribution from the PICs. As COVID-19 emphasised, in the short term that aid was irreplaceable.

Towards Permanency?

In this century international migration has become more significant throughout the Pacific, more diverse in terms of destinations and occupations, more valuable to national economies through remittances and more likely to be permanent. Migration has followed more flexible labour markets and cheaper travel, but reflected legal restrictions on access and permanency. In the wider context of Pacific migration, temporary mobility for employment remains exceptional. Moreover, governments in the smaller Pacific island states especially have become increasingly involved in negotiating migration strategies, as remittances have become more significant for national and household economies. Responding to scarce international opportunities has been particularly difficult in smaller, more remote states, whereas in the nineteenth century their inhabitants usually took the lead.

The shift towards uneven development and urban bias in the SWP (Chap. 7) has been paralleled and accentuated by the shift towards longer-term opportunities—marked by more return migration, longer contract periods and diversification into other sectors. More recent employment in caring, meat works and tourism took migration beyond agriculture as the Australian national economy moved towards services and the national population aged, increasing the need for carers and reducing the proportion of people of working age. Temporary labour migration has thus acquired wider social and geographical significance within Australia, resulting in an active guestworker presence in some small towns and tourist resorts. In this century Australia has sought more skilled workers (with other migration streams dedicated to them). In the PICs, some of these workers have been trained through Australia Pacific Training Coalition (APTC). Boosting the APTC would extend the range of PIC worker categories and enable the PICs to benefit from potentially superior and more enduring remittance flows from longer-term migrants, but could also contribute to emerging skill shortages in many PICs, typified in the health sector. This would offer potential for more permanent migration (and hence greater earnings), the preferred goal for many Pacific islanders from the Polynesian states. Contemporary seasonal guestworker schemes are something of a halfway house. Again, parallels exist, remarkably with the distant past. As Jim Johnson (1967, p. 110) observed of seasonal migration from mid-nineteenth-century Ireland to England: 'a movement of this kind contains within itself the seeds of its own destruction, since it introduces its participants to the benefits of a cash income and familiarizes them with the higher material standard of living possible elsewhere'. In short, as expectations are regularly revised upwards, it invariably leads on to more permanent migration, the situation that characterised Irish migration and, a century later, marked Turkish migration to Germany.

Significantly, permanence reduces the potential for exploitation and vulnerability. Just and equitable temporary work programmes would, over time, enable workers to gain permanence and citizenship. That, of course, would spell the end of traditional guestworker schemes which require that workers return home (Lenard & Straehle, 2011). In any case, repeated return migration has come close to achieving that. The end of migrants being 'permanently temporary—living and working in Australia … but kept at arm's length and unable to settle' (Mares, 2012; Walsh, 2014) would contribute to a renewed diasporic development. As Stephen Howes (2019) has suggested, 'Circular migration—a concept often used to promote the SWP and PLS—sounds good, until you remember that in the Pacific migrants are being forced to circulate in and out of economies that simply do not generate enough jobs'. When arguments over the virtues of the SWP are reduced to simple economics, it is undeniable that sending and receiving states, and participating households, gain from the scheme in material terms, so that greater permanence and a greater scale of migration might just possibly be more beneficial to the workers.

Being temporary was at the root of many of the problems that workers experienced in an unfamiliar cultural context. Providing avenues to citizenship, and therefore to a much longer stay, could simultaneously prevent returning workers from becoming a 'permanent temporary' workforce, increase their rights (Reilly, 2011) and make it a little easier to address power imbalances. The relationship between permanent residence and freedom, however, is not straightforward however much greater agency would come with permanency. Writing about Filipino hospitality workers in Canada with access to permanent residence, Geraldina Polanco (2019, p. 12) argues that unfreedom 'is not only a question of labour rights and vulnerabilities, but also centrally about (racialized) social exclusion and belonging to the nation'. It is likely that Pacific islanders, racialised and considered 'well-suited' to unskilled agricultural work (Stead, 2019a, 2021; Nishitani & Lee, 2019), would face similar issues as do Filipinos in Canada, where power imbalances exist and only hard-won marginal reform seems possible. No shift towards permanence will be easy since the greater rights that would come with permanence would raise costs for employers.

Nonetheless, while greater permanence does not necessarily result in superior working and other conditions, it does offer the prospect of family reunions that might reduce social costs, ease integration into regional areas and small towns, as refugees have managed to do, and offer prospects of fairer wages and conditions. If the PLS expands, the probability of permanence and more graduates from APTC programmes arriving, mobility is more likely to be associated with skills and with less remote residence. Permanence might be achieved through a 'Pacific Integration Visa' with places allocated within the existing migration cap, justified by commitment to the 'Pacific family' frequently rhetorically expressed in the context of the 'step-up' (Howes, 2019), however much political will remains lacking.

As 'seasonal' migrants stay longer, with seasons extending into years through the PLS scheme, their value to Australia will have increased: already with tentative roots, routes and routines, and with skills evidently lacking in the regions where they live, and able to be regular employees rather than part of a rigid government programme, they offer the potential to bring new life to regional communities (Rose, 2020). The value of greater permanency and greater integration to Australia soon became evident in the COVID-19 labour shortages; Wes Morgan thus argued that 'During the COVID-19 crisis, while international travel is restricted, Australia is presented with an opportunity to cement the long-term relationship with its Pacific family and build a more integrated regional economy' (Morgan, 2020; Howes, 2019). That aligns with how repeated return migration has effectively turned seasonal migration for some into a career.

Pathways to permanence were favoured by various workers we met. Some of the ni-Vanuatu women working in Victoria in 2012 believed that their unhappy workplace experiences were directly linked to their temporary status. Edna suggested,

> The others are all permanent staff, so they always have first priority. We've just come for six months and then we're leaving.

Joshua agreed that the men were similarly treated poorly as compared to other more permanent staff—even though the ni-Vanuatu workers were reliable.

> The way he [employer] speaks to us at work isn't very nice. He's always telling us to go back to Vanuatu [...] OK, I can go back. But whenever he's in trouble and there aren't enough workers, he comes running to us. He's lucky he has us too because we're fast workers.

The workers agreed that they would like the opportunity to stay in Australia a little longer, and to be able to bring their children and families with them while they worked. Alice and Keith had two children in Port Vila, a five-year-old and a two-year-old. They had left their children with Keith's parents while they came to work in Australia for six months. Alice explained,

> We've come to work here to earn money so that we can develop our land in Vanuatu. But we'd really like it if we could stay in Australia for a small time and have our children attend school here. That way we could work until we had met all of our goals and finished our plans for our life in Vanuatu. Then we'd move back to Vanuatu again.

Vanessa was working in Australia for the first time; however, her husband was also engaged in seasonal work and was away in New Zealand for the third time. Their young daughter was staying with Vanessa's parents on Ambrym. Vanessa confided,

> It's a little bit hard to leave your child behind. It would be good if there was a way our children could travel with us. Yes, if you have one or two children, you could bring them to Australia while you were working and they would be close by, that would be good.

Much the same was true in New Zealand. As one RSE worker in New Zealand said, as early as 2010:

> So if we lived here for five years and were good citizens and we didn't have a bad record or anything it would be good if we could get permanent resident status. If we are good workers, we are experienced and we fit into the industry and we are helping the economy then we should be allowed PR after five years. I have been here three years now and have a good record. We want the government to do this for the good workers like us. If there are no problems with the workers and they need us to keep returning why don't they give us permanent residency so that we can stay in the industry and keep working in it? (Quoted in Williams, 2010, p. 108)

By 2021, such aspirations were no closer to fruition in either New Zealand or Australia. Were greater permanence to be achieved it would put guestworkers on a similar footing to skilled workers for whom the transition to permanence has been relatively easy (since their labour was more highly regarded) and contribute to strengthening Australia's relations with the PICs.

Much as blackbirding literally changed the face of parts of Queensland, where South Sea Islanders eventually made long-term homes, a shift towards permanence may have a similar impact on other parts of rural and regional Australia—hitherto some of the more Anglo-Australian parts of the country—as the demand for agricultural workers, meat workers, carers and others creates a more permanent population (a process that has already taken many refugees to small Australian towns) that also counteracts the out-migration of young people. In Australia, as in other OECD countries, that is part of a broader trend that includes skilled migrants, given incentives or directed to go to rural areas (Dufty-Jones, 2014; Boese & Moran, 2021). But political support for anything other than circular migration has remained lacking: unskilled Pacific seasonal migrant workers were not on a track to citizenship.

Eternal Demand?

Underpinning analysis and assessment of the SWP is some consideration of its sustainability. Is it merely a temporary phenomenon to serve immediate localised agricultural (and other) labour requirements in Australia, or even serve soft power and strategic needs (see below) that coincidentally meet the needs of PICs? If so will such circumstances endure and enable long-term continuity? The broadening out of the PLS, and the quest for new skills, suggests expansion rather than contraction, as one component of the internationalisation of the Australian workforce. That too follows the constant industry push over more than two decades for labour—provoking periodic media headlines over rotting fruit—that suggests either a lack of alternatives or a reluctance to pay wages that would attract local interest. Despite intermittent arguments and parliamentary enquiries, issues relating to lazy Australians, welfare dependence, dole bludgers, remote places and poor conditions remain unresolved; inducements to Australians have consistently failed, even during COVID-19 when they were ramped up (towards fatuousness in some circumstances), providing enough evidence that a labour shortage existed that was unlikely to be remedied by plausible modifications to wages and conditions. No examples exist of farmers deliberately reverting to employing local workers after a period of employing guestworkers (or WHMs) without a significant structural change such as mechanisation. An implausible radical restructuring of a complex and diverse industry is no more likely in Australia than in other OECD countries. The new age of seasonal labour migration is likely to expand rather than contract.

Mechanisation, automation and genetic changes may make a contribution to decreasing, but not replacing, the need for human labour. Scientists have been optimistic over fruit picking, as one has declared for New Zealand orchards:

> Of course, we will still have the essential pickers around, but we can enhance productivity. We plan to implement some Internet of Things (IoT) devices to the picking buckets that will give information such as the location and the weight that is being carried. It will eliminate the need to walk to a fixed bin, as it will be coming to them. … The mobile bin robots can also be used throughout the whole year to automate spraying through the orchards and vineyards with pesticides and fertilizers. It can reduce costs and the severity of impacts of labour shortages. Full automation is not looking promising at all. The idea of fully eliminating workers hasn't been practical in the past 70 years… hundreds of prototypes of full automation have proven impractical and unreliable (Fresh Plaza, 10 June 2020)

Significant labour-saving mechanisation is not imminent. Dwarf varieties of stone fruit have been introduced; salads can be mechanically harvested, cleaned and packed; and cherries and avocadoes packed mechanically, but many soft fruits are fragile. Robots can pick some vegetables and fruits—but cannot easily determine ripeness and are not always gentle enough, although they can work up to 20 hours a day (and in problematic weather). Technological possibilities remain expensive and under-developed.

Australia might find a cheaper source of migrant labour (perhaps in south Asia, through increased support for an Asian agricultural visa [Chap. 7]). WHMs have been valuable in the past, and despite farmers concerns that they are unreliable and too flexible, they are likely to eventually again become part of the workforce. In a race to the bottom, the SWP could be opened up to workers from nearby Asian countries such as Indonesia, Vietnam and Bangladesh—and development trends (and climate change) in those countries would suggest possible interest. After all, the small Asian state of Timor-Leste is part of the scheme and many Asian workers are familiar with agricultural work. Promoting the migration of workers has long been employed by various Asian governments, and there, as in the Caribbean and Ireland, migration has not been a development panacea but remains a temporarily permanent policy and practice. Equally, more places could be given to WHMs from Asia and elsewhere. The argument for not opening up is threefold: first, numbers are small and demand can be presently satisfied from the Pacific; second, to switch to Asia would threaten Australia's strategic interests; and, third, taking a few thousand workers from Asian states would have no significant impact on their economic status (Howes, 2018). Nonetheless were particular political and environmental changes to occur greater pressure would most likely be felt.

Conversely, some PICs may no longer seek seasonal employment or flows might reverse. Both the two small Pacific New Zealand dependencies of the Cook Islands and Niue were bringing in workers from Indonesia (more than 200 in the Cook Islands), rather than turn to nearby PICs, for various semi-skilled activities (PacNews, 17 July 2019; Radio New Zealand, 28 July 2016), an indication of the unavailability of any such surpluses in the region. Early in 2018, the Fiji government was attracting both praise and criticism over plans to introduce Asian seasonal workers to cut sugar cane and undertake construction work. More remarkably, late in 2019, Fiji introduced a group of 19 ni-Vanuatu men to become cane cutters, when no local workers could be found (RNZ, 22 November 2019). Samoa too was experiencing shortages of agricultural labour (Chap. 9). While that too reflected unattractive wages and conditions, it suggested a lack of obviously surplus labour. Small parts of the Pacific labour market were shifting direction.

Since the 1980s, workers in western economies have experienced rising inequality and growing economic polarisation has left many people having to accept poorly paid jobs, short-term contracts, casual employment and a gig economy: broadly a precariat that faces insecurity. It has taken a pandemic for this to become obvious, and undervalued workers in several sectors (in such 'humble' and often 'non-standard' jobs as cleaners, shelf stackers, delivery drivers and carers) to be recognised as invaluable. Just as there is little indication that the wages and conditions of those 'menial' workers, invaluable in the pandemic, will substantially change, so there is even less indication that the status of agricultural workers will change, nor even that they will achieve greater agency. While the pandemic has shown that usually conservative governments have intervened in labour markets to an unprecedented degree—when recent history has demonstrated growing distaste for intervention—regulating and enforcing superior conditions for agricultural workers may prove to be two steps too far. That alone will ensure that the Pacific labour market will remain open. While COVID-19 incidentally provided a pause for thought about future directions, seasonal workers as key workers, other than in rhetoric, have been valued no more highly and rewarded accordingly. Despite Michael McCormack's possibility of falling in love in the vines, the stigma of seemingly unskilled agricultural employment remains, and any new labour market, domestic or international, is far from obvious.

The Future PICs: A Kind of Blue

Continuity of guestworker programmes depends on the PIC interest in and continued need for involvement. While PICs prefer development at home and some greater degree of self-sufficiency, where dependence on labour migration may not be needed, prospects for locally based strategies remain limited. A widespread shift away from agriculture has occurred despite pockets of intensification centred around local market access. Environmental changes are putting further pressure on the coral island states especially, and the margins of settlement in the Pacific are gradually contracting. Extending from Polynesia to Micronesia and Melanesia, deterritorialisation is becoming the norm, as the demographic balance shifts outwards (Connell & Corbett, 2016). The value of remittances has shaped a substantial outward urge—following aid fatigue and stagnant economic growth—that has drawn in PICs, like Vanuatu, hitherto less involved with international migration. The relevance of MIRAB has extended from the smallest states to Fiji and parts of Melanesia as migration grows in importance, while the SWP is a politically expedient means of dealing with youth discontent by providing a 'choice' to leave. Much as in the Caribbean, an outwards orientation seems unlikely to change. In the context of globalisation and technological advancements, Pacific people are more mobile than ever, travel more frequently and cover greater distances.

Beyond tourism and agriculture, deep sea mining and fisheries are increasingly seen as key components in an emerging 'blue economy' which, reflecting the greater presence of Small Island Developing States (SIDS) in international organisations, has gained popularity in donor circles in recent years. At the 48th Pacific Island Forum in 2017 leaders launched *The Blue Pacific* 'as a new narrative that calls for inspired leadership and a long-term Forum foreign policy commitment to act as one "Blue Continent"' (Keen et al., 2018). Such narratives empower Pacific islanders in providing a vision for overcoming structural challenges, which might combine exploiting economic niches, but also migration opportunities and subsequent remittances. While regional organisations have touted the idea of a future blue economy and claimed the Pacific as the region of the future, unless fishing delivers even greater rewards or mining seabed minerals becomes economically and environmentally feasible (both of which deliver financial rewards rather than employment growth), the future economies of the region will not be greatly different from the present ones.

Sustainable development is as challenging now as at any time in the past. As global economic shifts and climate change continue on their present trajectory, the development and maintenance of sustainable livelihoods will become more difficult and increase the need and demand for migration and resettlement. The PICs already face multiple challenges to creating sustainable development and employment growth, while coping with accelerated environmental change and its rise as a security threat. Papua New Guinea (PNG) and Fiji have the greatest potential to generate revenue from mining and most PICs can revitalise export agriculture. Fiji and Vanuatu especially, but also Tonga and Samoa, have experienced the loss of the valuable tourism sector in the wake of COVID-19. The PICs are increasingly dependent on remittances, which in turn are dependent on migration opportunities, but these too may be limited in future years. The contours of future aid delivery are hard to envisage, in an era of changing global security and resource concerns, but aid fatigue is less evident for small states, and location is one partial guarantor of continued external support.

The focus on international migration from the South Pacific has hitherto primarily centred on New Zealand and Australia (and, unlike Micronesia, to a much lesser extent the United States). Might there be superior (i.e. greater) opportunities for long-term and permanent migration to metropolitan countries? Might such opportunities extend to other countries, including the United States and Canada, where there has already been PIC migration? Are there opportunities to migrate to the United Kingdom, once a significant Pacific colonial power? Might Japan or even China be destinations? Possible opportunities may even exist within the Pacific region. The Commonwealth of the Northern Marianas at different times has employed migrant labour, in sectors ranging from agriculture to the casino industry. It might be possible for governments of countries like Nauru and Kiribati, fellow Micronesians, to negotiate for some such employment opportunities to be reserved for PICs. The World Bank's *At Home and Away* (2006) suggested Malaysia and South Korea as possible labour markets, yet Asia has a relative abundance of cheap labour. Conceivably migrants from the PICs may find other destinations, perhaps following the pattern of migration from Micronesia to the United States. Pacific islanders from all the independent PICs became eligible for seasonal employment in the United States from 2011, being able to acquire H-2A (agricultural work) and H-2B (non-agricultural work) visas, but there is no evidence that any have taken advantage of this, and American recruiters have not visited the Pacific. A few Solomon Islander workers had gone to Canada by 2018 to work in the Seasonal Agricultural Worker Program (SAWP) and in hospitality, and Canada was actively seeking out more skilled and unskilled workers (*Island Sun*, 7 November 2018). A year later more than 50 Solomon Islanders were employed in Canada mainly in agriculture but also in tourism and hospitality through the Canadian International Training and Education Corp (CITREC). That is a small but suggestive start. There may be more scope than is usually assumed for new mobility in the Asia-Pacific region, but limited immediate likelihood of that becoming significant or preferred in the PICs.

A Strategic Arena

Regardless of PIC aspirations, both the RSE and SWP have largely been dictated by metropolitan requirements in Australia and New Zealand. That will not change. Australia has always been concerned by events in its perceived 'backyard' though, as its establishment of the SWP after the RSE suggests, if rarely with the same enthusiasm or commitment as New

Zealand. Migration too was never part of any commitment. Indeed, Australia has largely ignored and discouraged migration from the Pacific islands whereas New Zealand has recognised its own position, socially and geographically, as a Pacific island state and absorbed Pacific islanders into the nation, and, visibly, into the political system. Many people of Pacific islander origin, resident in Australia, have arrived via New Zealand. Australia, contemplating at the turn of the century an 'arc of instability' in its Melanesian neighbours, but with a preference for a future in Asia, has been more likely to see PICs as problematic rather than partners, or simply to forget that they existed, an amnesia particularly reserved for PNG (Dorney, 2016). Unsurprisingly then Australia was drawn somewhat reluctantly into developing a seasonal workers scheme, that might play a small part in regional stability.

Almost two decades on, the rise of China has presented a new political environment, and a geographical expansion, of rather more than an 'arc of instability' and unwanted competition for hegemony, at a time when climate change has put new Pacific pressures on Australia. Diplomatic trends (as several PICs have broken longstanding ties with Taiwan to recognise China), a growing physical presence (evident in diverse shipping movements) or economic shifts with increased trade, aid (soft and hard loans, infrastructure provision and port construction) and investment (especially in the mining sector as in PNG) all part of its maritime Belt and Road initiative, have prompted PICs to 'look north' with greater frequency (Szadziewski, 2020). Rather later, the arrival of COVID brought rivalry through 'mask' and then 'vaccine diplomacy'. China's more obvious presence has raised regional diplomatic stakes, stimulating greater Australian interest (the 'step-up'), New Zealand interest (the 'reset'), renewed American interest (but particularly in Guam and the Micronesian Compact states) and even the return of the UK to the region (the 'Pacific Uplift') after a hiatus of several decades. France also expanded its interests in the region, becoming through New Caledonia and French Polynesia increasingly involved in regional development agencies such as the Forum Secretariat (where in earlier decades it was seen by several PICs as a pariah nation, the 'last colonial power'). At the same time, secessionism and violent repression in West Papua were increasing and the island of Bougainville had voted in 2019 for an independence that neither PNG nor Australia welcomed. Quite simply, in the 2010s, the Pacific became of greater global and regional interest than at any time since the Second World War. The 'blue continent' became an increasingly contested space.

Such developments prompted Australia and New Zealand to seek to renew and develop their presumed status as trusted partners of PICs, to step up and step in through extended aid programmes. Concern and interests could easily become overbearing. Bureaucrats and politicians, including the former Prime Minister, Kevin Rudd, even canvassed the idea of Australia offering citizenship and residency rights to smaller Pacific island states, such as Tuvalu, Nauru and Kiribati, within a loose Compact of Free Association, analogous to that between the United States and the Micronesian states, or that between Australia and Norfolk Island, whereby they would give up sovereignty to obtain economic support, employment opportunities and defence protection from Australia (Colton, 2017). That call was quickly rejected by the Prime Minister of Tuvalu as 'imperial thinking' and a recolonisation of the Pacific incompatible with the interests of the island states. Nonetheless, the negotiation of PACER-Plus and the reference to 'economic integration' on the 2021 Department of Foreign Affairs and Trade (DFAT) website indicate the pressure on PICs, while the United States has been encouraged to extend its Free Association arrangements for the Micronesian states to Nauru, Kiribati and Tuvalu (Gray, 2021).

Australia remains the largest provider of development assistance in the region, yet, because of its failure to develop policies that adequately challenge climate change, it has been strongly criticised by the PICs that see New Zealand as a more reliable and sympathetic partner. Indeed, globally Australia is a less generous aid donor than most OECD countries and also ranks poorly in aid allocated for climate change adaptation (Wood et al., 2021). At the end of 2020, the Fijian Prime Minister, Frank Bainimarama, rejected the notion that 'Fijians and our Pacific Island sisters and brothers be some sacrificial canary for coal-burning countries' (quoted in *Sydney Morning Herald*, 12 December 2020). Australia is deemed to have taken inadequate response (Bradshaw, 2019). The urgency, ethics and recognition of climate change migration as a 'human rights' issue are of critical importance to PICs where people are now at risk of climate change displacement, already on a small scale in some coastal areas. Not only does this have obvious moral overtones, but it is an element of international security (Wyett, 2014). That has undercut Australia's step up and its aim of strategic denial.

Aid was confounded with labour migration, in Michael McCormack's view that the PICs will survive because 'their workers come here and pick our fruit' made while the Pacific Islands Forum, attended by the Prime Minister, in the atoll state of Tuvalu, was focused on climate change. As Victoria Stead (2019b) phrased it, the workers are not in Australia 'as an expression of our benevolence, they are here because they are vital to the productivity and profits of the industry … many working under conditions that would not pass muster under Australian industrial law'. The views of McCormack conflated national security and economic issues with a crude distaste for difference and a preference for a reserve army of labour rather than any mitigation of climate change in vulnerable countries. To consolidate PIC perspectives several island states donated aid to Australia in the midst of the 'climate-change' driven east coast bushfire disasters of 2019–2020. Vanuatu alone pledged A$250,000 to support relief and recovery efforts while Fiji and PNG sent military engineering forces: a reciprocity that mocked the words of the Deputy Prime Minister.

Both an expansionist China and climate change have provided PICs with leverage, which they have increasingly sought to use collectively and to assert greater agency (Morgan, 2018). While Australia could find it difficult to match the largesse of China in the region, it has soft power advantages in education, language and Christianity, and one further obvious advantage that of PIC access to its labour market. It costs little to further liberalise and expand the scheme, as is happening, and gain a small strategic 'win': a limited response to powerful on climate change. The SWP and PLS took their place in regional strategy and a more 'personal' engagement with the Pacific 'family' to which the Australian Prime Minister often referred. Ironically, much as with bushfire relief, the SWP scheme constituted a donation of aid from the PICs to Australia, where the value of the agricultural output of migrant workers was greater than the remittances that travelled home.

While the SWP contributes to growth and development, it does rather more than send (hard-earned) Australian dollars to much poorer states. Perhaps even perversely, the PICs send young men and women—often in the prime of their working lives—to ensure that such a basic industry as food production is able to continue. These workers have been nurtured, educated and kept healthy in, and at the cost of, the PICs, and agents and employers select the best of them for overseas employment: youthful home-made workers without their costs of production, willing to work cheaply in sometimes unpleasant jobs (and thus temporarily unavailable for their own national development), with a strong work ethic and who go home when their contribution is over (or return again to use now enhanced skills to make a superior contribution). Were it not for pressures from Australian farmers, this 'reverse aid' programme might never have existed or been restored so quickly after the onset of COVID-19. Here, in a region where strategic matters have become more important than ever, the triple win can very easily be reinterpreted.

A Widening Sphere?

International migration has deferred and mitigated, but not resolved, issues of poverty and development. The combination of scarce resources, weak economies, overburdened bureaucracies, urban unemployment, fractured social networks and uneven development challenges notions of sustainable development. In conventional economic terms most PICs are likely to remain weak for the foreseeable future, become increasingly dependent on the wider world and require some degree of external support. International migration constitutes one increasingly less hesitant solution: an expanding and unsatisfied outward urge, a necessary bottom-up globalisation that will always be both uneven and somewhat unsatisfying. Demand for opportunities is unlikely to diminish but continue to exceed the supply. Without substantial and improbable changes in Australian agriculture, especially in a socio-economic context of employment that would challenge powerful domestic economic and political interests and increase the price of food in Australia, a demand for cheap and pliable labour from PICs will continue. What was true 15 years ago seems no less valid now: 'as long as considerable economic challenges face island states, as their population growth rates remain above world averages, as development prospects are few, as the possibility of declining aid becomes more apparent, and as expectations rise, the ability to migrate will be crucial' (Connell & Brown, 2005, p. 55). By default it is valuable, while in a post-COVID world environmental change is likely to enhance all these trends.

Despite government intervention and regulation, throughout the PICs most migration is of islanders making their own decisions, and usually either moving into already overstretched urban centres or to cities overseas. Pacific people have used multiple strategies to achieve mobility: kinship, lotteries, New Zealand's Pacific Access Category, national agreements, educational scholarships, temporary work schemes and even illegal overstaying. Simultaneously government interventions in favour of migration have increased and been oriented to metropolitan needs.

While the diversity of migration possibilities has increased, and skilled migration expanded, the legal rights and entitlements of Pacific island migrants have been unchanged—including the right to stay—exemplifying stratification, inequality and persistent precarity. Temporary agricultural workers are denied permanency but encouraged to return, to take advantage of their acquired skills, becoming 'permanently temporary', both in being away more than at home and in never securing the rights of permanent migrants, a situation that does nothing for equity, challenges the etymology of 'guest' workers and questions the morality of hosts. Indeed, as Cindy Hahamovitch (2008, p. 770) has suggested, 'The greatest problem with the term, of course, is that guestworkers have so rarely been treated as guests'.

Demand for migration is therefore likely to become even more important in the future for a combination of social, economic and environmental reasons. Whatever forms of migration develop it is necessary to create an attractive political, social and economic environment that encourages migrants to remain in contact and be involved in their home countries and islands, to return and to remit and, where feasible, to make investments there. National strategies in the PICs are very slowly evolving in this direction. Yet, as migratory numbers and aspirations have grown, reaction to migration has become particularly strident, evident in the populism of Brexit and Donald Trump's wall against Mexico, and a lack of solidarity and empathy, highlighting the political ground-swell against globalisation, marked by draconian policies towards refugees. In this context temporary guestworkers, or what Claudia Tazreiter et al. (2016) call 'less desired non-citizens', are valuable and permanency is a threat.

Migration has never been easy and there has rarely been a good time to be a migrant while the considerable extent of global opposition to migration seems unlikely to decline in the near future, however beneficial migrants may be. Managing migration to the benefit of all involved—countries, people and economic enterprises—is one of the pressing issues of our time. Borders are hardening and legal restrictions increasing, at the same time as both greater but selective demand for migrants, where flexible labour markets offer neither stable nor well-paid jobs. It is in this context that the present and future of Pacific migration exists. Migration cannot be a development panacea and a judicious combination of policies is needed to secure a 'migration-development nexus' (Storey & Steinmayer, 2011). This must be based on reciprocal relationships between countries and places of origin and destination. On the margins especially, few opportunities exist for PICs and their residents to establish some degree of 'migration with dignity'—where migration satisfies a range of needs and can be suitably planned and managed—but the politics of migration will become increasingly sensitive and critical. At present, it continues to favour Australia.

Migration and development issues have become more challenging after COVID-19 and in the face of climate change, despite uncertainties about their impacts. If PICs can develop national economies where employment opportunities are growing, population growth is slowed, and harmful environmental change mitigated and reduced, the 'rush to migration' will be rather less urgent and less likely to end in precarity. That scarcely seems imminent. While inevitable COVID uncertainty raises questions about what a 'new normal' will be like, the rush to get guestworkers back into Australia (and New Zealand) suggests that it will be little different from the old familiar normal, and that in terms of migration and development very little will change (Connell & Campbell, 2021). Ultimately, there is a distinct irony in workers from Vanuatu still migrating to Australia a century and a half after blackbirding began and still caught up in an asymmetrical, ephemeral and occasionally exploitative relationship as development appears elusive at home. The 'new age' of migration has prominent antecedents.

Bibliography

Boese, M., & Moran, A. (2021). The regional migration-development Nexus in Australia: What migration? Whose development? *Frontiers in Sociology, 6*, 1–13.

Bradshaw, S. (2019). *Save Tuvalu, save the world: The climate crisis and the Pacific*. Oxfam Australia.

Colton, G. (2017, January). Time for Australia to Forge Free Compact Agreements in the Pacific. *The Interpreter*.

Connell, J., & Brown, R. (2005). *Remittances in the Pacific: An overview*. ADB.

Connell, J., & Campbell, Y. (2021). Aftermath. Towards a 'New Normal'. In Y. Campbell & J. Connell (Eds.), *COVID in the Islands: The Pacific and the Caribbean* (pp. 517–528). Palgrave Macmillan.

Connell, J., & Corbett, J. (2016). Deterritorialisation. Reconceptualising development in the Pacific Islands. *Global Society, 30*(4), 583–604.

Dorney, S. (2016). *The embarrassed colonialist*. Lowy Institute and Penguin Random House.

Dufty-Jones, R. (2014). Rural economies in the 'Age of Migration': Perspectives from OECD countries. *Geography Compass, 8*(6), 368–380.

Gray, A. (2021, May 7). How the US can protect the sovereignty of the smallest Pacific Islands. *The Diplomat*.

Hahamovitch, C. (2008). 'The Worst Job in the World': Reform, revolution and the secret rebellion in Florida's cane fields. *Journal of Peasant Studies, 35*(4), 770–800.

Hao'uli, E. (2013). Triple wins or Trojan horse? Examining the recognised Seasonal Employer Scheme under a TWAIL lens. *New Zealand Yearbook of International Law, 11*, 183–220.

Howes, S. (2018, May 18). *Asia or the Pacific: Where will we source our agricultural labour?* ANU DevPolicyBlog.

Howes, S. (2019, December 12). *Time for a permanent Australian step-up in Pacific labour mobility*. ANU DevPolicyBlog.

Johnson, J. (1967). Harvest migration from nineteenth-century Ireland. *Transactions of the Institute of British Geographers, 41*, 97–112.

Keen, M., Schwarz, A.-M., & Wini-Simeon, L. (2018). Towards defining the blue economy: Practical lessons from Pacific Ocean governance. *Marine Policy, 88*, 333–341.

Lenard, P., & Straehle, C. (2011). Temporary labour migration, global redistribution, and democratic justice. *Politics, Philosophy and Economics, 11*(2), 206–230.

Mares, P. (2012). *Temporary migration and its implications for Australia*. mimeo.
Morgan, W. (2018). Back on the map: Pacific Islands in a new era of strategic competition. USP SGDIA Working Paper No. 5, Suva.
Morgan, W. (2020, November 26). *Pacer plus: A third view*. ANU DevPolicyBlog.
Nishitani, M., & Lee, H. (2019). Becoming 'Overstayers': The coloniality of citizenship and the resilience of Pacific Farm Workers. In V. Stead & J. Altman (Eds.), *Labour lines and colonial power: Indigenous and Pacific Islander Labour Mobility in Australia* (pp. 159–184). ANU Press.
Phillips, N. (2009). Migration as development strategy? The new political economy of dispossession and inequality in the Americas. *Review of International Political Economy, 16*(2), 231–259.
Polanco, G. (2019). Migration regimes and the production of (labor) unfreedom. *Journal of Asian American Studies, 22*(1), 11–30.
Ramasamy, S., Krishnan, V., Bedford, R., & Bedford, C. (2008). The recognised seasonal employer policy: Seeking the elusive triple wins for development through international migration. *Pacific Economic Bulletin, 23*(3), 171–186.
Reilly, A. (2011). The ethics of seasonal labour migration. *Griffith Law Review, 20*(1), 127–152.
Rodman, M. (1993). Keeping options open: Copra and fish in rural Vanuatu. In T. Harding & B. Wallace (Eds.), *Contemporary Pacific societies* (pp. 171–184). Prentice Hall.
Rogaly, B. (2021). Commentary: Agricultural racial capitalism and rural migrant workers. *Journal of Rural Studies, 88*, 527–531.
Rose, M. (2020, January 28). *The Pacific Labour Scheme needs a pathway to permanency*. : ANU DevPolicy Blog.
Smith, A. (2015). Troubling 'Project Canada': The Caribbean and the making of 'unfree migrant labor'. *Canadian Journal of Latin American and Caribbean Studies, 40*(2), 274–293.
Stead, V. (2019a). Money trees, development dreams, and colonial legacies in contemporary Pasifika Horticultural Labour. In V. Stead & J. Altman (Eds.), *Labour lines and colonial power: Indigenous and Pacific Islander Labour Mobility in Australia* (pp. 133–158). ANU Press.
Stead, V. (2019b, August 18). Picking fruit is work, not benevolence, and doesn't absolve Australia of climate responsibility, *Guardian*.
Stead, V. (2021). Precarity's reach: Intersections of history, life, and labour in the Australian horticultural industry. *Journal of the Royal Anthropological Institute, 27*(2), 303–320.

Storey, D., & Steinmayer, V. (2011). Mobility as a development strategy: The case of the Pacific Islands. *Asia-Pacific Population Journal, 26*(4), 57–72.

Szadziewski, H. (2020). Converging anticipatory geographies in Oceania: The Belt and Road Initiative and Look North in Fiji. *Political Geography, 77*, 102–119.

Tazreiter, C., Weber, L., Pickering, S., Segrave, M., & McKernan, H. (2016). *Fluid Security in the Asia Pacific* (pp. 103–127). Macmillan.

Villafuerte, L. (2018). *Seasons of change: Ni-Vanuatu and the Recognised Seasonal Employers (RSE) scheme*. PhD thesis, University of Auckland.

Walsh, J. (2014). From nations of immigrants to states of transience: Temporary migration in Canada and Australia. *International Sociology, 29*(6), 584–606.

Williams, A. (2010). *The experiences of the recognised seasonal employer scheme and Ni-Vanuatu temporary workers in Te Puke, New Zealand.* Master of Planning thesis, University of Otago, .

Wood, T., Dornan, M., & Muller, S. (2021). *Change and continuity in Australian aid: What the aid flows show*. Development Policy Centre.

World Bank. (2006). *At home and away. Expanding job opportunities for Pacific Islanders through labour mobility*. World Bank.

Wyett, K. (2014). Escaping a rising tide: Sea level rise and migration in Kiribati. *Asia and the Pacific Policy Studies, 1*, 171–185.

Bibliography

Abbott, D. (2007). Poverty and pro-poor policies for Pacific Island countries. *Asia-Pacific Population Journal, 22*(3), 59–74.

Abbott, D., & Pollard, S. (2004). *Hardship and poverty in the Pacific: Strengthening poverty analysis and strategies in the Pacific*. Asian Development Bank.

Abbott, D., & Pollard, S. (2020, March 2 and 3). Mired in MIRAB: Migration and remittances. ANU DevPolicyBlog.

Able, S. (2010). Letter. *SPC Fisheries Newsletter, 133*, 27–28.

Adams, R. (1990). Plantation Labour in Vanuatu. In C. Moore, J. Leckie, & D. Munro (Eds.), *Labour in the South Pacific* (pp. 140–143). James Cook University.

Addinsall, C., Rihai, N., & Nasse, A. (2020). Agroecology and sustainable livelihoods as a framework to empower rural Ni-Vanuatu women. In L. Singh-Peterson & M. Carnegie (Eds.), *Integrating gender in agricultural development. Learnings from South Pacific contexts* (pp. 105–120). Bingley.

Agar, C., & Manolchev, C. (2020). Migrant labour as space: Rhythmanalysing the agri-food industry. *Organization, 27*(2), 252–271.

AIDAB. (1987). *Australia's relations with the South Pacific*. International Development Issues No. 2.

Albert, S., Bronen, R., Tooler, N., Leon, J., et al. (2018). Heading for the hills: Climate-driven community relocations in the Solomon Islands and Alaska

K. Petrou, J. Connell, *Pacific Islands Guestworkers in Australia*,
https://doi.org/10.1007/978-981-19-5387-3

provide insight for a 1.5°C future. *Regional Environmental Change, 18*(8), 2261–2272.

Allen, B., Bourke, M., & Gibson, J. (2005). Poor rural places in Papua New Guinea. *Asia Pacific Viewpoint, 46*(2), 201–217.

Allen, M. (1968). The establishment of Christianity and cash cropping in a New Hebridean community. *Journal of Pacific History, 3*, 25–46.

Allen, M. (2020). Framing food security in the Pacific Islands: Resilience in Malo, Vanuatu. In J. Connell & K. Lowitt (Eds.), *Food security in Small Island states* (pp. 91–110). Springer.

Anderson, B. (2010). Migration, immigration controls and the fashioning of precarious workers. *Work, Employment and Society, 4*(2), 300–317.

Anderson, J. (2021). Managing labour migration in Malaysia: Foreign workers and the challenges of 'control' beyond liberal democracies. *Third World Quarterly, 42*(1), 86–104.

Anderson, J., & Franck, A. (2019). The public and the private in Guestworker Schemes: Examples from Malaysia and the US. *Journal of Ethnic and Migration Studies, 45*(7), 1207–1223.

Asian Development Bank. (2003). *Priorities of the people: Hardship in Vanuatu*. ADB.

Asian Development Bank. (2009). *Kiribati Social and Economic Report 2008: Managing development risk*. ADB.

Australia. (1978). *Australia and the South Pacific*. Senate Standing Committee on Foreign Affairs and Defence.

Australia. (1979). *Australia and the Third World*. Senate Standing Committee on Foreign Affairs and Defence.

Australia. (2006). *Perspectives on the future of the harvest labour force*. Senate Standing Committee on Employment, Workplace Relations and Education.

Australian Government. (2003). A Pacific Engaged. Australia's relations with Papua New Guinea and the Island states of the Southwest Pacific. .

Australian Government. (2019). *Report of the Migrant Workers' Taskforce*. Commonwealth of Australia.

Australian Government. (2020). *Growing Australia*. House of Representatives Standing Committee on Agriculture and Water Resources.

Australian Government. (2021). *Harvest trail. Work your way around Australia*. Department of Education, Skills and Employment.

Australian Government. (n.d.). *Working and living in Australia: Pre-departure guidebook for seasonal workers*. AGPS.

Axelsson, L., & Hedberg, C. (2018). Emerging topologies of transnational employment: 'Posting' Thai workers in Sweden's wild berry industry beyond regulatory reach. *Geoforum, 89*, 1–10.

Ayson, R. (2007). The 'arc of instability' and Australia's strategic policy. *Australian Journal of International Affairs, 61*(2), 215–231.

Baak, P. (1999). About enslaved ex-slaves, uncaptured contract coolies and Unfreed Freedmen. *Modern Asian Studies, 33*(1), 121–157.

Bailey, R. (2013). *Ni-Vanuatu in the Recognised Seasonal Employer Scheme: Impacts at home and away*. ANU SSGM Discussion Paper 2013/4.

Bailey, R. (2014). *Working the Vines: Seasonal migration, money, and development in New Zealand and Ambrym, Vanuatu*. PhD thesis, University of Otago, Dunedin.

Bailey, R. (2018a, April 9). RSE workers ten years on. ANU DevPolicyBlog.

Bailey, R. (2018b, July 12). *Blacklisting seasonal workers*. ANU DevPolicyBlog.

Bailey, R. (2019a, April 26). *Pacific workers in rural Victoria*. ANU DevPolicyBlog.

Bailey, R. (2019b). *New Zealand's Recognised Employer Scheme (RSE): 10 year longitudinal case study*. ANU Department of Pacific Affairs.

Bailey, R. (2019c). *Limiting possible exploitation in transportation services for seasonal workers*. ANU InBrief 2019/12.

Bailey, R. (2019d). *Ni-Vanuatu 'Making Business' through Seasonal Worker Programs*. ANU InBrief 2019/13.

Bailey, R. (2021). Mobilising pathways through cross-border labour schemes, DPA Seminar Series, 2 September 2021, ANU [online].

Bailey, R., & Ng Shiu, R. (2016). *New responses to natural disasters through Seasonal Labour Mobility Programs*. ANU InBrief 2016/12.

Bailey, R., & Rereman, J. (2019, January 8). Vanuatu's seasonal workers: Where are they from? ANU DevPolicyBlog.

Bailey, R., & Sorensen, M. (2019, January 31). Labour mobility and medical mobility: Lessons from Vanuatu. ANU DevPolicyBlog.

Bailey, R., & Wells, S. (2016). *Temporary Labour Schemes, the new avenue of social mobility: Case study Vanuatu*. ANU InBrief 2016/21.

Bakewell, O. (2008). 'Keeping them in their place': The ambivalent relationship between development and migration in Africa. *Third World Quarterly, 29*(7), 1341–1358.

Ball, R. (2010). Australia's Pacific Seasonal Worker Pilot Scheme and its interface with the Australian horticultural labour market: Is it time to refine the policy? *Pacific Economic Bulletin, 25*(1), 114–130.

Banivanua Mar, T. (2019). 'Boyd's Blacks': Labour and the making of settler lands in Australia and the Pacific. In V. Stead & J. Altman (Eds.), *Labour lines and colonial power: Indigenous and Pacific Islander labour mobility in Australia* (pp. 57–73). ANU Press.

Banivanua-Mar, T. (2007). *Violence and colonial dialogue. The Australian-Pacific Indentured Labour Trade*. University of Hawai'i Press.

Barclay, K. (2010). Impacts of tuna industries on coastal communities in Pacific Island countries. *Marine Policy, 34*, 406–413.

Barlass, T. (2021, November 14). *National park to be renamed over slave link* (p. 2). Sun-Herald.

Barnes, N. (2013). Is health a labour, citizenship or human right? Mexican seasonal agricultural workers in Leamington. *Canada. Global Public Health, 8*(6), 654–669.

Barry, K. (2020). Momentarily immobile: Backpacking, farm work, and hostels in Bundaberg, Australia. *Geographical Research, 59*(1), 46–55.

Basok, T. (2002). *Tortillas and tomatoes: Transmigrant Mexican Harvesters in Canada*. McGill-Queen's University Press.

Basok, T. (2003). Mexican seasonal migration to Canada and development: A community-based comparison. *International Migration, 41*(2), 3–26.

Bastin, R. (1985). Weasisi mobility: A committed rural proletariat? In M. Chapman & R. Prothero (Eds.), *Circulation in population movement. Substance and concepts from the Melanesian Case* (pp. 175–190). Routledge and Kegan Paul.

Bauder, H. (2005). Landscape and scale in media representations: The construction of Offshore Farm Labour in Ontario, Canada. *Cultural Geographies, 12*(1), 41–58.

Bauder, H. (2008). Foreign farm workers in Ontario (Canada): Exclusionary discourse in the newsprint media. *Journal of Peasant Studies, 35*, 100–118.

Beckford, C. (2016). The experiences of Caribbean Migrant Farmworkers in Ontario, Canada. *Social and Economic Studies, 65*(1), 153–188.

Bedford, C. (2013). *Picking winners?: New Zealand's recognised seasonal employer (RSE) policy and its impacts on employers, Pacific workers and their island-based communities.* PhD thesis, The University of Adelaide.

Bedford, C. (2020, November 24). *New Zealand's seasonal labour shortage: Why local workers aren't enough*. ANU DevPolicyBlog.

Bedford, C. (2021, March 24). *Supporting NZ's seasonal workers to remit and save*. ANU DevPolicyPlog.

Bedford, C., Bedford, R. & Nunns, H. (2020). *RSE Impact Study: Pacific stream report*. Bedford Consulting.

Bedford, R. (1973a). A transition in circular mobility: Population movement in the New Hebrides, 1800–1970. In H. Brookfield (Ed.), *The Pacific in transition* (pp. 187–228). Arnold.

Bedford, R. (1973b). *New Hebridean mobility: A study of circular migration*. ANU Department of Human Geography Publication HG/9.

Bedford, R., & Bedford, C. (2017). *RSE earnings and remittance surveys: Samoans, Tongans and Ni-Vanuatu employed in Hawke's Bay, the Bay of Plenty and Marlborough*. Ministry of Business, Innovation and Employment.

Bedford, R., Bedford, C., & Ho, E. (2010). Engaging with New Zealand's recognised seasonal employer work policy: The case of Tuvalu. *Asian and Pacific Migration Journal, 19*, 421–445.

Bedford, R., Bedford, C., Wall, J., & Young, M. (2017). Managed temporary labour migration of pacific islanders to Australia and New Zealand in the early twenty-first century. *Australian Geographer, 48*(1), 37–57.

Bedford, R., & Hugo, G. (2011). *Population movement in the Pacific: A perspective on future prospects*. Department of Labour.

Bell, J., Johnson, J., & Hobday, A. (2011). *Vulnerability of tropical Pacific fisheries and aquaculture to climate change*. SPC.

Bell, J., Taylor, M., Amos, M., & Andrew, N. (2016). *Climate change and Pacific Island Food Systems*. CCAFS and CTA.

BellchambersBarrett. (2020, December 9). *Pacific labour mobility accommodation review: Key findings and recommendations*. BellchambersBarrett.

Berg, L., & Farbenblum, B. (2017). *Wage theft in Australia: Findings of the national temporary migrant work survey*. Migrant Worker Justice Initiative.

Berg, L., & Farbenblum, B. (2020). *International students and wage theft in Australia*. Migrant Worker Justice Initiative.

Bertram, G. (1986). 'Sustainable development' in Pacific micro-economies. *World Development, 14*(7), 809–822.

Bertram, G., & Watters, R. (1985). The MIRAB economy and South Pacific microstates. *Pacific Viewpoint, 26*(3), 497–519.

Binford, A. (2019). Assessing temporary foreign worker programs through the prism of Canada's Seasonal Agricultural Worker Program: Can they be reformed or should they be eliminated? *Dialectical Anthropology, 43*(4), 347–366.

Binford, L. (2002). Social and economic contradictions of rural migrant contract labor between Tlaxcala. *Mexico and Canada. Culture & Agriculture, 24*(2), 1–19.

Binford, L. (2003). Migrant remittances and (under) development in Mexico. *Critique of Anthropology, 23*(3), 305–336.

Binford, L. (2009). From fields of power to fields of sweat: The dual process of constructing temporary migrant labour in Mexico and Canada. *Third World Quarterly, 30*(3), 503–517.

Binford, L. (2013). *Tomorrow we're all going to the harvest: Temporary foreign worker programs and neoliberal political economy*. University of Texas Press.

Blanco, C. (2009). *Temporary migration, remittances and development: Early impressions of the Australian Pacific seasonal worker pilot scheme.* Honours thesis, University of Sydney, School of Geosciences.

Boese, M. (2015). The roles of employers in the regional settlement of recently arrived migrants and refugees. *Journal of Sociology, 51*(2), 401–416.

Boese, M., & Moran, A. (2021). The regional migration-development Nexus in Australia: What migration? Whose development? *Frontiers in Sociology, 6*, 1–13.

Böhning, W. (1984). *Studies in international labour migration.* Macmillan.

Bonnemaison, J. (1985). Territorial control and mobility within ni-Vanuatu societies. In M. Chapman & R. Prothero (Eds.), *Circulation in population movement: Substance and concepts from the Melanesian case* (pp. 57–79). Routledge and Kegan Paul.

Bonnemaison, J. (1994). *The tree and the Canoe. History and ethnography of Tanna.* University of Hawaii Press.

Borovnik, M. (2007). Labor circulation and changes among seafarers' families and communities in Kiribati. *Asian and Pacific Migration Journal, 16*(2), 225–249.

Borovnik, M. (2011). Occupational health and safety of merchant seafarers from Kiribati and Tuvalu. *Asia Pacific Viewpoint, 52*(3), 333–346.

Bourke, R. (2019). Subsistence food production in Melanesia. In E. Hirsch & W. Rollason (Eds.), *The Melanesian World* (pp. 143–163). Routledge.

Bradshaw, S. (2019). *Save Tuvalu, save the world: The climate crisis and the Pacific.* Oxfam Australia.

Brass, T., & Bernstein, H. (1992). Introduction: Proletarianisation and deproletarianisation on the colonial frontier. *Journal of Peasant Studies, 19*(3), 1–40.

Breman, J. (1990). *Taming the coolie beast. Plantation society and the colonial order in Southeast Asia.* Oxford University Press.

Brickenstein, C. (2014). *Foreign Seasonal Workers in Agriculture; a comparison of regulations and practices in the Pacific and in Europe.* PhD thesis, Macquarie University, Sydney.

Brickenstein, C. (2017). The development impact of seasonal worker schemes: A comparison of European and Pacific experiences. *Migration and Development, 6*(1), 60–82.

Bridi, R. (2015). Migrant workers in Ontario's tobacco belt: An examination of workplace dynamics. *Human Geography, 8*, 54–67.

Bridi, R. (2022). Spaces of alienation in the tobacco fields: The case of migrant workers in Ontario, Canada. *GeoJournal, 87*, 131–146.

Brookes, S. (2018, May 23). Agri Labour Australia suspended from government scheme amid underpayment, unsafe work claims. *ABC News.* https://www.

abc.net.au/news/2018-05-23/government-suspends-labour-hire-firm-amid-underpayment-claims/9788742.

Brookfield, H. (1968). 'The money that grows on trees': The consequences of an innovation within a man-environment system. *Australian Geographical Studies, 6*(2), 97–119.

Brookfield, H. (1972). *Colonialism, development and independence. The case of the Melanesian Islands in the South Pacific*. Cambridge University Press.

Brown, M. (2018, July 26). RSE football cup unites overseas vineyard labourers. *Stuff.co.NZ*. https://www.stuff.co.nz/sport/football/105736746/work-hard-play-hard-rse-football-cup-unites-vineyard-labourers.

Brown, R., & Connell, J. (2015). Migration and remittances: A multidisciplinary synthesis. In J. Connell & R. Brown (Eds.), *Migration and remittances* (pp. xiiv–lxviii). Edward Elgar.

Brown, R., Connell, J., Jimenez, E., & Leeves, G. (2006). Cents and sensibility: The economic benefits of remittances, in World Bank (2006b). In *At home and away* (pp. 47–99). World Bank.

Brown, R., Connell, J., & Jimenez-Soto, E. (2014). Migrants' remittances, poverty and social protection in the South Pacific: Fiji and Tonga. *Population, Space and Place, 20*(5), 434–454.

Cahn, M. (2008). Indigenous entrepreneurship, culture and micro-enterprise in the Pacific Islands: Case studies from Samoa. *Entrepreneurship & Regional Development: An International Journal, 20*(1), 1–18.

Callick, R. (2010, March 2). Pacific scheme withers on the vine. *The Australian*.

Callister, P., Badkar, J., & Williams, J. (2009a). Paid caregivers & domestic workers: Some policy issues in meeting future demand in New Zealand. *Policy Quarterly, 5*(3), 38–43.

Callister, P., Badkar, J., & Williams, J. (2009b). The future supply of caregivers for the elderly: Should Australia and New Zealand look to the Pacific? *Pacific Economic Bulletin, 24*(3), 162–172.

Calnitsky, N. (2016). Grape vines and orchard lines. *Left History: An Interdisciplinary Journal of Historical Inquiry and Debate, 20*(1), 11–35.

Cameron, E. (2011). *Social development outcomes of participation in the New Zealand Recognised Seasonal Employer (RSE) scheme for ni-Vanuatu seasonal migrant workers*. Master's thesis, Massey University, Palmerston North.

Campbell, I. (2019). Harvest labour markets in Australia: Alleged labour shortages and employer demand for temporary migrant workers. *Journal of Australian Political Economy, 84*, 46–88.

Campbell, I., Tranfaglia, M., Tham, J., & Boese, M. (2019). Precarious work and the reluctance to complain: Italian temporary migrant workers in Australia. *Labour and Industry, 29*(1), 98–117.

Carmo, R., & Hedberg, C. (2019). Translocal mobility systems: Social inequalities and flows in the wild berry industry. *Geoforum, 99*, 102–110.

Carnegie, M., & Jeffries, L. (2021). Pacific Labour Scheme: Preliminary examination into the implications and impacts of family accompaniment presented at Pacific Migration Research Workshop, 17th June 2021, ANU [online].

Castles, S. (1986). The guest-worker in Europe—An obituary. *International Migration Review, 20*(4), 761–778.

Castles, S. (1988). Temporary migrant workers—Economic and social aspects, in committee to advise on Australia's Immigration Policies. In *Immigration. A commitment to Australia* (pp. 1–10). Australian Government Publishing Service.

Castles, S. (1998). Globalization and migration: Some pressing contradictions. *International Social Science Journal, 50*(2), 179–186.

Castles, S. (2006). Guestworkers in Europe. A resurrection? *International Migration Review, 40*(4), 741–766.

Castles, S., & Ozkul, D. (2014). Circular migration: Triple win, or a new label for temporary migration? In G. Battistella (Ed.), *Global and Asian perspectives on international migration* (pp. 27–49). Springer.

Cavanough, E., & Wherrett, C. (2020). *Blue Harvest. Wage theft & other labour infringements in the NSW Mid-North coasts's 2019/20 berry harvest.* The McKell Institute.

Cave, D. (2012). *Digital Islands. How the Pacific's ICT revolution is transforming the region.* Lowy Institute.

Caxaj, S., & Diaz, L. (2018). Migrant workers' (non)belonging in rural British Columbia, Canada: Storied experiences of Marginal Living. *International Journal of Migration, Health and Social Care, 14*(2), 208–220.

Cecil, R., & Ebanks, G. (1991). The human condition of West Indian Migrant Farm labour in Southwestern Ontario. *International Migration, 29*(3), 389–405.

Cecil, R., & Ebanks, G. (1992). The Caribbean Migrant Farm Worker Programme in Ontario: Seasonal expansion of West Indian Economic Spaces. *International Migration, 30*(1), 19–37.

Centenera, N. (2016, December 1). *Mallala hospital becomes seasonal worker home* [Online]. Two Wells, South Australia: Two Wells and Districts Echo. Retrieved from https://www.twowellsecho.com.au/2016/12/01/mallala-hospital-becomes-seasonal-worker-home/

Chacko, E., & Price, M. (2021). (Un)settled sojourners in cities: The scalar and temporal dimensions of migrant precarity. *Journal of Ethnic and Migration Studies, 47*(20), 4597–4614.

Chambers, A. (1975/1984). *Nanumea.* : ANU Development Studies Centre.

Chand, S. (2006). Labour mobility for sustainable livelihoods in Pacific Island States. In M. Powles (Ed.), *Pacific futures* (pp. 127–141). Pandanus Books.

Chand, S. (2008). Biceps or the brain: Investing in our most valuable resource. *Islands Business, 34*(6), 40–41.

Chand, S. (2021). Luve ni Yali: Lost son of a seasonal worker. *Islands Business, 47*(5), 16.

Charan, D., Kur, M., & Singh, P. (2017). Customary land and climate change induces relocation—A case study of Vunidogoloa Village, Vanua Levu, Fiji. In W. Filho (Ed.), *Climate change adaptation in Pacific countries* (pp. 19–34). Springer.

Chattier, P. (2019). Beyond development impact: Gender and care in the Pacific Seasonal Worker Programme. *Gender and Development, 27*(1), 49–65.

Cheer, J., & Reeves, K. (2013). Roots tourism: Blackbirding and the South Sea Islander diaspora. *Tourism Analysis, 18*, 245–257.

Chin, R. (2007). *The guest worker question in postwar Germany.* Cambridge University Press.

Choithani, C., van Duijne, R., & Nijman, N. (2021). Changing livelihoods at India's rural–urban transition. *World Development, 146*, 105617.

Clemens, M., Graham, C., & Howes, S. (2015). Skill development and regional mobility: Lessons from the Australia-Pacific Technical College. *Journal of Development Studies, 51*(11), 1502–1517.

Clibborn, S., & Wright, C. (2018). Employer theft of temporary migrant workers' wages in Australia: Why has the State Failed to Act. *Economic and Labour Relations Review, 29*(2), 207–227.

Cohen, A. (2019). 'Slavery hasn't ended, it has just become modernized': Border imperialism and the lived realities of migrant farmworkers in British Columbia, Canada. *ACME: An International Journal for Critical Geographies, 18*(1), 130–148.

Cohen, A., & Hjalmarson, E. (2020). Quiet struggles: Migrant farmworkers, informal labor, and everyday resistance in Canada. *International Journal of Comparative Sociology, 61*(2–3), 141–158.

Colton, G. (2017, January). Time for Australia to Forge Free Compact Agreements in the Pacific. *The Interpreter.*

Connell, J. (1983a). *Migration, employment and development in the South Pacific. Country Report No. 7. Kiribati.* SPC and ILO.

Connell, J. (1983b). *Migration, employment and development in the South Pacific. Country Report No. 18. Tonga*. SPC and ILO.
Connell, J. (1985). Cocoa, copper and cash: Terminal, temporary and circular mobility in Siwai, North Solomons. In M. Chapman & R. Prothero (Eds.), *Circulation in population movement. Substance and concepts from the Melanesian Case* (pp. 119–148). Routledge and Kegan Paul.
Connell, J. (2003). Losing ground? Tuvalu, the Greenhouse effect and the garbage can. *Asia Pacific Viewpoint, 444*, 89–107.
Connell, J. (2006a). Migration, dependency and inequality in the Pacific: Old wine in bigger bottles? In S. Firth (Ed.), *Globalisation and governance in the Pacific Islands* (pp. 59–106). ANU E-Press.
Connell, J. (2006b). Nauru: The first Pacific failed state? *The Round Table, 95*(383), 47–63.
Connell, J. (2007). Towards free trade in the Pacific? The genesis of the 'Kava-Biscuit War' between Fiji and Vanuatu. *Geographical Research, 45*, 1–12.
Connell, J. (2008). Niue: Embracing a culture of migration. *Journal of Ethnic and Migration Studies, 34*, 1021–1040.
Connell, J. (2009). *The global care chain. From the Pacific to the World*. Routledge.
Connell, J. (2010). From blackbirds to guestworkers in the South Pacific. Plus ça change…? *The Economic and Labour Relations Review, 20*(2), 111–122.
Connell, J. (2012). Population resettlement in the Pacific: Lessons from a Hazardous History. *Australian Geographer, 43*, 127–142.
Connell, J. (2013). *Islands at risk*. Edward Elgar.
Connell, J. (2015). Vulnerable Islands: Climate change, tectonic change, and changing livelihoods in the Western Pacific. *The Contemporary Pacific, 27*(1), 1–36.
Connell, J. (2016). Last days in the Carteret Islands? Climate change, livelihoods and migration on coral atolls. *Asia Pacific Viewpoint, 57*(1), 3–15.
Connell, J. (2020). Lost roots? Fading food security in Micronesia. In J. Connell & K. Lowitt (Eds.), *Food security in Small Island states* (pp. 57–76). Springer.
Connell, J., & Brown, R. (2004). The remittances of migrant Tongan and Samoan Nurses in Australia. *Human Resources for Health, 2*(2), 2.
Connell, J., & Brown, R. (2005). *Remittances in the Pacific: An overview*. ADB.
Connell, J., & Campbell, Y. (2021). Aftermath. Towards a 'New Normal'. In Y. Campbell & J. Connell (Eds.), *COVID in the Islands: The Pacific and the Caribbean* (pp. 517–528). Palgrave Macmillan.
Connell, J., & Conway, D. (2000). Migration and remittances in island microstates: A comparative perspective on the South Pacific and the Caribbean. *International Journal of Urban and Regional Research, 24*(1), 52–78.

Connell, J., & Corbett, J. (2016). Deterritorialisation. Reconceptualising development in the Pacific Islands. *Global Society, 30*(4), 583–604.

Connell, J., & Hammond, J. (2009). *Iso Iso*: The first ni-Vanuatu guest workers in New Zealand. *Pacific Economic Bulletin, 24*(1), 83–95.

Connell, J., & Lea, J. (2002). *Urbanisation in the Island Pacific. Towards sustainable development*. Routledge.

Connell, J., & Lutkehaus, N. (2017a). Escaping Zaria's fire? The volcano resettlement problem of Manam Island, Papua New Guinea. *Asia Pacific Viewpoint, 58*(1), 14–26.

Connell, J., & Lutkehaus, N. (2017b). Environmental refugees? A tale of two resettlement projects in Papua New Guinea. *Australian Geographer, 48*(1), 79–95.

Connell, J., & Tabucanon, G. (2016). From Banaba to Rabi: A Pacific model for resettlement? In S. Price & J. Singer (Eds.), *Global implications of development, disasters and climate change* (pp. 91–107). Routledge.

Consterdine, E., & Samuk, S. (2018). Temporary migration programmes: The cause or antidote of migrant worker exploitation in UK agriculture. *Journal of International Migration and Integration, 19*, 1005–1020.

Cooper, E. (2020, December 2). 'Squatter-like' property owned by Burnie Mayor Steve Kons used to house Tongan workers. *ABC News*. https://www.abc.net.au/news/2020-02-12/burnie-seasonal-worker-accommodation-concern-from-union/11959286.

Corlett, E. (2019, December 12). RSE Scheme transformed the New Zealand fruit growing industry. *Radio New Zealand*. https://www.rnz.co.nz/national/programmes/eyewitness/audio/2018726025/.

Corrado, A., de Castro, C., & Perrotta, D. (Eds.). (2016). *Migration and agriculture. Mobility and change in the Mediterranean area*. Routledge.

Corris, P. (1970). Kaisulia of Ada Gege: A strongman in the Solomon Islands. In J. Davidson & D. Scarr (Eds.), *Pacific Islands portraits* (pp. 253–265). ANU Press.

Corris, P. (1973). *Passage, port and plantation. A history of Solomon Islands Labour Migration 1870–1914*. Melbourne University Press.

Costa, D., & Martin, P. (2018). *Temporary Labor Migration Programs*. Economic Policy Institute.

Cox, J. (2009). The money pit: An analysis of Nauru's phosphate mining policy. *Pacific Economic Bulletin, 24*(1), 174–186.

Craven, L. (2013). *The relationship between temporary migration and vulnerability to climate change in rural Vanuatu*. Honours thesis, University of Sydney Department of Government.

Craven, L. (2015). Migration-affected change and vulnerability in rural Vanuatu. *Asia Pacific Viewpoint, 56*(2), 223–236.

Craven, L., & Gartaula, H. (2015). Conceptualising the migration-food security Nexus: Lessons from Nepal and Vanuatu. *Australian Geographer, 46*(4), 455–471.

Cullen, R. (2012). Empire, Indian indentured labor and the colony: The Debate over 'coolie' Labor in New South Wales, 1836–1838. *History Australia, 9*, 84–109.

Cummings, M. (2013). Imagining transnational futures in Vanuatu. In A. Quayson & G. Daswani (Eds.), *A companion to diaspora and transnationalism* (pp. 381–396). Hoboken, New Jersey.

Cummings, M. (2016). Uncertain belongings: Relationships, money and returned migrant workers in Port Vila, Vanuatu. *Journal of Pacific Studies, 35*(3), 21–33.

Curry, G. (2005). Doing "Business" in Papua New Guinea: The social embeddedness of small business enterprises. *Journal of Small Business & Entrepreneurship, 18*(2), 231–246.

Curtain, R. (1980). The structure of internal migration in Papua New Guinea. *Pacific Viewpoint, 21*(1), 42–61.

Curtain, R. (2018). What can Papua New Guinea do to lift its numbers in the seasonal worker programs of Australia and New Zealand? ANU Development Policy Centre Discussion Paper 70.

Curtain, R., Dornan, M., Doyle, J., & Howes, S. (2016). *Pacific possible. Labour mobility: The ten million dollar prize*. World Bank and ANU.

Curtain, R., Dornan, M., Howes, S., & Sherrell, H. (2018). Pacific seasonal workers: Learning from the contrasting temporary migration outcomes in Australian and New Zealand Horticulture. *Asia and the Pacific Policy Studies, 5*, 462–480.

Curtain, R., & Howes, S. (2018, July 13). *Making the SWP more employer friendly: A work in progress*. ANU DevPolicyBlog.

Curtain, R., & Howes, S. (2019, June 28). *Tiny Tonga's SWP success*. ANU DevPolicyBlog.

Curtain, R., & Howes, S. (2020). *Governance of the seasonal worker programme in Australia and sending countries*. ANU Development Policy Centre.

Curtain, R., & Howes, S. (2021a). *Worsening employment outcomes for Pacific technical graduate job-seekers*. ANU Development Policy Centre Discussion Paper No. 91.

Curtain, R., & Howes, S. (2021b). *Helping APTC trades graduates to migrate to Australia under the TSS*. ANU Development Policy Centre Policy Brief No. 20.

Curtain, R., & Sherrell, H. (2017, February 2). *How recruitment and selection can shape seasonal work programs: Comparing Fiji and Papua*. ANU: DevPolicy Blog.

D'Arcy, M. (2011). *Away from subsistence: Food security and resource use on Paama, Vanuatu*. Honours thesis, University of Sydney School of Geosciences.

Dao, A., Green, M., & Kass, T. (2020, May 16). We feed you. *The Saturday Paper*. https://www.thesaturdaypaper.com.au/news/politics/2020/05/16/we-feed-you/15895512009836?fbclid=IwAR1X5r411aDQfb397ODGaQey8a3YtJ_hdIBAGnPLm0d4w5GXp9hE1XjDvn4.

Davidson, H., & Earl, C. (2020, October 18). The fruit pickers: Inside Australia's seasonal worker program, *The Guardian*.

Davidson, P. (2020, September 22). 'If JobSeeker was cut, the unemployed would be picking fruit'? Why that's not true. *The Conversation*. https://theconversation.com/if-jobseeker-was-cut-the-unemployed-would-be-picking-fruit-why-thats-not-true-145951

Davies, L. (2019). A regulated labour trade across the Torres Strait: Papuan and New Guinean Domestic Labourers in Australia. In V. Stead & J. Altman (Eds.), *Labour lines and colonial power: Indigenous and Pacific Islander Labour mobility in Australia* (pp. 75–102). ANU Press.

De Bres, J., Campbell, R., & Harris, P. (1974). *Migrant labour in the Pacific*. CORSO.

De Castro, C., Reigada, A., & Gadea, E. (2020). The devaluation of female labour in fruit and vegetable packaging plants in Spanish Mediterranean agriculture. *Organization, 27*(2), 232–250.

De Genova, N. (2009). Conflicts of mobility, and the mobility of conflict: Rightlessness, presence, subjectivity, freedom. *Subjectivity, 29*, 445–466.

de Haas, H. (2005). International migration, remittances & development: Myths and facts. *Third World Quarterly, 26*(8), 1269–1284.

de Haas, H. (2007). *Remittances, migration and social development: A conceptual review of the literature*. United Nations Research Institute for Social Development Social Policy & Development Programme Paper Number 34.

de Haas, H. (2010). Migration and development: A theoretical perspective. *International Migration Review, 44*(1), 227–264.

de Haas, H. (2012). The migration and development pendulum: A critical view on research and policy. *International Migration, 50*(3), 8–25.

de Haas, H., Castles, S., & Miller, M. (2020). *The age of Migration: International population movements in the modern world*. London: Red Globe Press.

de Moor, N. (2011). Temporary labour migration for victims of natural disasters: The Colombia-Spain Model. In M. Leighton, X. Shen, & K. Warner (Eds.), *Climate change and migration: Rethinking polices for adaptation and disaster risk reduction* (pp. 92–103). UNU-EHS.

Dean, A., Green, D., & Nunn, P. (2017). Too much sail for a small craft? Donor requirements, scale and capacity discourses in Kiribati. In E. Stratford (Ed.), *Island Geographies: Essays and conclusions* (pp. 54–77). Routledge.

Deeming, C., & Gubhaju, B. (2015). The mis-measurement of extreme global poverty: A case study in the Pacific Islands. *Journal of Sociology, 51*(3), 689–706.

Dennis, J. (2003). *Pacific island seafarers: A study of the economic and social implications of seafaring on dependents and communities*. South Pacific Commission.

Department of Agriculture & Rural Development and Department of Labour & Employment Services. (2020). *Framework for the reintegration of agricultural workers in Labour Mobility Programs 2021–2023*. Vanuatu Government.

DESE. (2021). Department of Education, Skills and Employment (dese.gov.au).

Docker, E. (1970). *The Blackbirders. The recruiting of South Seas Labour for Queensland* (pp. 1863–1907). Angus and Robertson.

Doherty, B. (2017, August 3). Modern day slavery in focus: Hungry, poor, exploited: Alarm over Australia's import of farm workers. *The Guardian*. https://www.theguardian.com/global-development/2017/aug/03/hungry-poor-exploited-alarm-over-australias-import-offarm-workers.

Dornan, M. (2018, March 16). *Vanuatu grapples with seasonal worker success*. ANU DevPolicyBlog.

Dornan, M., Howes, S., & Curtain, R. (2018, September 13). *Is a new visa for agricultural workers needed?* ANU DevPolicyBlog.

Dorney, S. (2016). *The embarrassed colonialist*. Lowy Institute and Penguin Random House.

Downes, A., & Clarke, K. (2007). *The Canadian seasonal agricultural workers program: The experience of Barbados*. Paper presented at the CSAWP conference, University of the West Indies, Barbados.

Doyle, J., & Howes, S. (2015). *Australia's Seasonal Worker Program: Demand-side constraints and suggested reforms*. World Bank.

Doyle, J., & Sharma, M. (2017). *Expanding employment pathways for Pacific Islanders in the Australian and New Zealand aged care sectors*. World Bank.

Dreby, J. (2010). *Divided by borders: Mexican migrants and their children*. University of California Press.

Dufty, N., Martin, P., & Zhao, S. (2019). *Demand for farm workers: ABARES farm surveys 2018*. ABARES.

Dufty-Jones, R. (2014). Rural economies in the 'Age of Migration': Perspectives from OECD countries. *Geography Compass, 8*(6), 368–380.

Dun, O., & Klocker, N. (2017). The migration of horticultural knowledge: Pacific Island seasonal workers in rural Australia—A missed opportunity? *Australian Geographer, 48*(1), 27–36.

Dun, O., McMichael, C., McNamara, K., & Farbotko, C. (2020). Investing in home: Development outcomes and climate change adaptation for seasonal workers living between Solomon Islands and Australia. *Migration and Development,* [In press].

Edwards, R. (2019). Thinking about brain drain. : ANU DevPolicyBlog, 3 September, 19 September, 14 October.

Ellerman, D. (2005). Labour migration: A developmental path or a low-level trap? *Development in Practice, 15*(5), 616–630.

Emberson-Bain, A. (1994). *Labour and Gold in Fiji.* Cambridge University Press.

Employment Services Unit. (2019). *Migrant Worker Disciplinary Policy*. Vanuatu Department of Labour.

Enoka, A. (2019). *Under the Gaze: A study of the portrayal by the New Zealand print media of Pacific Island workers in the Recognised Seasonal Employer (RSE) scheme, 2007–2012*, PhD thesis, Massey University, Palmerston North.

Eriksson, M., & Tollefsen, A. (2013). Disciplining temporary workers. Thai seasonal labourers and the Swedish berry industry. In M. Geiger & A. Pécoud (Eds.), *Disciplining the transnational mobility of people* (pp. 185–206). Palgrave Macmillan.

Fair Work Ombudsman. (2018). *Harvest Trail Inquiry*. AGPS.

Fair Work Ombudsman. (2019, April 23). Agri-Labour to pay migrant workers $50,000. Australian Government Fair Work Ombudsman. https://www.fairwork.gov.au/about-us/news-and-media-releases/2019-media-releases/april-2019/20190423-agri-labour-eu-media-release#twitter.

Faist, T. (2008). Migrants as transnational development agents: An inquiry into the newest round of the migration-development nexus. *Population, Space and Place, 41*, 21–42.

Falefou, T. (2017). *TOKU TIA: Tuvalu and the impacts of climate change*. PhD thesis, University of Waikato, Hamilton.

Farbenblum, B., & Berg, L. (2018). *Wage theft in silence.* Migrant Worker Justice Initiative.

Fatupaito, A., Utuva, L., Tauave, S., Alofipo, A., Meleisia, M., Schoeffel, P., Arthur, T., & Alexeyeff, K. (2021). Samoa's New Labour Trade. *Journal of Samoan Studies, 11*(1), 51–63.

Feist, H., Tan, G., McDougall, K., & Hugo, G. (2014). *Enabling rural migrant settlement: A case study of the Limestone Coast*. Australian Population and Migration Research Centre, University of Adelaide.

Fels, A., & Cousins, D. (2019). The migrant workers' taskforce and the Australian government's response to migrant worker wage exploitation. *Journal of Australian Political Economy, 84*, 13–45.

Ferguson, A. (2020, February 22). Scourge of wage fraud demands action, *Sydney Morning Herald*, 2.

Finau, G., Kant, R., Logan, S., Prasad, A., Tarai, J., & Cox, J. (2014). Social media and e-democracy in Fiji, Solomon Islands and Vanuatu. 20th Americas Conference on Information Systems, Georgia, mimeo.

Firth, S., & Munro, D. (1990). Compagnie et Consulat: Loies germaniques et emploi des travailleurs sur les plantations de Samoa, 1864–1914. *Journal de la Société des Océanistes, 91*, 115–134.

Fitzpatrick, P. (1980). *Law and State in Papua New Guinea*. Academic Press.

Flinn, J. (1997). Private houses, public sharing: Pollapese migrants coping with change. In J. Rensel & M. Rodman (Eds.), *Home in the Islands* (pp. 132–150). University of Hawai'i Press.

Fonseka, D. (2020) *Boss allegedly assaulted seasonal worker* [Online]. New Zealand: Newsroom. Retrieved April 17, 2021.

Frater, M. (1922). *Midst Volcanic Fires*. James Clarke.

Gailey, C. (1992). A good man is hard to find. *Critique of Anthropology, 12*, 47–74.

Galbraith, J. (1979). *The nature of mass poverty*. Harvard University Press.

Gangopadhyay, P. (2018, February 27). What Australia can learn from Fiji in reducing the working poor. *The Conversation*.

Garae, L. (2018a, February 1). SWP agent and workers donate Vt90,000 to assist disabled people. *The Daily Post*. http://dailypost.vu/news/swp-agent-and-workers-donate-vt-to-assist-disabled-people/article_cb7b3128-d3e3-507c-884b-da6ff3807667.html.

Garae, L. (2018b, September 21). Persons with disabilities enter SWP. *The Daily Post*. http://dailypost.vu/news/persons-with-disabilities-enter-swp/article_ced9d615-a8a6-517a-941d-371c4153b6a1.html

Garrapa, A. M. (2016). The citrus fruit crisis. Value chains and "just in time" migrants in Rosarno (Italy) and Valencia (Spain). In A. Corrado, C. Castro, & D. Perotta (Eds.), *Migration and agriculture: Mobility and change in the Mediterranean area* (pp. 111–127). Routledge.

Geddes, W. (1988). *South Pacific Emigration Study*. Deakin University.

Gibbon, P., & Riisgaard, L. (2014). A new system of labour management in African largescale agriculture? *Journal of Agrarian Change, 14*(1), 94–128.

Gibson, J. (2015). Circular migration, remittances and inequality in Vanuatu. *New Zealand Population Review, 41*, 153–167.

Gibson, J., & McKenzie, D. (2011). Australia's PSWPS: Development impacts in the first two years. *Asia Pacific Viewpoint, 52*(3), 361–370.

Gibson, J., & McKenzie, D. (2014). The development impact of a best practice seasonal worker policy. *Review of Economics and Statistics, 96*(2), 229–243.

Gibson, J., McKenzie, D., & Rohorua, H. (2008). How pro-poor is the selection of seasonal migrant workers from Tonga under New Zealand's Recognised Seasonal Employer Program? *Pacific Economic Bulletin, 23*(3), 187–204.

Gil, P. (2016, July 15). Tanna farms. Bringing back the coconut. *Island Life.*

Gleeson, P. (2020a, June 18). From Louisiana to Queensland: How American slave owners started again in Australia, *The Conversation.*

Gleeson, P. (2020b, August 16). Pickers can't be choosers. *Sunday Telegraph*, 106.

Górny, A., & Kaczmarczyk, P. (2018). A known but uncertain path: The role of foreign labour in Polish agriculture. *Journal of Rural Studies, 64*, 177–188.

Gosling, L. (2020, September 24). Australia needs a new Pacific compact. ANU DevPolicyBlog.

Government of Vanuatu. (2019). *National Labour Mobility Policy*. Government of Vanuatu.

Graves, A. (1986). Colonialism and indentured labour migration in the Western Pacific, 1840–1915. In P. Emmer (Ed.), *Colonialism and migration: Indentured labour before and after slavery* (pp. 237–259). Martinus Nijhoff.

Graves, A. (1993). *Cane and labour. The political economy of the Queensland Sugar Industry 1862–1906*. Edinburgh University Press.

Gray, A. (2021, May 7). How the US can protect the sovereignty of the smallest Pacific Islands. *The Diplomat.*

Griffiths, D. (2010). Working through the Pandemic: What crises tell us about economics. *Anthropology Now, 12*(2), 18–25.

Haberkorn, G. (1989). *Port Vila: Transit station or final stop?* Australian National University.

Haberkorn, G. (1992). Temporary versus permanent population mobility in Melanesia: A case study from Vanuatu. *International Migration Review, 26*(3), 806–842.

Hahamovitch, C. (2008). 'The Worst Job in the World': Reform, revolution and the secret rebellion in Florida's cane fields. *Journal of Peasant Studies, 35*(4), 770–800.

Hahamovitch, C. (2011). *No man's land. Jamaican Guestworkers in America and the Global History of Deportable Labor*. Princeton University Press.

Hammond, J., & Connell, J. (2009). The new blackbirds? Vanuatu guestworkers in New Zealand. *New Zealand Geographer, 65*(3), 201–210.

Hanich, Q., Wabnitz, C., Ota, Y., Amos, M., Donato-Hunt, C., & Hunt, A. (2018). Small-scale fisheries under climate change in the Pacific Islands region. *Marine Policy, 88*, 279–284.

Hanson, J., & Bell, M. (2007). Harvest trails in Australia: Patterns of seasonal migration in the fruit and vegetable industry. *Journal of Rural Studies, 23*, 101–117.

Hao'uli, E. (2013). Triple wins or Trojan horse? Examining the recognised Seasonal Employer Scheme under a TWAIL lens. *New Zealand Yearbook of International Law, 11*, 183–220.

Harari, F. (2021, June 5). Help wanted. *Weekend Australian Magazine*, 12–16.

Hartman, T. (2008). States, markets, and other unexceptional communities: Informal Romanian labour in a Spanish agricultural zone. *Journal of the Royal Anthropological Institute, 14*, 496–514.

Hau'ofa, E. (1993). Our Sea of Islands. In E. Waddell, V. Naidu, & E. Hau'ofa (Eds.), *A New Oceania: Rediscovering Our Sea of Islands* (pp. 2–16). University of the South Pacific.

Hay, D., & Howes, S. (2012). *Australia's Pacific seasonal worker Pilot Scheme: Why has take-up been so low?* ANU Development Policy Centre Discussion Paper No. 17.

Hayes, G. (1982). Migration, metascience and development policy in island Polynesia. *The Contemporary Pacific, 3*(1), 1–58.

Hedberg, C. (2013). 'Grapes of Wrath'? Power spatialities and aspects of labour in the Wild Berry Global Commodity Chain. *Competition and Change, 17*(1), 57–74.

Hedberg, C., & Haandrikman, K. (2014). Repopulation of the Swedish countryside: Globalisation by international migration. *Journal of Rural Studies, 34*, 128–138.

Hennebry, J. (2014). Falling through the cracks: Migrant workers and the Global Social Protection Floor Initiative. *Global Social Policy, 14*(3), 369–388.

Hennebry, J., & Preibisch, K. (2012). A model for managed migration? Re-examining best practices in Canada's Seasonal Agricultural Worker Program. *International Migration, 50*(21), e19–e40.

Hennebry, J., & Williams, G. (2015). Making vulnerability visible: Medical repatriation and Canada's migrant agricultural workers. *Canadian Medical Association Journal, 187*(6), 391–392.

Hennebry, J., McLaughlin, J., & Preibisch, K. (2016). Out of the loop: (In) access to health care for migrant workers in Canada. *Journal of International Migration and Integration, 17*, 521–538.

Hermann, E., & Kempf, W. (2017). Climate change and the imagining of migration: Emerging discourses on Kiribati's Land Purchase in Fiji. *The Contemporary Pacific, 29*(2), 231–263.

Hermant, N. (2020, November 6). Workers from Pacific Islands paying hundreds to live in cramped quarters in regional Australia. *ABC News*. Retrieved April 10, 2021, from https://www.abc.net.au/news/2020-06-11/pacific-island-scheme-wages-deducted-high-rent-inverell/12336278

Hjalmarson, K. (2016). Race, labour and the postmodern plantation: Jamaican Migrant Farmworkers in Canada's Seasonal Agricultural Worker Program. MA thesis, University of British Columbia, Okanagan.

Hobbis, G. (2019). New Media, New Melanesia? In E. Hirsch & W. Rollason (Eds.), *The Melanesian World* (pp. 546–560). Routledge.

Hoggart, K., & Mendoza, C. (1999). African immigrant workers in Spanish agriculture. *Sociologia Ruralis, 39*(4), 538–562.

Holani, A. (2017). *Enhancing the development impact of labour mobility in Pacific Island countries: The role of the seasonal worker program and the PACER plus*. PhD thesis, University of Adelaide.

Holmes, S. (2013). *Fresh fruit, broken bodies: Migrant farmworkers in the United States*. University of California Press.

Hooper, A., & Huntsman, J. (1973). A demographic history of the Tokelau Islands. *Journal of the Polynesian Society, 82*(4), 366–411.

Horvath, K. (2014). Securitisation, economisation and the political constitution of temporary migration: The making of the Austrian seasonal workers scheme. *Migration Letters, 11*(2), 154–170.

Houghton, K. (2019). *The future of regional jobs*. Regional Australia Institute.

Howe, A., Reilly, J., Clibborn, S., van den Broek, D., & Wright, C. (2020). Slicing and dicing work in the Australian horticulture industry: Labour market segmentation within the temporary migrant workforce. *Federal Law Review, 48*(2), 247–271.

Howe, J. (2021, November 5). Pay ruling puts an end to slave labour on farms. *Sydney Moring Herald*.

Howe, J., Clibborn, S., Reilly, A., van den Broek, D., & Wright, C. (2019). *Towards a durable future: Tackling labour challenges in the Australian Horticulture Industry*. University of Adelaide.

Howe, J., & Reilly, A. (2018). A tale of two Visas: Interrogating the substitution effect between Pacific seasonal workers and backpackers in addressing horticultural labour supply challenges and worker exploitation. *Australian Journal of Labour Law, 31.* University of Adelaide Law Research Paper No. 2019–38.

Howe, K. (1974). Review of Peter Corris. *Passage, Port and Plantation, Journal of the Polynesian Society, 84*(3), 382–384.

Howe, K. (1984). *Where the waves fall. A New South Sea Islands History from First Settlement to Colonial Rule.* Allen and Unwin.

Howes, S. (2018a, February 23). The Pacific Labour Scheme: No families allowed? ANU DevPolicyBlog.

Howes, S. (2018b, August 6 and 8). ANZ Seasonal labour programs are not too reliant on return workers. ANU DevPolicyBlog.

Howes, S. (2018c, May 18). *Asia or the Pacific: Where will we source our agricultural labour?* ANU DevPolicyBlog.

Howes, S. (2018d, April 3). *Satisfied seasonal workers.* ANU DevPolicyBlog.

Howes, S. (2019a, August 23). *Why 'term limits' on return workers in the SWP is a bad idea.* ANU DevPolicyBlog.

Howes, S. (2019b, December 12). *Time for a permanent Australian step-up in Pacific labour mobility.* ANU DevPolicyBlog.

Howes, S. (2021, July 23). *Foreign workers on Australian farms: Not if but how.* ANU DevPolicyBlog.

Howes, S., & Curtain, R. (2019, July 3–5). Pacific labour mobility: Success and sending country governance. *Pacific Update 2019.* The University of the South Pacific, Suva, mimeo.

Howes, S., & Curtain, R. (2021, June 17). ASEAN ag visa makes no sense. ANU DevPolicyBlog.

Howes, S., & Lawton, H. (2019, July 29). *The Pacific Labour Scheme: Is it a flop?* ANU DevPolicyBlog.

Howes, S., & Orton, B. (2020, January 21). For Tonga, Australian labour mobility more important than aid and trade combined. ANU DevPolicyBlog.

Howlett, D. (1985). Australia's next boat people? The need for a concessionary migration policy for Kiribati and Tuvalu, unpublished mimeo.

Hughes, H. (2003). Helping the islands to help themselves. *Quadrant, 47*(7), 1–3.

Hughes, H., & Sodhi, G. (2006). *Should Australia and New Zealand open their doors to guest workers from the Pacific? Costs and benefits.* Centre for Independent Studies Policy Monograph 72.

Hugo, G. (1982). Circular migration in Indonesia. *Population and Development Review, 8*(1), 59–83.

Hugo, G. (2009). Best practice in temporary labour migration for development: A perspective from Asia and the Pacific. *International Migration, 47*(5), 23–74.
Hugo, G. (2013). *What we know about circular migration and enhanced mobility.* Migration Policy Institute Policy Brief No. 7.
Hugo, G. (2014). Immigrant settlement in regional Australia: Patterns and processes. In R. Dufty-Jones & J. Connell (Eds.), *Rural change in Australia. Population, economy, environment* (pp. 57–82). Ashgate.
International Labour Organisation. (2012). *Pacific seasonal migration to New Zealand and Australia: Literature review.* ILO.
International Labour Organisation. (2015). *Pre-departure training for Nauru seasonal workers: Trainer's manual.* ILO.
Jackson Report (Committee to Review the Australian Overseas Aid Program). (1984). *Report of the committee to review the Australian Overseas Aid Program.* Australian Government Publishing Service.
Jenkin, S. (2015, March 5). *Issues with the Seasonal Workers Program: An employer's perspective.* ANU DevPolicyBlog.
Johnson, J. (1967). Harvest migration from nineteenth-century Ireland. *Transactions of the Institute of British Geographers, 41*, 97–112.
Joint Standing Committee on Foreign Affairs, Defence & Trade. (2017). *Hidden in Plain Sight: An enquiry into establishing a modern slavery act in Australia.* Parliament of the Commonwealth of Australia.
Jolly, M. (1987). The forgotten women: A history of migrant labour and gender relations in Vanuatu. *Oceania, 58*(2), 119–139.
Kagan, S. (2015, January 20). Making the case for preferential access to labour markets for Kiribati and Tuvalu migrants. ANU DevPolicyBlog.
Kagan, S. (2016a). 'On the ship, you can do anything': The impact of international cruiseship employment for I-Kiribati women. *Pacific Studies, 36*(1), 35–51.
Kagan, S. (2016b). The role of remittances in risk management and resilience in Tuvalu: Evidence and potential policy responses. In A. Milan, B. Schraven, K. Warner, & N. Cascone (Eds.), *Migration, risk management and climate change: Evidence and policy responses* (pp. 175–191). Springer.
Kaiapam, S. (2021). Impacts of labour migration on migrants and their families: A case study of Vanuatu. MA thesis, University of the South Pacific.
Kapur, D. (2004). *Remittances: The new development Mantra?* UNCTAD G24 Discussion Paper No. 29.

Kasimis, C., Papadopoulos, A., & Pappas, C. (2010). Gaining from rural migrants: Migrant employment strategies and socioeconomic implications for rural labour markets. *Sociologia Ruralis, 50*(3), 258–276.

Kautoke-Holani, A. (2017). Labour mobility in the PACER Plus. *Asia and the Pacific Policy Studies, 5*(1), 90–101.

Keen, M., Schwarz, A.-M., & Wini-Simeon, L. (2018). Towards defining the blue economy: Practical lessons from Pacific Ocean governance. *Marine Policy, 88*, 333–341.

Keith-Reid, R. (1984, July 10). Australia's new vision of islands. *Islands Business*, 66.

Kilby, P. (2008). Migrant labour, and the neo-liberal development paradigm: Balancing the contradictions in the Australian aid program. *Australian Journal of International Affairs, 62*(3), 345–356.

Kinch, J., Vitukawalu, B., Nalasi, U., Waqainabete, P., & Bermudes, M. (2019). Socioeconomic aspects of Oyster harvesting in the Rewa river delta area, Fiji. *SPC Fisheries Newsletter, 159*, 45–54.

King, R. (2018). Is migration a form of development aid given by poor to rich countries? *Journal of Intercultural Studies, 39*(2), 114–128.

Kite, S. (1975). Overseas employment in New Zealand for Tongan workers. Noumea: Unpublished paper, Regional Conference on Manpower Planning, Labour Recruitment and Migration Policies.

Knudson, K. (1981). Adaptational persistence in the Gilbert Islands. In R. Force & B. Bishop (Eds.), *Persistence and exchange* (pp. 91–99). Honolulu.

Koczberski, G., & Curry, G. (2005). Making a living: Land pressures and changing livelihood strategies among oil palm settlers in Papua New Guinea. *Agricultural Systems, 85*, 324–339.

Komaisavai, P., & Magick, S. (2019, July). Keeping our seafarers safe. *Islands Business, 45*(7), 10–13.

Kotsios, N. (2020, November). Picking up the pieces. *AgJournal*, 6–7.

Kraemer, D. (2013). *Planting roots, making place: An ethnography of young men in port Vila, Vanuatu*. PhD thesis, London School of Economics.

Kumar, R. (2012). *Migration and development: The development impacts of short-term labour migration of Ni-Vanuatu workers to New Zealand*. MA thesis, University of the South Pacific.

Kusakabe, K., & Myae, A. (2019). Precarity and vulnerability: Rubber plantations in Northern Laos and Northern Shan State, Myanmar. *Journal of Contemporary Asia, 49*(4), 586–601.

Lal, B. (2004). *Girmitiyas. The Origins of the Fiji Indians*. Fiji Institute of Applied Studies.

Larson, E. (1981). The effects of a plantation economy on a Polynesian population. In K. Michaelson (Ed.), *And the poor get children. Radical perspectives on population dynamics* (pp. 180–193). Monthly Review Press.

Lawton, H. (2021, May 27). Pacific Labour Scheme: Expanding while borders are closed. ANU DevPolicyBlog.

Le De, L., Gaillard, J.-C., Friesen, W., & Matautia-Smith, F. (2015). Remittances in the face of disasters: A case study of rural Samoa. *Environment, Development and Sustainability, 17*, 653–672.

Lee, T. (2015, May 2). Pacific Seasonal Worker Program: Vital employment for Vanuatuans, assured labour for farmers, *ABC Landline*.

Lees-Galloway, I. (2017). *Government continues to fail migrant workers*. Labour New Zealand.

Leeton Shire Council. (2020). *Seasonal Worker Programme (SWP): Understanding and recommendations for Leeton Shire Council*. Leeton Shire.

Lenard, P., & Straehle, C. (2010). Temporary labour migration: Exploitation, tool of development, or both? *Policy and Society, 29*, 283–294.

Lenard, P., & Straehle, C. (2011). Temporary labour migration, global redistribution, and democratic justice. *Politics, Philosophy and Economics, 11*(2), 206–230.

Lepon, D. (2010). *'It does not go as well as it could'. The views of Melanesian migrant farm workers of the cultural, economic and social benefits and costs of seasonal work in New Zealand*. Master of Social Sciences thesis, University of Waikato.

Levick, W., & Bedford, R. (1988). Fiji labour migration to New Zealand in the 1980s. *New Zealand Geographer, 44*(2), 14–21.

Levitt, P. (1996). *Social remittances: A conceptual tool for understanding migration and development*. Harvard University Working Paper Series (96.04).

Lindner, K., & Kathmann, T. (2014). Mobility partnerships and circular migration. In J. Gertel & S. Sippel (Eds.), *Seasonal workers in Mediterranean agriculture: The social costs of eating fresh* (pp. 121–129). Routledge.

Lipset, D. (2011). The tides: Masculinity and climate change in coastal Papua New Guinea. *Journal of the Royal Anthropological Institute, 17*(1), 20–43.

LMAP. (2017a). *LMAP Vanuatu reintegration and tracer study report for management*. LMAP.

LMAP. (2017b). *Approved employer survey and focus group discussion*. CARDNO.

Locke, S. (2017, November 1). How labour hire villains make millions from Australia's 'modern slavery' market. *Australia: ABC News*. https://www.abc.net.au/news/rural/2017-11-01/exploitation-or-slavery-of-tongans-malaysians-in-horticulture/9102676.

Loy-Wilson, S. (2017). Coolie Alibis: Seizing gold from Chinese Miners in New South Wales. *International Labor and Working Class History, 91*, 28–45.

Lozanski, K., & Baumgartner, K. (2021). Local gastronomy, transnational labour: Farm-to-table tourism and migrant agricultural workers in Niagara-on-the Lake, Canada. *Tourism Geographies*, (in press).

Luthria, M. (2008). Seasonal migration for development? Evaluating New Zealand's RSE Program. *Pacific Economic Bulletin, 23*(3), 165–170.

Macdonald, B. (1982). *Cinderellas of the empire*. ANU Press.

Maclellan, N. (2008a). Seasonal workers for Australia: Lessons for New Zealand. *Farm Policy Journal, 5*(3), 43–53.

Maclellan, N. (2008b). *Workers for all Seasons? Issues from New Zealand's Recognised Seasonal Employer (RSE) program*. Swinburne Institute for Social Research.

Maclellan, N., & Mares, P. (2006). *Remittances and labour mobility in the Pacific*. Swinburne Institute for Social Research.

Maclellan, N., & Mares, P. (2007). Labour mobility in the Pacific: Creating seasonal work programs in Australia. In S. Firth (Ed.), *Globalisation and governance in the Pacific Islands* (pp. 137–172). ANU E-Press.

Macpherson, C. (1981). Guest-worker movements and their consequences for donor and recipient countries: A case study. In G. Jones & H. Richter (Eds.), *Population mobility and development: Southeast Asia and the Pacific* (pp. 257–277). ANU Development Studies Centre Monograph No. 27.

Maebuta, H., & Maebuta, J. (2009). Generating livelihoods: A study of urban squatter settlements in Solomon Islands. *Pacific Economic Bulletin, 24*(3), 118–131.

Maher, G., & Cawley, M. (2016). Short-term labour migration: Brazilian migrants in Ireland. *Population, Space and Place, 22*, 23–35.

Manning, F. (1986, January 10). West Indians complete another season on Ontario farms. *Caribbean Chronicle,* 1589.

Marazita, J. (2021, December 7). Pacific nurses in the desert: Kiribati brain drain is outback Australia's gain, *The Guardian*.

Mares, P. (2007). Objections to Pacific Seasonal Work Programs in rural Australia. *Public Policy, 2*(1), 68–87.

Mares, P. (2012). *Temporary migration and its implications for Australia*. mimeo.

Martin, P. (1991). *The unfinished story: Turkish labour migration to Western Europe*. ILO.

Martinez, J. (2005). The end of indenture? Asian workers in the Australian Pearling Industry, 1901–1972. *International Labor and Working Class History, 67*, 125–147.

Massey, D., Arango, J., Hugo, G., Kouaouci, A., Pellegrino, A., & Taylor, E. (1993). Theories of international migration: A review and appraisal. *Population and Development Review, 19*(3), 431–466.

Maude, H. E. (1981). Slavers in paradise. The Peruvian Labour Trade in Polynesia, 1862–1864. : ANU Press.

Mayell, S. (2016). *Up-rooted lives, deep-rooted memories: Stress and resilience among Jamaican agricultural workers in Southern Ontario.* MA thesis, McMaster University, Hamilton.

McAdam, J. (2014). Historical cross-border relocations in the Pacific: Lessons for planned relocations in the context of climate change. *Journal of Pacific History, 49*, 301–332.

McAdam, J. (2015). The emerging New Zealand jurisprudence on climate change, disasters and displacement. *Migration Studies, 3*(1), 131–142.

McCall, G. (1976). European impact on Easter Island: Response, recruitment and the Polynesian experience in Peru. *Journal of Pacific History, 11*, 90–105.

McCarthy, K. (2017, December 1). *The seasonal worker program: A personal story*. ANU DevPolicyBlog.

McDonald, H. (2008, April 28). Rich and poor share a harvest of plenty. *The Australian*, 1–2.

McDonald, H. (2021, May 28). Bitter Harvest. *Inside Story*.

McGlone, T. (2021, April 11). Seasonal work on Australia's farms: 'No one wants to do this sort of work'. *Guardian*.

McKenzie, D., Garcia Martinez, P., & Winters, A. (2008). *Who is coming to New Zealand under the new Recognised Seasonal Employer program?* World Bank.

McKenzie, D., & Gibson, J. (2010). *The development impact of a best practice seasonal worker policy*. The World Bank.

McKinnon, A. (2019, June). Blackbirds. Australia had a slave trade?, *The Monthly*, 44–55.

McLaughlin, J., & Hennebry, J. (2013). Pathways to precarity: Structural vulnerabilities and lived consequences for migrant farmworkers in Canada. In L. Goldring & P. Landolt (Eds.), *Producing and negotiating non-citizenship: Precarious legal status in Canada* (pp. 175–194). University of Toronto Press.

McMichael, C., Katonivualiku, M., & Powell, T. (2019). Planned relocation and everyday agency in low-lying coastal villages in Fiji. *Geographical Journal, 185*(3), 325–337.

Meléndez, E. (2017). *Sponsored migration. The state and Puerto Rican Migration to the United States*. Ohio University Press.

Mésini, B. (2014). The transnational recruitment of temporary Latino workers in European agriculture. In J. Gertel & S. Sippel (Eds.), *Seasonal workers in Mediterranean agriculture: The social costs of eating fresh* (pp. 71–82). Routledge.

Millbank, A. (2006). *A seasonal guest-worker program for Australia?* Parliament of Australia Research Brief No. 16.

Mimura, N., Nurse, L., McLean, R., Agard, J. et al. (2007). Small islands. In M. Parry, O. Canziani, J. Palutikof, P. van der Linden & C. Hanson (Eds.), *Climate change 2007: Impacts, adaptation and vulnerability* (pp. 687–716). Cambridge: Cambridge University Press.

Mitchell, J. (2004). 'Killing time' in a postcolonial town: Young people and settlements in Port Vila, Vanuatu. In V. Lockwood (Ed.), *Globalization and culture change in the Pacific Islands* (pp. 358–376). Upper Saddle River: Pearson.

Mitchell, J. (2011). 'Operation restore public hope': Youth and the magic of modernity in Vanuatu. *Oceania, 81*, 36–50.

Moala-Tupou, L. (2016). The social impacts of seasonal migration on left-behind children: An exploratory study from Lifuka, Tonga. Master of Social Science thesis, University of Waikato, Hamilton.

Moore, C. (1985). Kanaka. *A History of Melanesian Mackay*. Institute of Papua New Guinea Studies, Port Moresby.

Moraes-Gorecki, V. (1994). 'Black Italians' in the sugar fields of North Queensland: A reflection on labour inclusion and cultural exclusion in tropical Australia. *Australian Journal of Anthropology, 5*(3), 306–319.

Morgan, W. (2018a). Back on the map: Pacific Islands in a new era of strategic competition. USP SGDIA Working Paper No. 5, Suva.

Morgan, W. (2018b). Much lost, little gained? Contemporary Trade Agreements in the Pacific Islands. *Journal of Pacific History, 53*(3), 268–286.

Morgan, W. (2020, November 26). *Pacer plus: A third view*. ANU DevPolicyBlog.

Morris, S. (2008, April 8). Fertile soil for guest workers, *Australian Financial Review*, 69.

Munro, D. (1992). Gilbert and Ellice Islanders on Queensland Canefields, 1894–1899. *Journal of the Royal Historical Society of Queensland, 14*(11), 449–465.

Murray, W. (2000). Neoliberal Globalisation, 'exotic' agro-exports and local change in the Pacific islands: A study of the Fijian kava sector. *Singapore Journal of Tropical Geography, 21*, 355–373.

Narain, D., Rogoiruwai, K., & Singh, T. (2020). *Trade and poverty: Some considerations for Pacific Island countries*. Griffith Asia Institute Research Working Paper No. 15.

Neef, A., Bege, L., Boruff, B., Pauli, N., Weber, E., & Varea, R. (2018). Climate adaptation strategies in Fiji: The role of social norms. *World Development, 107*, 125–137.

Nef, D., Kotra, K., Stauffacher, M., Six, J., & Kruetli, P. (2022). Vulnerability to food insecurity in a telecoupled world: Insights from Vanuatu. *Frontiers in Sustainable Food Systems* 6:818586.

Negin, J., Coffman, J., Connell, J., & Short, S. (2016). Foreign-born aged care workers in Australia: A growing trend. *Australasian Journal on Ageing, 35*(4), E13–E17.

Nelson, H. (1976). *Black, white and gold. Goldmining in Papua New Guinea 1978–1930*. ANU Press.

Newbury, C. (1980). The Melanesian Labor Reserve. Some reflections on pacific labor markets in the nineteenth century. *Pacific Studies, 4*(1), 1–25.

Newman, A. (2013). The legal in/Security of temporary migrant agricultural work: Case studies from Canada and Australia. *Deakin Law Review, 18*(2), 361–389.

Nishitani, M., & Lee, H. (2017). Invisible Islanders: Precarious work and Pacific settlers in rural Australia. *Pacific Studies, 40*(3), 430–449.

Nishitani, M., & Lee, H. (2019). Becoming 'Overstayers': The coloniality of citizenship and the resilience of Pacific Farm Workers. In V. Stead & J. Altman (Eds.), *Labour lines and colonial power: Indigenous and Pacific Islander Labour Mobility in Australia* (pp. 159–184). ANU Press.

O'Brien, L. (2013). *'Migrating with dignity'. A study of the Kiribati-Australia Nursing Initiative (KANI)*. MA thesis, University of Kansas.

Oceania Football Center. (2020). Seasonal working: A tough reality for football players from Vanuatu. https://oceaniafootballcenter.org/2020/06/21/.

Opeskin, B., & MacDermott, T. (2009). Resources, population and migration in the Pacific: Connecting islands and rims. *Asia Pacific Viewpoint, 50*(3), 353–373.

Orton, B. (2020, August 19). *Working holiday makers not a long-term solution for horticulture*. ANU DevPolicyBlog.

Overton, J., Murray, W., Prinsen, G., Ulu, T., & Wrighton, N. (2018). *Aid, ownership and development : The inverse sovereignty effect in the Pacific Islands*. Routledge.

Pabon, L., Umapathi, N., & Waqavonovono, E. (2012). How geographically concentrated is poverty in Fiji? *Asia Pacific Viewpoint, 53*(2), 205–217.

Pacific Periscope. (2018, February 21). What next for senior RSE workers? *Vanuatu Independent.* https://vanuatuindependent.com/2018/02/21/next-senior-rse-workers/

Paciulan, M., & Preibisch, K. (2013). Navigating the productive/reproductive split: Latin American transnational mothers and fathers in Canada's temporary migration programs. *Transnational Social Review, 3*(2), 173–192.

Papadopoulos, A., Fratsea, L., & Mavrommatis, G. (2018). Governing migrant labour in an intensive agricultural area in Greece: Precarity, political mobilization and migrant agency in the fields of Manolada. *Journal of Rural Studies, 64*, 200–209.

Parliament of the Commonwealth of Australia. Joint Committee on Foreign Affairs Defence and Trade. (1989). *Australia's relations with the South Pacific.* Australian Government Publishing Service.

Parreñas, R. (2005a). *Children of global migration. Transnational families and gendered woes.* Stanford University Press.

Parreñas, R. (2005b). Long distance intimacy: Class, gender and intergenerational relations between mothers and children in Filipino transnational families. *Global Networks, 5*(4), 317–336.

Paton, F. (1903). *Lomai of Lenakel. A hero of the New Hebrides.* Hodder and Stoughton.

Paton, J. (1891). In G. John & D. D. Paton (Eds.), *Missionary to the new Hebrides. An autobiography*. Hodder and Stoughton.

Pearce, T., Currenti, R., Mateiwai, A., & Doran, B. (2018). Adaptation to climate change and freshwater resources in Vusama village, Viti Levu, Fiji. *Regional Environmental Change, 18*, 501–510.

Peattie, M. (1988). *Nan'yo. The rise and fall of the Japanese in Micronesia, 1885–1945.* University of Hawaii Press.

Perry, A. (2012). Barely legal: Racism and migrant farm labour in the context of Canadian multiculturalism. *Citizenship Studies, 16*(2), 189–201.

Petrou, K. (2020). *If everyone returned, the Island Would Sink: Urbanisation and migration in Vanuatu*. Berghahn.

Petrou, K., & Connell, J. (2018). 'We don't feel free at all': Temporary ni-Vanuatu workers in the Riverina, Australia. *Rural Society, 27*(1), 66–79.

Petrou, K., & Connell, J. (2020). Overcoming precarity? Social media, agency and Ni-Vanuatu seasonal workers in Australia. *Journal of Australian Political Economy, 84*, 116–146.

Petrou, K., Dun, O., Farbotko, C., & Kitara, T. (2021). Pacific labour mobility on pause: Consequences of temporary immobility during the pandemic. In Y. Campbell & J. Connell (Eds.), *COVID in the Islands: A comparative perspective on the Caribbean and the Pacific* (pp. 299–319). London: Palgrave Macmillan.

Phillips, N. (2009). Migration as development strategy? The new political economy of dispossession and inequality in the Americas. *Review of International Political Economy, 16*(2), 231–259.

PIFS. (2013). *Framework for Pacific regionalism.* Suva: Pacific Islands Forum Secretariat.

Piper, N. (2010). All quiet on the Eastern front? Temporary contract migration in Asia revisited from a development perspective. *Policy and Society, 29*(4), 399–411.

Plewa, P., & Miller, M. (2005). Postwar and post-cold war generations of European Temporary Foreign Worker Policies: Implications from Spain. *Migraciones Internacionales, 3*(2), 58–83.

Poirine, B. (1998). Should we hate or love MIRAB? *The Contemporary Pacific, 10*(1), 65–105.

Polanco, G. (2019). Migration regimes and the production of (labor) unfreedom. *Journal of Asian American Studies, 22*(1), 11–30.

Polkowska, D. (2019). New dimensions of precarious work among migrants: A qualitative meta-analysis of current research. *Asian and Pacific Migration Journal, 28*(4), 440–456.

Polkowska, D., & Filipek, K. (2020). Grateful precarious worker? Ukrainian migrants in Poland. *Review of Radical Political Economics, 52*(3), 564–581.

Potter, M., & Hamilton, J. (2014). Picking on vulnerable migrants: Precarity and the mushroom industry in Northern Ireland. *Work, Employment and Society, 28*(3), 390–406.

Preibisch, K. (2007). Local produce, foreign labor: Labor mobility programs and global trade competitiveness in Canada. *Rural Sociology, 72*, 418–449.

Preibisch, K. (2010). Pick-your-own labor: Migrant workers and flexibility in Canadian agriculture. *International Migration Review, 44*(2), 404–441.

Preibisch, K., & Binford, L. (2007). Interrogating racialized global labour supply: An exploration of the racial/national replacement of foreign agricultural workers in Canada. *Canadian Review of Sociology and Anthropology, 44*(1), 5–36.

Preibisch, K., & Grez, E. (2010). The other side of el otro lado: Mexican migrant women and labor flexibility in Canadian agriculture. *Signs: Journal of Women in Culture and Society, 35*(2), 289–316.

Preibisch, K., & Grez, E. (2013). Between hearts and pockets: Locating the outcomes of transnational homemaking practices among Mexican women in Canada's temporary migration programmes'. *Citizenship Studies, 17*(6–7), 785–802.

Preibisch, K., & Hennebry, J. (2013). Buy local, hire global: Temporary migration in Canadian agriculture. In C. Straehle & P. Lenard (Eds.), *Legislating inequality: Canada's temporary migrant worker program* (pp. 49–72). McGill-Queen's University Press.

Radio New Zealand. (2019a, February 7). Canada extends labour scheme to all Solomons provinces. *RadioNZ*. https://www.radionz.co.nz/international/pacific-news/381913/canada-extends-labour-scheme-to-all-solomons-provinces

Radio New Zealand. (2019b, February 15). Fiji to address shortage of caregivers for elderly demand. *RadioNZ*. https://www.radionz.co.nz/international/pacific-news/382573/fiji-to-address-shortage-of-caregivers-for-elderly-demand

Radio New Zealand. (2019c, April 8). Extra-marital affairs issue for Samoa seasonal workers 'manageable'. *RadioNZ*. https://www.radionz.co.nz/international/pacific-news/386617/extra-marital-affairs-issue-for-samoa-seasonal-workers-manageable

Radio New Zealand. (2019d, April 23). Vanuatu rogue agents threatened with jail. *RadioNZ*. https://www.radionz.co.nz/international/pacific-news/387565/vanuatu-rogue-agents-threatened-with-jail

Rai, P. (2018). The labor of social change: Seasonal labor migration and social change in rural western India. *Geoforum, 92*, 171–180.

Rajah, R., Dayant, A., & Pryke, J. (2019). *Ocean of debt? Belt and road and debt diplomacy in the Pacific*. Lowy Institute.

Ramasamy, S., Krishnan, V., Bedford, R., & Bedford, C. (2008). The recognised seasonal employer policy: Seeking the elusive triple wins for development through international migration. *Pacific Economic Bulletin, 23*(3), 171–186.

Rasmussen, A. (2015). *In the absence of the Gift. New forms of value and Personhood in a Papua New Guinea Community*. Berghahn.

Reid-Musson, E. (2017). Grown close to home™: Migrant farmworker (Im) mobilities and unfreedom on Canadian family farms. *Annals of the American Association of Geographers, 107*(3), 716–730.

Reilly, A. (2011). The ethics of seasonal labour migration. *Griffith Law Review, 20*(1), 127–152.

Reilly, A., Howe, J., van den Broek, D., & Wright, C. F. (2018). Working holiday makers in Australian horticulture: Labour market effect, exploitation and avenues for reform. *Griffith Law Review, 27*(1), 99–130.

Rensel, J. (1997). From Thatch to cement: Social implications of housing change on Rotuma. In J. Rensel & M. Rodman (Eds.), *Home in the Islands. Housing and social change in the Pacific* (pp. 27–54). University of Hawai'i Press.

Republic of Vanuatu (2007). Seasonal Employment Act. Government of Vanuatu.

Research New Zealand. (2017). *Recognised Seasonal Employers Survey 2017—Working Report*. Research New Zealand.

Research New Zealand. (2018). *Recognised Seasonal Employers Survey 2018—Working Report*. Research New Zealand.

Rice, S. (2021, November 10). Australia's 'slave trade' disgrace. *The Australian*, 4.

Rio, K. (2011). Policing the holy nation: The state and righteous violence in Vanuatu. *Oceania, 81*, 51–71.

Robertson, H. (1902). *Erromanga. The Martyr Isle*. Hodder and Stoughton.

Robillard, C., McLaughlin, J., Cole, D., Vasilevska, B., & Gendron, R. (2018). 'Caught in the Same Webs'—Service providers' insights on gender-based and structural violence among female temporary foreign workers in Canada. *Journal of International Migration and Integration, 19*, 583–606.

Rockell, D. G. (2015). *Pacific Island labour programmes in New Zealand: An aid to Pacific Island development?: A critical lens on the recognised seasonal employer policy.* PhD thesis, Massey University, Palmerston North.

Rodgers, G., & Rodgers, J. (1989). *Precarious Jobs in Labour Marker Regulation: The growth of atypical employment in Western Europe*. ILO.

Rodman, M. (1993). Keeping options open: Copra and fish in rural Vanuatu. In T. Harding & B. Wallace (Eds.), *Contemporary Pacific societies* (pp. 171–184). Prentice Hall.

Rogaly, B. (2008). Intensification of workplace regimes in British horticulture: The role of migrant workers. *Population, Space and Place, 14*, 497–510.

Rogaly, B. (2021). Commentary: Agricultural racial capitalism and rural migrant workers. *Journal of Rural Studies, 88*, 527–531.

Rohorua, H. T., Gibson, J., McKenzie, D., & Martinez, P. G. (2009). How do Pacific island households and communities cope with seasonally absent members? *Pacific Economic Bulletin, 24*(3), 19–38.

Rose, M. (2019, October 11). Seasonal workers and rising oceans: McCormack needs to understand reality. ANU DevPolicyBlog.

Rose, M. (2020, January 28). *The Pacific Labour Scheme needs a pathway to permanency*. ANU DevPolicy Blog.

Rosewarne, S. (2010). Globalisation and the commodification of labour: Temporary labour migration. *Economic and Labour Relations Review, 20*(2), 99–110.

Ruehr, S. (2022). Beyond the vulnerability/resilience dichotomy: Perceptions of and responses to the climate crisis on Emau, Vanuatu. *Island Studies Journal, 17*(1), 157–176.

Russell, R. (2003). Jamaican workers' participation in CSAWP and development-consequences in the workers' rural home communities. In *Canada's Seasonal Agricultural Workers' Program (CSAWP) as a model best practices in the employment of Caribbean and Mexican farm workers* (p. 82). : North-South Institute.

Russell, L. (2009). *Stayin' alive. A report on urban livelihoods in Honiara, Solomon islands*. Sydney: APHEDA.

Rye, J. (2018). Labour migrants and rural change: The 'mobility transformation' of Hitra/Frøya, Norway, 2005–2015. *Journal of Rural Studies, 64*, 189–199.

Rye, J., & Andrzejewska, J. (2010). The structural disempowerment of Eastern European migrant farm workers in Norwegian agriculture. *Journal of Rural Studies, 26*, 41–51.

Rye, J., & O'Reilly, K. (Eds.). (2021). *International labour migration to Europe's rural regions*. Routledge.

Rye, J., & Scott, S. (2018). International labour migration and food production in Rural Europe: A review of the evidence. *Sociologia Ruralis, 58*(4), 928–952.

Rye, J., & Slettebak, M. (2020). The new geography of labour migration: EU11 migrants in rural Norway. *Journal of Rural Studies, 75*, 125–131.

Sampaio, D., & Carvalho, R. (2016). From the Al-Maghrib to the Al-Gharb: An anatomy of the recruitment and labour incorporation of Moroccan agricultural workers. In A. Corrado, C. Castro, & D. Perotta (Eds.), *Migration and agriculture: Mobility and change in the Mediterranean area* (pp. 231–245). Routledge.

Saunders, K. (1982). *Workers in bondage. The origins and bases of unfree labour in Queensland 1824–1916*. University of Queensland Press.

Scarr, D. (1967). Introduction. In W. E. Giles (Ed.), *A Cruise in a Queensland labour vessel to the South seas* (pp. 1–32). ANU Press.

Scarr, D. (1970). Recruits and recruiters: A portrait of the labour trade. In J. Davidson & D. Scarr (Eds.), *Pacific Islands portraits* (pp. 225–251). ANU Press.

Schneiders, B. (2018, May 18). Key farm scheme badly rorted, migrant workers paid $8 an hour. *Sydney Morning Herald*. https://www.smh.com.au/business/workplace/key-farm-scheme-badly-exploited-migrant-workers-paid-8-an-hour-20180518-p4zg53.html

Schneiders, B. (2019, August 1). Exploited Pacific migrant farm workers settle significant case. *Sydney Morning Herald*. https://www.smh.com.au/business/workplace/exploited-pacific-migrant-farm-workers-settle-landmark-case-20190801-p52cvi.html.

Scott, J. (1985). *Weapons of the weak: Everyday forms of peasant resistance*. Yale University Press.

Scott, S. (2015). Making the case for Temporary Migrant Worker Programmes: Evidence from the UK's rural guestworker ('SAWS') scheme. *Journal of Rural Studies, 40*, 1–15.

Sen, A. (2001). *Development as Freedom*. Oxford University Press.

Seo, S., & Skelton, T. (2017). Regulatory migration regimes and the production of space: The case of Nepalese workers in South Korea. *Geoforum, 78*, 159–168.

Shah, A., & Lerche, J. (2020). Migration and the invisible economies of care: Production, social reproduction and seasonal migrant labour in India. *Transactions of the Institute of British Geographers, 45*, 719–734.

Sharma, N. (2001). On being *not* Canadian: The social organisation of 'migrant workers' in Canada. *Canadian Review of Sociology and Anthropology, 38*(4), 415–439.

Sharp, T. (2013). Baias, Bisnis, and Betel nut: The place of traders in the making of a Melanesian market. In F. McCormack & K. Barclay (Eds.), *Engaging with capitalism: Cases from Oceania* (pp. 227–256). Emerald.

Sharp, T., & Busse, M. (2019). Cash crops and markets. In E. Hirsch & W. Rollason (Eds.), *The Melanesian World* (pp. 194–222). Routledge.

Shen, S., & Binns, T. (2012). Pathways, motivations and challenges: Contemporary Tuvaluan migration to New Zealand. *GeoJournal, 77*, 63–82.

Shen, S., & Gemenne, F. (2011). Contrasted views on environmental change and migration: The case of Tuvaluan migration to New Zealand. *International Migration, 49*, 224–242.

Shine, K. (2005, March 5). Fruit rots as workers pick and choose. *The Australian*, 5.

Shineberg, D. (1967). *They came for sandalwood: A study of the sandalwood trade in the South-West Pacific 1830–1865*. Melbourne University Press.

Shineberg, D. (1999). *The people trade. Pacific Island Laborers and New Caledonia, 1865–1930*. University of Hawai'i Press.

Silverman, S., & Hari, A. (2016). Troubling the fields: Choice, consent, and coercion of Canada's seasonal agricultural workers. *International Migration, 54*(5), 91–104.

Smith, A. (2015a). Troubling 'Project Canada': The Caribbean and the making of 'unfree migrant labor'. *Canadian Journal of Latin American and Caribbean Studies, 40*(2), 274–293.

Smith, R. (2015b, November 12). The 'hidden abodes' of temporary migration programs. FocaalBlog. www.focaalblog.com/2015/11/12/rachel-smith-the-hidden'abodes-of-temporary-migration-programs.

Smith, R. (2016). *The goal of the good house: Seasonal work and seeking a good life in Lamen and Lamen Bay, Epi, Vanuatu.* Unpublished PhD Thesis, University of Manchester, Manchester.

Smith, R. (2018). Changing standards of living: The paradoxes of building a good life in rural Vanuatu. In C. Gregory & J. Altman (Eds.), *The quest for the good life in precarious times: Ethnographic perspectives on the domestic moral economy* (pp. 33–55). ANU Press.

Smith, R. (2019). Be our guest/worker: Reciprocal dependency and expressions of hospitality in ni-Vanuatu overseas labour migration. *Journal of the Royal Anthropological Institute, 25*(1), 1–19.

Smith, R. (2021). Declarations of self-reliance: Alternative visions of dependency. *Citizenship and Development in Vanuatu. Oceania, 91*(2), 236–256.

Speiser, F. (1913). *Two years with the natives of the Western Pacific*. Mills and Boon.

Standing, G. (2011). *The Precariat, The New Dangerous Class*. London: Bloomsbury Academic.

Stapleton, J. (2008, June 21). Fruit rots on the vine for want of seasonal workers. *The Australian*, 9.

Stead, V. (2019a). Money trees, development dreams, and colonial legacies in contemporary Pasifika Horticultural Labour. In V. Stead & J. Altman (Eds.), *Labour lines and colonial power: Indigenous and Pacific Islander Labour Mobility in Australia* (pp. 133–158). ANU Press.

Stead, V. (2019b, August 18). Picking fruit is work, not benevolence, and doesn't absolve Australia of climate responsibility, *Guardian*.

Stead, V. (2021a). Precarity's reach: Intersections of history, life, and labour in the Australian horticultural industry. *Journal of the Royal Anthropological Institute, 27*(2), 303–320.

Stead, V. (2021b, November 10). Australia needs better working conditions, not shaming, for Pacific Islander farm workers. *The Conversation*.

Stenbacka, S. (2019). Responsibilities, caring practices and agriculture: Farmers' perspectives on recruitment and employer–employee relationships. *Sociologia Ruralis, 59*(2), 255–274.

Stewart-Withers, R. (2011). Contesting a Third World development category: Female-headed households in Samoa. *Women's Studies International Forum, 34*, 171–184.

Stilwell, F. (2003). Refugees in a region: Afghans in young, NSW. *Urban Policy and Research, 21*(3), 235–248.

Stoler, A. (1985). *Capitalism and confrontation in Sumatra's Plantation Belt, 1870–1979*. Yale University Press.

Storey, D., & Hunter, S. (2010). Kiribati: An environmental 'perfect storm'. *Australian Geographer, 47*, 161–181.

Storey, D., & Steinmayer, V. (2011). Mobility as a development strategy: The case of the Pacific Islands. *Asia-Pacific Population Journal, 26*(4), 57–72.

Strauss, K., & McGrath, S. (2017). Temporary migration, precarious employment and unfree labour relations: Exploring the 'continuum of exploitation' in Canada's temporary foreign worker program. *Geoforum, 78*, 199–208.

Sullivan, K. (2020, September 30). Coronavirus restrictions could lead to 26,000-person shortfall for coming harvests, report says [Online]. *Australia: ABC News*. https://www.abc.net.au/news/2020-09-30/coronavirus-farm-worker-shortage-coming-harvests/12714694.

Sullivan, K., & Hynninen, E. (2020, February 14). Seasonal worker program faces fresh scrutiny after 70 workers found living in one house. *Australia: ABC News*. https://www.abc.net.au/news/2020-02-14/seasonal-worker-program-under-scrutiny-after-70-people-in-house/11960818.

Sunam, R., Barney, K., & McCarthy, J. (2021). Transnational labour migration and livelihoods in rural Asia: Tracing patterns of agrarian and forest change. *Geoforum, 118*, 1–13.

Surak, K. (2015). Guestworker regimes globally: A historical comparison. In L. Talani & S. McMahon (Eds.), *Handbook of the International Political Economy of Migration* (pp. 167–184). Edward Elgar.

Szadziewski, H. (2020). Converging anticipatory geographies in Oceania: The Belt and Road Initiative and Look North in Fiji. *Political Geography, 77*, 102–119.

Tabucanon, G., & Opeskin, B. (2011). The resettlement of the Nauruans in Australia: An early case of failed environmental migration. *Journal of Pacific History, 46*, 337–356.

Tan, Y., & Lester, L. (2012). Labour market and economic impacts of international working holiday temporary migrants to Australia. *Population, Space and Place, 18*(3), 359–383.

TASWA. (2016). Issues and concerns by Tongan seasonal workers regarding exploitation, Sydney (mimeo).

Taylor, J. (1999). The new economics of labour migration and the role of remittances in the migration process. *International Migration, 37*(1), 63–88.

Taylor, M., McGregor, A., & Dawson, B. (2016). *Vulnerability of Pacific Island agriculture and forestry to climate change.* Pacific Community and Australian Aid.

Tazreiter, C., Weber, L., Pickering, S., Segrave, M., & McKernan, H. (2016). *Fluid Security in the Asia Pacific* (pp. 103–127). Macmillan.

Terry, W. (2016). Solving seasonality in tourism? Labour shortages and guest worker programmes in the USA. *Area, 48*(1), 111–118.

Thomas, E. (2019). *'Forgetting' farmworkers: Transnational experiences of black Jamaicans 'retired' from Canada's seasonal agricultural worker program.* PhD thesis, York University, Toronto.

Thomas-Hope, E. (2017). Migration, small farming and food security in the Caribbean: Jamaica and St. Vincent and the Grenadines. *International Migration, 55*(4), 35–47.

Thompson, B. (2018, March 23). Australians avoiding farm work despite abundant jobs, award rates. *Australian Financial Review.* https://www.afr.com/companies/agriculture/australians-avoiding-farm-work-despite-abundant-jobs-award-rates-20180323-h0xv9f

Thomson, D. (2018, October 18). Forced labour is not the answer for horticulture. *The North Queensland Register*, 19.

Thurnwald, R. (1932). The psychology of acculturation. *American Anthropologist, 34*(4), 557–569.

TNS. (2011). *Final evaluation of the Pacific Seasonal Worker Pilot Scheme.* TNS.

Toa, E. (2016, November 15). Vanuatu workers win praise at prestigious resort. *The Vanuatu Independent.* https://vanuatuindependent.com/2016/11/15/vanuatu-workers-win-praise-prestigious-resort/

Toatu, T. (1980). Seamen and cultural change in Kiribati. *Pacific Perspectives, 8*, 31–32.

Tonga Ministry of Finance. (2011). *Second national millennium development goals report.* Nuku'alofa: Ministry of Finance.

Trundle, A., Barth, B., & McEvoy, D. (2018). Leveraging endogenous climate resilience: Urban adaptation in Pacific Small Island developing states. *Environment and Urbanization, 31*(1), 53–74.

Turhan, E., Zografos, C., & Kallis, G. (2015). Adaptation as biopolitics: Why state policies in Turkey do not reduce the vulnerability of seasonal agricultural workers to climate change. *Global Environmental Change, 31*, 296–306.

Uera, M. (2019). *A Pacific/ Tongan perspective. Pacific Update 2019*. USP.

Uibu, K. (2020, September 15). There are no human rights here. *ABC News*. https://www.abc.net.au/news/2020-09-15/backpacker-farm-workers-speak-of-wage-exploitation/12545294?nw=0.

Underhill, E., Groutsis, D., van den Broek, D., & Rimmer, M. (2015). Organising across borders: Mobilising temporary migrant labour in Australian food production. *Journal of Industrial Relations, 62*(2), 278–303.

Underhill, E., & Rimmer, M. (2015). Itinerant foreign harvest workers in Australia: The impact of precarious employment on occupational safety and health. *Policy and Practice in Health and Safety, 13*(2), 25–46.

Underhill, E., & Rimmer, M. (2016). Layered vulnerability: Temporary migrants in Australian horticulture. *Journal of Industrial Relations, 58*(5), 608–626.

Underhill-Sem, Y., & Marsters, E. (2017). *Labour mobility in the Pacific: A systematic literature review of development impacts*. New Zealand Institute for Pacific Research.

UNDP. (2013). *Vanuatu Hardship and Poverty Report*. Vanuatu National Statistics Office and UNDP Pacific Centre.

Union Aid Abroad APHEDA. (2017). *Temporary labour mobility in Australia and the Pacific Islands*. Union Aid Abroad APHEDA.

Union Aid Abroad APHEDA. (2020). Inquiry into strengthening Australia's relationships with countries in the Pacific region: Submission 20. Submission to Joint Standing Committee on Foreign Affairs, Defence and Trade—Foreign Affairs and Aid Sub-Committee: Canberra.

Unions NSW. (2020). *Wage theft. The shadow market*. Unions NSW.

Unions NSW. (2021). *Wage theft. The shadow market. Part Two: The horticultural industry*. Unions NSW.

Vainikolo, L. (2021). Small and isolated Vava'u, Tonga: From weakness to strength during COVID-19? In Y. Campbell & J. Connell (Eds.), *COVID in the Islands: The Pacific and the Caribbean* (pp. 207–218). Palgrave Macmillan.

van den Broek, D., Wright, C., Howe, J., & Reilly, A. (2021). Pro-market governance, migration status and worker vulnerability: The case of Australian horticulture. *Economic and Industrial Democracy, 42*(4), 1305–1325.

Vanuatu National Statistics Office. (2012). *Vanuatu Household Income and Expenditure Survey 2010.* Vanuatu Government.

Vanuatu National Statistics Office. (2017). *2016 Post-TC Pam Mini-Census Report: Basic Table Volume 1.* Vanuatu Government.

Vanuatu National Statistics Office. (2021a). *Wellbeing in Vanuatu. 2019–20 NSDP Baseline Survey.* Vanuatu Government.

Vanuatu National Statistics Office. (2021b). *Hardship in Vanuatu: 2019–20 NSDP Baseline Survey.* Vanuatu Government.

Villadóniga, J. (2020). Unstable places. Modern-day slavery of the 21st century in the Strawberry sector in Huelva in Spain. *Migraciones Internacionales, 11*(4), 1.

Villafuerte, L. (2018). *Seasons of change: Ni-Vanuatu and the Recognised Seasonal Employers (RSE) scheme.* PhD thesis, University of Auckland.

Viviani, N. (1970). *Nauru. Phosphate and Political Progress.* ANU Press.

Voigt-Graf, C. (2003). Fijian teachers on the move: Causes, implications and policies. *Asia Pacific Viewpoint, 44*, 163–174.

Wade, J. (2020, September 1). *Licensed agents: The key to Vanuatu's SWP success.* DevPolicy Blog.

Wainiqolo, I. (2019, December). Labour productivity and youth unemployment in Fiji. *Pacific Economic Monitor*, 7–8.

Walsh, J. (2014). From nations of immigrants to states of transience: Temporary migration in Canada and Australia. *International Sociology, 29*(6), 584–606.

Wang, Z., & Connell, J. (2021). Taiwanese working holiday makers in rural and regional Australia: Temporary transnational identities and employment challenges. *Australian Geographer, 52*(2), 191–208.

Ward, M. (1971). *The role of investment in the development of Fiji.* Cambridge University Press.

Ward, R. (1990). Contract labor recruitment from the highlands of Papua New Guinea, 1950–1974. *International Migration Review, 24*, 273–296.

Watters, R. with Banibati, K. (1977/1984). Abemama. ANU Development Studies Centre.

Webb, J. (2020a). Kiribati economic survey: Oceans of opportunity. *Asia Pacific Policy Studies, 7*, 5–26.

Webb, J. (2020b, December). Samoa and Tonga: Opportunities in the storm?, *Pacific Economic Monitor*, 19–22.

Weiler, A., Sexsmith, K., & Minkoff-Zern, L.-A. (2020). Parallel precarity: A comparison of U.S. and Canadian Agricultural Guestworker Programs. *International Journal of the Sociology of Agriculture and Food, 26*(2), 143–163.

Wessen, A., Hooper, A., Huntsman, J., Prior, I., & Salmond, C. (1992). *Migration and health in a small society. The case of Tokelau*. Clarendon Press.

West, P. (2016). *Dispossession and the environment: Rhetoric and inequality in Papua New Guinea*. Columbia University Press.

Wheeler, T. (2020, February 5). *South West farm workers take part in Aus Day soccer tournament.* Western Australia: Manjimup-Bridgetown Times. https://www.swtimes.com.au/news/manjimup-bridgetown-times/south-west-farm-workers-take-part-in-aus-day-soccer-tournament-ng-b881447078z?fbclid=IwAR0lJDs0BLSFocKvhI9cTyCeAnH7btO5IfYyJEVs9zRDZH-CLaaZsN8IxHR8.

Wickramasekera, P. (2011). *Circular migration: A triple win or a dead end?* ILO.

Wigglesworth, A. (2018, June 12). Why Timorese workers love the Seasonal Worker Program. ANU DevPolicy Blog.

Wigglesworth, A., & Boavida dos Santos, A. (2016). *Timorese workers in Australia: A study of seasonal work experiences, skills and future options*. Labour Mobility Assistance Program.

Williams, A. (2010). *The experiences of the recognised seasonal employer scheme and Ni-Vanuatu temporary workers in Te Puke, New Zealand.* Master of Planning thesis, University of Otago.

Wilson, J. (1966). *Economic survey of the new hebrides*. HMSO.

Wilson, K. (2013). Wan Laki Aelan? Diverse development strategies on Aniwa, Vanuatu. *Asia Pacific Viewpoint, 54*(2), 246–263.

Wood, C., & McCoy, T. (1985). Migration, remittances and development: A study of Caribbean cane cutters in Florida. *International Migration Review, 19*(2), 251–277.

Wood, T., Dornan, M., & Muller, S. (2021). *Change and continuity in Australian aid: What the aid flows show*. Development Policy Centre.

World Bank. (2006a). *Global economic prospects. Economic implications of remittances and migration*. World Bank.

World Bank. (2006b). *At home and away. Expanding job opportunities for Pacific Islanders through labour mobility*. World Bank.

World Bank. (2014). *Vanuatu Socio-Economic Atlas*. World Bank.

World Bank. (2017). *Maximizing the development impacts from temporary migration*. World Bank.

World Bank. (2018a). *The social impacts of seasonal migration: Lessons from Australia's Seasonal Worker Programme for Pacific Islanders*. World Bank.

World Bank. (2018b). *Expanding employment pathways for Pacific Islanders in the Australian and New Zealand aged care sectors*. World Bank.

World Bank. (2021). *Pacific labor mobility, migration and remittances in Times of COVID-19*. World Bank.

Wright, C., & Clibborn, S. (2020). A guest-worker state? The declining power and agency of migrant labour in Australia. *Economic and Labour Relations Review, 31*(1), 34–58.

Wright, C., Groutsis, D., & van den Broek, D. (2017). Employer sponsored temporary labour migration schemes in Australia, Canada and Sweden: Enhancing efficiency, compromising fairness? *Journal of Ethnic and Migration Studies, 43*(11), 1854–1872.

Wu, A. (2019). *Timorese participants in the Australian seasonal workers programme: Migration, remittances and development*. PhD thesis, University of Sydney.

Wyett, K. (2014). Escaping a rising tide: Sea level rise and migration in Kiribati. *Asia and the Pacific Policy Studies, 1*, 171–185.

Ye, J., & Yeoh, B. (2018). Disciplining deserving subjects through social assistance: Migration and the diversification of precarity in Singapore. *Annals of the American Association of Geographers, 108*(2), 478–485.

Zann, S., & Zann, L. (2008). Poverty in paradise? Issues in poverty and development in Fijian fishing villages. *SPC Women in Fisheries Information Bulletin, 18*, 36–41.

Zhao, S., Binks, B., Kruger, H., Xia, C., & Stenekes, N. (2018). *What difference does labour choice make to farm productivity and profitability in the Australian horticulture industry? A comparison between seasonal workers and working holiday makers*. ABARES.

Index

K. Petrou, J. Connell, *Pacific Islands Guestworkers in Australia*,
https://doi.org/10.1007/978-981-19-5387-3

N

O

P

FSC
www.fsc.org
MIX
Papier | Fördert
gute Waldnutzung
FSC® C083411

Kolibri 360 GmbH
Ferdinand-Jühlke-Straße 7
99095 Erfurt, Deutschland
produktsicherheit@kolibri360.de